Principles of Aquaponics Food Production Systems: Volume 1

Principles of Aquaponics Food Production Systems: Volume 1

Stephnie Glenn

Callisto Reference

www.callistoreference.com

Callisto Reference,
118-35 Queens Blvd., Suite 400,
Forest Hills, NY 11375, USA

Visit us on the World Wide Web at:
www.callistoreference.com

ISBN: 978-1-64116-742-0 (Hardback)

Cataloging-in-Publication Data

Principles of aquaponics food production systems : volume 1 / Stephnie Glenn.
p. cm.
Includes bibliographical references and index.
ISBN 978-1-64116-742-0
1. Aquaponics. 2. Aquaculture. 3. Hydroponics. 4. Food industry and trade--Appropriate technology. I. Glenn, Stephnie.

SB126.5 .E44 2023
631.585--dc23

Table of Contents

Preface

Aquaponic food production systems integrate two production systems, namely, aquaculture and hydroponics in a single production system. Aquaculture is related to growing fish in a closed environment whereas hydroponics involves growing plants in a soil-less environment. Aquaponic systems offer an alternate technology solution for some of the major problems in the world such as population growth, urbanization, water shortages, land and soil degradation, environmental pollution, world hunger, and climate change. There are three major components of the aquaponics freshwater systems which include freshwater aquatic animals (such as fish), nitrifying bacteria, and plants. All these three living entities are dependent on each other for their survival. Bacteria consume the fish waste and generate nutrients which are used by the plants. As plants grow, nutrients are removed from the water, making it clean for the fish. Aquaponic food production systems are an intensive yet sustainable food production system. Moreover, aquaponics do not require soil and are therefore not susceptible to soil-borne diseases. This book outlines the principles of aquaponics food production systems in detail. It will provide comprehensive knowledge to the readers.

All of the data presented henceforth, was collaborated in the wake of recent advancements in the field. The aim of this book is to present the diversified developments from across the globe in a comprehensible manner. The opinions expressed in each chapter belong solely to the contributing authors. Their interpretations of the topics are the integral part of this book, which I have carefully compiled for a better understanding of the readers.

At the end, I would like to thank all those who dedicated their time and efforts for the successful completion of this book. I also wish to convey my gratitude towards my friends and family who supported me at every step.

Stephnie Glenn

Part I
Limited Resources and Associated Framework Conditions

1

Aquaponics and Food Production: The Global Issue

Simon Goddek, Alyssa Joyce, Benz Kotzen, and Maria Dos-Santos

Abstract As the world's population grows, the demands for increased food production expand, and as the stresses on resources such as land, water and nutrients become ever greater, there is an urgent need to find alternative, sustainable and reliable methods to provide this food. The current strategies for supplying more produce are neither ecologically sound nor address the issues of the circular economy of reducing waste whilst meeting the WHO's Millennium Development Goals of eradicating hunger and poverty by 2015. Aquaponics, a technology that integrates aquaculture and hydroponics, provides part of the solution. Although aquaponics has developed considerably over recent decades, there are a number of key issues that still need to be fully addressed, including the development of energy-efficient systems with optimized nutrient recycling and suitable pathogen controls. There is also a key issue of achieving profitability, which includes effective value chains and efficient supply chain management. Legislation, licensing and policy are also keys to the success of future aquaponics, as are the issues of education and research, which are discussed across this book.

Keywords Aquaponics · Agriculture · Planetary boundaries · Food supply chain · Phosphorus

---

S. Goddek (✉)
Mathematical and Statistical Methods (Biometris), Wageningen University, Wageningen, The Netherlands
e-mail: simon.goddek@wur.nl; simon@goddek.nl

A. Joyce
Department of Marine Science, University of Gothenburg, Gothenburg, Sweden
e-mail: alyssa.joyce@gu.se

B. Kotzen
School of Design, University of Greenwich, London, UK
e-mail: b.kotzen@greenwich.ac.uk

M. Dos-Santos
ESCS-IPL, DINÂMIA'CET, ISCTE-Institute University of Lisbon, Lisbon, Portugal
e-mail: mjpls@iscte-iul.pt

1.1 Introduction

Food production relies on the availability of resources, such as land, freshwater, fossil energy and nutrients (Conijn et al. 2018), and current consumption or degradation of these resources exceeds their global regeneration rate (Van Vuuren et al. 2010). The concept of planetary boundaries (Fig. 1.1) aims to define the environmental limits within which humanity can safely operate with regard to scarce resources (Rockström et al. 2009). Biochemical flow boundaries that limit food supply are more stringent than climate change (Steffen et al. 2015). In addition to nutrient recycling, dietary changes and waste prevention are integrally necessary to transform current production (Conijn et al. 2018; Kahiluoto et al. 2014). Thus, a major global challenge is to shift the growth-based economic model towards a

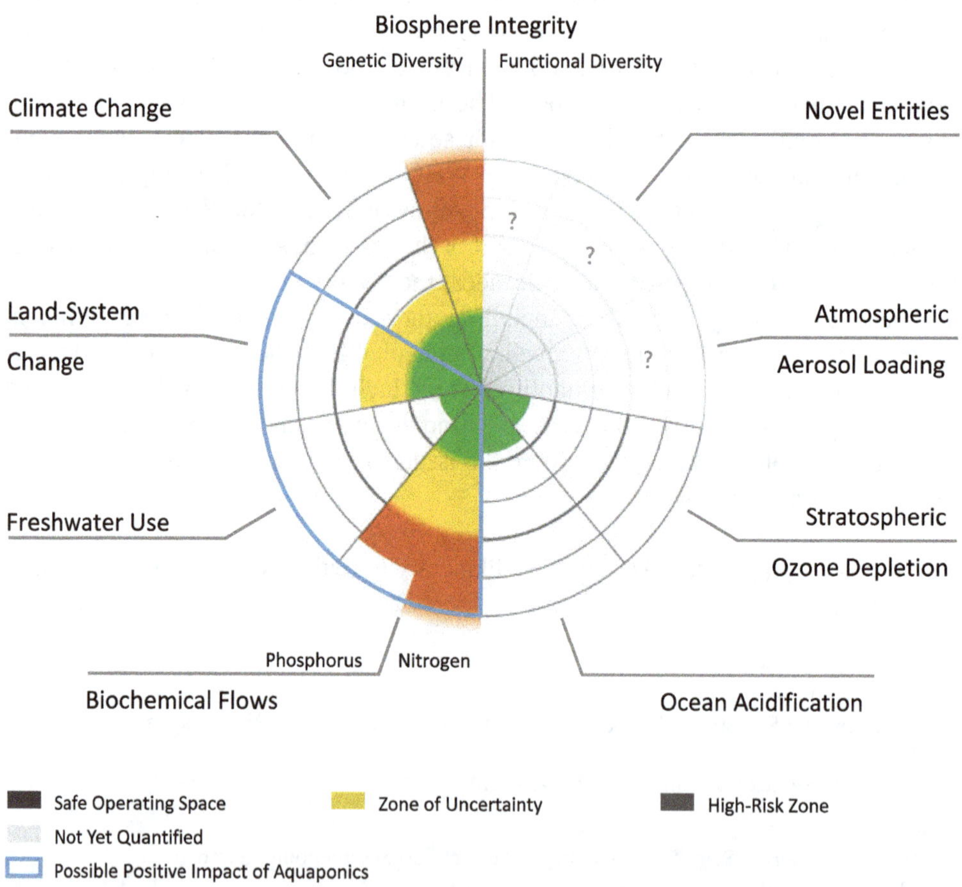

Fig. 1.1 Current status of the control variables for seven of the planetary boundaries as described by Steffen et al. (2015). The green zone is the safe operating space, the yellow represents the zone of uncertainty (increasing risk), the red is a high-risk zone, and the grey zone boundaries are those that have not yet been quantified. The variables outlined in blue (i.e. land-system change, freshwater use and biochemical flows) indicate the planetary boundaries that aquaponics can have a positive impact on

balanced eco-economic paradigm that replaces infinite growth with sustainable development (Manelli 2016). In order to maintain a balanced paradigm, innovative and more ecologically sound cropping systems are required, such that trade-offs between immediate human needs can be balanced whilst maintaining the capacity of the biosphere to provide the required goods and services (Ehrlich and Harte 2015).

In this context, aquaponics has been identified as a farming approach that, through nutrient and waste recycling, can aid in addressing both planetary boundaries (Fig. 1.1) and sustainable development goals, particularly for arid regions or areas with nonarable soils (Goddek and Körner 2019; Appelbaum and Kotzen 2016; Kotzen and Appelbaum 2010). Aquaponics is also proposed as a solution for using marginal lands in urban areas for food production closer to markets. At one time largely a backyard technology (Bernstein 2011), aquaponics is now growing rapidly into industrial-scale production as technical improvements in design and practice allow for significantly increased output capacities and production efficiencies. One such area of evolution is in the field of coupled vs. decoupled aquaponics systems. Traditional designs for one-loop aquaponics systems comprise both aquaculture and hydroponics units between which water recirculates. In such traditional systems, it is necessary to make compromises to the conditions of both subsystems in terms of pH, temperature and nutrient concentrations (Goddek et al. 2015; Kloas et al. 2015) (see Chap. 7). A decoupled aquaponics system, however, can reduce the need for trade-offs by separating the components, thus allowing the conditions in each subsystem to be optimized. Utilization of sludge digesters is another key way of maximizing efficiency through the reuse of solid wastes (Emerenciano et al. 2017; Goddek et al. 2018; Monsees et al. 2015). Although many of the largest facilities worldwide are still in arid regions (i.e. Arabian Peninsula, Australia and sub-Saharan Africa), this technology is also being adopted elsewhere as design advances have increasingly made aquaponics not just a water-saving enterprise but also an efficient energy and nutrient recycling system.

1.2 Supply and Demand

The 2030 Agenda for Sustainable Development emphasizes the need to tackle global challenges, ranging from climate change to poverty, with sustainable food production a high priority (Brandi 2017; UN 2017). As reflected in the UN's Sustainable Development Goal 2 (UN 2017), one of the greatest challenges facing the world is how to ensure that a growing global population, projected to rise to around 10 billion by 2050, will be able to meet its nutritional needs. To feed an additional two billion people by 2050, food production will need to increase by 50% globally (FAO 2017). Whilst more food will need to be produced, there is a shrinking rural labour force because of increasing urbanization (dos Santos 2016). The global rural population has diminished from 66.4% to 46.1% in the period from 1960 to 2015 (FAO 2017). Whilst, in 2017, urban populations represented more than 54% of the total world population, nearly all future growth of the world's population will occur in urban areas, such that by 2050, 66% of the global population will live in cities (UN 2014). This increasing urbanization of cities is accompanied by a simultaneously growing network of infrastructure systems, including transportation networks.

To ensure global food security, total food production will need to increase by more than 70% in the coming decades to meet the Millennium Development Goals (FAO 2009), which include the 'eradication of extreme poverty and hunger' and also 'ensuring environmental sustainability'. At the same time, food production will inevitably face other challenges, such as climate change, pollution, loss of biodiversity, loss of pollinators and degradation of arable lands. These conditions require the adoption of rapid technological advances, more efficient and sustainable production methods and also more efficient and sustainable food supply chains, given that approximately a billion people are already chronically malnourished, whilst agricultural systems continue to degrade land, water and biodiversity at a global scale (Foley et al. 2011; Godfray et al. 2010).

Recent studies show that current trends in agricultural yield improvements will not be sufficient to meet projected global food demand by 2050, and these further suggest that an expansion of agricultural areas will be necessary (Bajželj et al. 2014). However, the widespread degradation of land in conjunction with other environmental problems appears to make this impossible. Agricultural land currently covers more than one-third of the world's land area, yet less than a third of it is arable (approximately 10%) (World Bank 2018). Over the last three decades, the availability of agricultural land has been slowly decreasing, as evidenced by more than 50% decrease from 1970 to 2013. The effects of the loss of arable land cannot be remedied by converting natural areas into farmland as this very often results in erosion as well as habitat loss. Ploughing results in the loss of topsoil through wind and water erosion, resulting in reduced soil fertility, increased fertilizer use and then eventually to land degradation. Soil losses from land can then end up in ponds, dams, lakes and rivers, causing damage to these habitats.

In short, the global population is rapidly growing, urbanizing and becoming wealthier. Consequently, dietary patterns are also changing, thus creating greater demands for greenhouse gas (GHG) intensive foods, such as meat and dairy products, with correspondingly greater land and resource requirements (Garnett 2011). But whilst global consumption is growing, the world's available resources, i.e. land, water and minerals, remain finite (Garnett 2011). When looking at the full life-cycle analysis of different food products, however, both Weber and Matthews (2008) and Engelhaupt (2008) suggest that dietary shifts can be a more effective means of lowering an average household's food-related climate footprint than 'buying local'. Therefore, instead of looking at the reduction of supply chains, it has been argued that a dietary shift away from meat and dairy products towards nutrition-oriented agriculture can be more effective in reducing energy and footprints (Engelhaupt 2008; Garnett 2011).

The complexity of demand-supply imbalances is compounded by deteriorating environmental conditions, which makes food production increasingly difficult and/or unpredictable in many regions of the world. Agricultural practices cannot only undermine planetary boundaries (Fig. 1.1) but also aggravate the persistence and propagation of zoonotic diseases and other health risks (Garnett 2011). All these factors result in the global food system losing its resilience and becoming increasingly unstable (Suweis et al. 2015).

The ambitious 2015 deadline of the WHO's Millennium Development Goals (MDGs) to eradicate hunger and poverty, to improve health and to ensure environmental sustainability has now passed, and it has become clear that providing nutritious food for the undernourished as well as for affluent populations is not a simple task. In summary, changes in climate, loss of land and diminution in land quality, increasingly complex food chains, urban growth, pollution and other adverse environmental conditions dictate that there is an urgent need to not only find new ways of growing nutritious food economically but also locate food production facilities closer to consumers. Delivering on the MDGs will require changes in practice, such as reducing waste, carbon and ecological footprints, and aquaponics is one of the solutions that has the potential to deliver on these goals.

1.3 Scientific and Technological Challenges in Aquaponics

Whilst aquaponics is seen to be one of the key food production technologies which 'could change our lives' (van Woensel et al. 2015), in terms of sustainable and efficient food production, aquaponics can be streamlined and become even more efficient. One of the key problems in conventional aquaponics systems is that the nutrients in the effluent produced by fish are different than the optimal nutrient solution for plants. Decoupled aquaponics systems (DAPS), which use water from the fish but do not return the water to the fish after the plants, can improve on traditional designs by introducing mineralization components and sludge bioreactors containing microbes that convert organic matter into bioavailable forms of key minerals, especially phosphorus, magnesium, iron, manganese and sulphur that are deficient in typical fish effluent. Contrary to mineralization components in one-loop systems, the bioreactor effluent in DAPS is only fed to the plant component instead of being diluted in the whole system. Thus, decoupled systems that utilize sludge digesters make it possible to optimize the recycling of organic wastes from fish as nutrients for plant growth (Goddek 2017; Goddek et al. 2018). The wastes in such systems mainly comprise fish sludge (i.e. faeces and uneaten feed that is not in solution) and thus cannot be delivered directly in a hydroponics system. Bioreactors (see Chap. 10) are therefore an important component that can turn otherwise unusable sludge into hydroponic fertilizers or reuse organic wastes such as stems and roots from the plant production component into biogas for heat and electricity generation or DAPS designs that also provide independently controlled water cycling for each unit, thus allowing separation of the systems (RAS, hydroponic and digesters) as required for the control of nutrient flows. Water moves between components in an energy and nutrient conserving loop, so that nutrient loads and flows in each subsystem can be monitored and regulated to better match downstream requirements. For instance, phosphorous (P) is an essential but exhaustible fossil resource that is mined for fertilizer, but world supplies are currently being depleted at an alarming rate. Using digesters in decoupled aquaponics systems allows microbes to convert the phosphorus in fish waste into orthophosphates that can be utilized by plants, with high recovery rates (Goddek et al. 2016, 2018).

Although decoupled systems are very effective at reclaiming nutrients, with near-zero nutrient loss, the scale of production in each of the units is important given that nutrient flows from one part of the system need to be matched with the downstream production potential of other components. Modelling software and Supervisory Control and Data Acquisition (SCADAS) data acquisition systems therefore become important to analyse and report the flow, dimensions, mass balances and tolerances of each unit, making it possible to predict physical and economic parameters (e.g. nutrient loads, optimal fish-plant pairings, flow rates and costs to maintain specific environmental parameters). In Chap. 11, we will look in more detail at systems theory as applied to aquaponics systems and demonstrate how modelling can resolve some of the issues of scale, whilst innovative technological solutions can increase efficiency and hence profitability of such systems. Scaling is important not only to predict the economic viability but also to predict production outputs based on available nutrient ratios.

Another important issue, which requires further development, is the use and reuse of energy. Aquaponics systems are energy and infrastructure intensive. Depending on received solar radiation, the use of solar PV, solar thermal heat sources and (solar) desalination may still not be economically feasible but could all be potentially integrated into aquaponics systems. In Chap. 12, we present information about innovative technical and operational possibilities that have the capacity to overcome the inherent limitations of such systems, including exciting new opportunities for implementing aquaponics systems in arid areas.

In Chap. 2, we also discuss in more detail the range of environmental challenges that aquaponics can help address. Pathogen control, for instance, is very important, and contained RAS systems have a number of environmental advantages for fish production, and one of the advantages of decoupled aquaponics systems is the ability to circulate water between the components and to utilize independent controls wherein it is easier to detect, isolate and decontaminate individual units when there are pathogen threats. Probiotics that are beneficial in fish culture also appear beneficial for plant production and can increase production efficiency when circulated within a closed system (Sirakov et al. 2016). Such challenges are further explored in Chap. 5, where we discuss in more detail how innovation in aquaponics can result in (a) increased space utilization efficiency (less cost and materials, maximizing land use); (b) reduced input resources, e.g. fishmeal, and reduced negative outputs, e.g. waste discharge; and (c) reduced use of antibiotics and pesticides in self-contained systems.

There are still several aquaponic topic areas that require more research in order to exploit the full potential of these systems. From a scientific perspective, topics such as nitrogen cycling (Chap. 9), aerobic and anaerobic remineralization (Chap. 10), water and nutrient efficiency (Chap. 8), optimized aquaponic fish diets (Chap. 13) and plant pathogens and control strategies (Chap. 14) are all high priorities.

In summary, the following scientific and technological challenges need to be addressed:

1. *Nutrients*: As we have discussed, systems utilizing sludge digesters make it possible to optimize the recycling of organic waste from fish into nutrients for plant growth, such designs allow for optimized reclamation and recycling of nutrients to create a near-zero nutrient loss from the system.
2. *Water*: The reuse of nutrient-depleted water from greenhouses can also be optimized for reuse back in the fish component utilizing condensers.
3. *Energy*: Solar-powered designs also improve energy savings, particularly if preheated water from solar heaters in the greenhouses can be recirculated back to fish tanks for reuse.

The ability to recycle water, nutrients and energy makes aquaponics a potentially unique solution to a number of environmental issues facing conventional agriculture. This is discussed in Chap. 2.

1.4 Economic and Social Challenges

From an economic perspective, there are a number of limitations inherent in aquaponics systems that make specific commercial designs more or less viable (Goddek et al. 2015; Vermeulen and Kamstra 2013). One of the key issues is that stand-alone, independent hydroponics and aquaculture systems are more productive than traditional one-loop aquaponics systems (Graber and Junge 2009), as they do not require trade-offs between the fish and plant components. Traditional, classic single-loop aquaponics requires a compromise between the fish and plant components when attempting to optimize water quality and nutrient levels that inherently differ for the two parts (e.g. desired pH ranges and nutrient requirements and concentrations). In traditional aquaponics systems, savings in fertilizer requirements for plants do not make up for the harvest shortfalls caused by suboptimal conditions in the respective subsystems (Delaide et al. 2016).

Optimizing growth conditions for both plants (Delaide et al. 2016; Goddek and Vermeulen 2018) and fish is the biggest challenge to profitability, and current results indicate that this can be better achieved in multi-loop decoupled aquaponics systems because they are based on independent recirculating loops that involve (1) fish, (2) plants and (3) bioreactors (anaerobic or aerobic) for sludge digestion and a unidirectional water (nutrient) flow, which can improve macro- and micro-nutrient recovery and bioavailability, as well as optimization of water consumption (Goddek and Keesman 2018). Current studies show that this type of system allows for the maintenance of specific microorganism populations within each compartment for better disease management, and they are more economically efficient in so much as the systems not only reduce waste outflow but also reutilize otherwise unusable sludge, converting it to valuable outputs (e.g. biogas and fertilizer).

Fig. 1.2 An aquaponics system seen as a black box scheme. We do not get to see inside the box, but we know the inputs, the outputs (i.e. fish and plants) and the waste

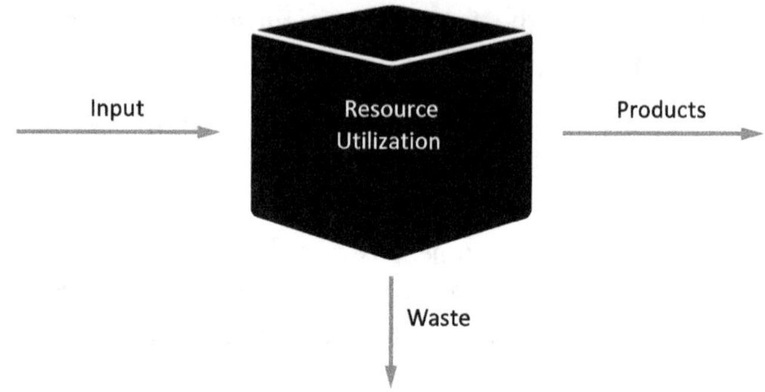

Independent, RAS systems and hydroponics units also have a wide range of operational challenges that are discussed in detail in Chaps. 3 and 4. Increasingly, technological advances have allowed for higher productivity ratios (Fig. 1.2), which can be defined as a fraction of the system's outputs (i.e. fish and plants) over the system's input (i.e. fish feed and/or additional fertilization, energy input for lighting, heating and pumping CO_2 dosing and biocontrols).

When considering the many challenges that aquaponics encounters, production problems can be broadly broken down into three specific themes: (1) system productivity, (2) effective value chains and (3) efficient supply chain management.

System Productivity Agricultural productivity is measured as the ratio of agricultural outputs to agricultural inputs. Traditional small-scale aquaponics systems were designed primarily to address environmental considerations such as water discharge, water inputs and nutrient recycling, but the focus in recent years has increasingly shifted towards economic feasibility in order to increase productivity for large-scale farming applications. However, this will require the productivity of aquaponics systems to be able to compete economically with independent, state-of-the-art hydroponics and aquaculture systems. If the concept of aquaponics is to be successfully applied at a large scale, the reuse of nutrients and energy must be optimized, but end markets must also be considered.

Effective Value Chains The value chains (added value) of agricultural products mainly arise from the processing of the produce such as the harvested vegetables, fruits and fish. For example, the selling price for pesto (i.e. red and green) can be more than ten times higher than that of the tomatoes, basil, olive oil and pine nuts. In addition, most processed food products have a longer shelf life, thus reducing spoilage. Evidently, fresh produce is important because nutritional values are mostly higher than those in the processed foods. However, producing fresh and high-quality produce is a real challenge and therefore a luxury in many regions of the world. Losses of nutrients during storage of fruit and vegetables are substantial if they are not canned or frozen quickly (Barrett 2007; Rickman et al. 2007). Therefore, for large-scale systems, food processing should at least be considered to balance out any fluctuations between supply and demand and reduce food waste. With respect to food waste reduction, vegetables that do not meet fresh produce standards, but are still of marketable quality, should be processed in order to reduce postharvest losses.

Although such criteria apply to all agricultural and fisheries products, value adding can substantially increase the profitability of the aquaponics farm, especially if products can reach niche markets.

Efficient Supply Chain Management In countries with well-developed transportation and refrigeration networks, fruit and vegetables can be imported from all around the world to meet consumer demands for fresh produce. But as mentioned previously, high-quality and fresh produce is a scarce commodity in many parts of the world, and the long-distance movement of goods – i.e. supply chain management – to meet high-end consumer demand is often criticized and justifiably so. Most urban dwellers around the world rely on the transport of foods over long distances to meet daily needs (Grewal and Grewal 2012). One of the major criticisms is thus the reliance on fossil fuels required to transport products over large distances (Barrett 2007). The issue of food miles directs focus on the distance that food is transported from the time of production to purchase by the end consumer (Mundler and Criner 2016). However, in terms of CO_2 emissions per tonne/km (tkm), one food mile for rail transportation (13.9 g CO_2/tkm) is not equal to one food mile of truck/road transportation, as truck transportation has more than 15 times greater environmental impact (McKinnon 2007). Therefore, transportation distance is not necessarily the only consideration, as the ecological footprint of vegetables grown on farms in rural areas is potentially less than the inputs required to grow food in greenhouses closer to urban centres.

Food miles are thus only a part of the picture. Food is transported long distances, but the greenhouse gas emissions associated with food production are dominated by the production phase (i.e. the impact of energy for heating, cooling and lighting) (Engelhaupt 2008; Weber and Matthews 2008). For example, Carlsson (1997) showed that tomatoes imported from Spain to Sweden in winter have a much lower carbon footprint than those locally grown in Sweden, since energy inputs to greenhouses in Sweden far outweigh the carbon footprint of transportation from Spain. When sourcing food, the transport of goods is not the only factor to take into consideration, as the freshness of products determines their nutritive value, taste and general appeal to consumers. By growing fresh food locally, many scholars agree that urban farming could help secure the supply of high-quality produce for urban populations of the future whilst also reducing food miles (Bon et al. 2010; dos Santos 2016; Hui 2011). Both areas will be discussed in more detail in Sect. 1.5.

From a consumer's perspective, urban aquaponics thus has advantages because of its environmental benefits due to short supply chains and since it meets consumer preferences for high-quality locally produced fresh food (Miličić et al. 2017). However, despite these advantages, there are a number of socio-economic concerns: The major issue involves urban property prices, as land is expensive and often considered too valuable for food production. Thus, purchasing urban land most likely makes it impossible to achieve a feasible expected return of investment. However, in shrinking cities, where populations are decreasing, unused space could be used for agricultural purpose (Bontje and Latten 2005; Schilling and Logan 2008) as is the case in Detroit in the United States (Mogk et al. 2010).

Additionally, there is a major issue of urban planning controls, where in many cities urban land is not designated for agricultural food production and aquaponics is seen to be a part of agriculture. Thus, in some cities aquaponic farming is not allowed. The time is ripe to engage with urban planners who need to be convinced of the benefits of urban farms, which are highly productive and produce fresh, healthy, local food in the midst of urban and suburban development.

1.5 The Future of Aquaponics

Technology has enabled agricultural productivity to grow exponentially in the last century, thus also supporting significant population growth. However, these changes also potentially undermine the capacity of ecosystems to sustain food production, to maintain freshwater and forest resources and to help regulate climate and air quality (Foley et al. 2005).

One of the most pressing challenges in innovative food production, and thus in aquaponics, is to address regulatory issues constraining the expansion of integrated technologies. A wide range of different agencies have jurisdiction over water, animal health, environmental protection and food safety, and their regulations are in some cases contradictory or are ill-suited for complex integrated systems (Joly et al. 2015). Regulations and legislation are currently one of the most confusing areas for producers and would-be entrepreneurs. Growers and investors need standards and guidelines for obtaining permits, loans and tax exemptions, yet the confusing overlap of responsibilities among regulatory agencies highlights the urgent need for better harmonization and consistent definitions. Regulatory frameworks are frequently confusing, and farm licensing as well as consumer certification remains problematic in many countries. The FAO (in 2015), the WHO (in 2017) and the EU (in 2016) all recently began harmonizing provisions for animal health/well-being and food safety within aquaponics systems and for export-import trade of aquaponic products. For instance, several countries involved in aquaponics are lobbying for explicit wording within the *Codex Alimentarius*, and a key focus within the EU, determined by the EU sponsored COST Action FA1305, the 'EU Aquaponics Hub', is currently on defining aquaponics as a clear and distinct entity. At present, regulations define production for both aquaculture and hydroponics, but have no provisions for merging of the two. This situation often creates excessive bureaucracy for producers who are required to license two separate operations or whose national legislation does not allow for co-culturing (Joly et al. 2015). The EU Aquaponics Hub, which has supported this publication (COST FA1305), defines aquaponics as 'a production system of aquatic organisms and plants where the majority (> 50%) of nutrients sustaining the optimal plant growth derives from waste originating from feeding the aquatic organisms' (see Chap. 7).

Consumer certification schemes also remain a difficult area for aquaponics producers in many parts of the world. For instance, in the United States and Australia, aquaponic products can be certified as organic, but not within the European Union.

From an economic perspective, aquaponics is in theory capable of increasing the overall value of fish farming or conventional hydroponics whilst also closing the food-water-energy cycle within a circular bio-based economy. In order to make small-scale aquaponics systems economically viable, aquaponics farmers generally have to operate in niche markets to obtain higher prices for products, so certification thus becomes very important.

The most pressing issues are whether aquaponics can become acceptable at the policy level. Food safety is a high priority for gaining public support, and although there is a much lowered pathogen risk in closed systems, thus implying less need for antimicrobials and pesticides, managing potential risks – or moreover managing perceptions of those risks, especially as they may affect food safety – is a high priority for government authorities and investors alike (Miličić et al. 2017). One concern that is often raised is the fear of pathogen transfer in sludge from fish to plants, but this is not substantiated in the literature (Chap. 6). As such, there is a need to allay any remaining food safety and biosecurity concerns through careful research and, where concerns may exist, to ascertain how it may be possible to manage these problems through improved system designs and/or regulatory frameworks.

Aquaponics is an emerging food production technology which has the ability to condense and compress production into spaces and places that would not normally be used for growing food. This not only means that it is exceptionally relevant in urban areas, where aquaponics can be placed on underutilized and unused places such as flat roofs, development sites, abandoned factories, housing estates and schools, but it provides a means both in the developed and developing world for people to take back part of the food production process by providing fresh local food to the market (van Gorcum et al. 2019). The integration of aquaponics with vertical farming and living wall technologies will, in time, most likely improve productivity by reducing the overall farming footprint with reduced land take and intensification.

The intense production methods in aquaponics rely on the knowledge of a combination of key factors which are highly suitable for use in teaching STEM (science, technology, engineering and maths) subjects in schools. Aquaponics provides the teacher and student with opportunities to explore the realm of complex systems, their design and management and a host of other subject areas, including environmental sciences, water chemistry, biology and animal welfare. Aquaponics is also being used in prisons/correctional facilities, such as at the San Francisco County Jail, to help inmates gain skills and experience in aquaculture and horticulture that they can use on their release. In the domestic context, there is a growing trend to design countertop systems that can grow herbs as well as small systems that can be located in offices, where exotic fish provide a calming effect, whilst plants, as part of living walls, similarly provide an aesthetic backdrop and clean the air.

Aquaponics is a farming technology advancing rapidly from its first exploits in the last years of the twentieth century and the first decades of the twenty-first century. But it still is an 'emerging technology and science topic' (Junge et al. 2017) which is subject to a considerable amount of 'hype'. When comparing the number of aquaculture, hydroponic and aquaponic peer-reviewed papers, aquaponic

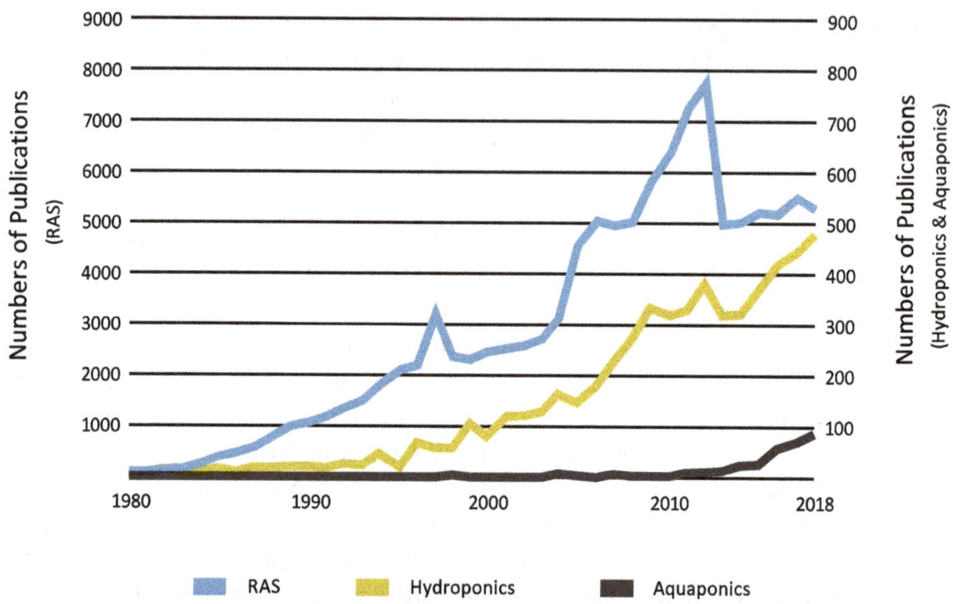

Fig. 1.3 The number of papers published on 'hydroponics', 'RAS' and 'aquaponics' from 1980 to 2018 (data were collected from the Scopus database on 30 January 2019). Please note that the scale for 'RAS' is one order of magnitude higher than that for 'hydroponics' and 'aquaponics'

papers are considerably lower (Fig. 1.3), but the numbers are rising and will continue to rise as aquaponics education, especially at university level, and general interest increases. A 'hype ratio' can be described as an indicator of the popularity of a subject in the public media relative to what is published in the academic press. This can, for example, be calculated by taking the search results in Google divided by the search results in Google Scholar. In the case of aquaponics, the hype ratio on 16 August 2016 was 1349, which is considerable when compared to the hype ratios of hydroponics (131) and recirculating aquaculture (17) (Junge et al. 2017). The sense one gets from this is that, indeed, aquaponics is an emerging technology but that there is enormous interest in the field which is likely to continue and increase over the next decades. The hype ratio, however, is likely to decline as more research is undertaken and scientific papers are published.

This book is aimed at the aquaponics researcher and practitioner, and it has been designed to discuss, explore and reveal the issues that aquaponics is addressing now and that will no doubt arise in the future. With such a broad spectrum of topics, it aims to provide a comprehensive but easily accessible overview of the rather novel scientific and commercial field of aquaponics. Apart from the production and technical side, this book has been designed to address trends in food supply and demand, as well as the various economic, environmental and social implications of this emerging technology. The book has been co-authored by numerous experts from around the world, but mostly from within the EU. Its 24 chapters cover the whole gamut of aquaponics areas and will provide a necessary textbook for all those interested in aquaponics and moving aquaponics forwards into the next decade.

References

Appelbaum S, Kotzen B (2016) Further investigations of aquaponics using brackish water resources of the Negev desert. Ecocycles 2:26. https://doi.org/10.19040/ecocycles.v2i2.53

Bajželj B, Richards KS, Allwood JM, Smith P, Dennis JS, Curmi E, Gilligan CA (2014) Importance of food-demand management for climate mitigation. Nat Clim Chang 4:924–929. https://doi.org/10.1038/nclimate2353

Barrett DM (2007) Maximizing the nutritional value of fruits and vegetables

Bernstein S (2011) Aquaponic gardening: a step-by-step guide to raising vegetables and fish together. New Society Publishers, Gabriola Island

Bon H, Parrot L, Moustier P (2010) Sustainable urban agriculture in developing countries. A review. Agron Sustain Dev 30:21–32. https://doi.org/10.1051/agro:2008062

Bontje M, Latten J (2005) Stable size, changing composition: recent migration dynamics of the Dutch large cities. Tijdschr Econ Soc Geogr 96:444–451. https://doi.org/10.1111/j.1467-9663.2005.00475.x

Brandi CA (2017) Sustainability standards and sustainable development –synergies and trade-offs of transnational governance. Sustain Dev 25:25–34. https://doi.org/10.1002/sd.1639

Carlsson A (1997) Greenhouse gas emissions in the life-cycle of carrots and tomatoes. Lund University, Lund

Conijn JG, Bindraban PS, Schröder JJ, Jongschaap REE (2018) Can our global food system meet food demand within planetary boundaries? Agric Ecosyst Environ 251:244–256. https://doi.org/10.1016/J.AGEE.2017.06.001

Delaide B, Goddek S, Gott J, Soyeurt H, Jijakli M (2016) Lettuce (*Lactuca sativa* L. var. Sucrine) growth performance in complemented aquaponic solution outperforms hydroponics. Water 8:467. https://doi.org/10.3390/w8100467

dos Santos MJPL (2016) Smart cities and urban areas–aquaponics as innovative urban agriculture. Urban For Urban Green 20:402. https://doi.org/10.1016/j.ufug.2016.10.004

Ehrlich PR, Harte J (2015) Opinion: to feed the world in 2050 will require a global revolution. Proc Natl Acad Sci U S A 112:14743–14744. https://doi.org/10.1073/pnas.1519841112

Emerenciano M, Carneiro P, Lapa M, Lapa K, Delaide B, Goddek S (2017) Mineralizacão de sólidos. Aquac Bras:21–26

Engelhaupt E (2008) Do food miles matter? Environ Sci Technol 42:3482

FAO (2009) The state of food and agriculture. FAO, Rome

FAO (2017) 2017 the state of food and agriculture leveraging food systems for inclusive rural transformation. FAO, Rome

Foley JA, Defries R, Asner GP, Barford C, Bonan G, Carpenter SR, Chapin FS, Coe MT, Daily GC, Gibbs HK, Helkowski JH, Holloway T, Howard EA, Kucharik CJ, Monfreda C, Patz JA, Prentice IC, Ramankutty N, Snyder PK (2005) Global consequences of land use. Science 309:570–574. https://doi.org/10.1126/science.1111772

Foley JA, Ramankutty N, Brauman KA, Cassidy ES, Gerber JS, Johnston M, Mueller ND, O'Connell C, Ray DK, West PC, Balzer C, Bennett EM, Carpenter SR, Hill J, Monfreda C, Polasky S, Rockström J, Sheehan J, Siebert S, Tilman D, Zaks DPM (2011) Solutions for a cultivated planet. Nature 478:337–342. https://doi.org/10.1038/nature10452

Garnett T (2011) Where are the best opportunities for reducing greenhouse gas emissions in the food system (including the food chain)? Food Policy 36:S23–S32. https://doi.org/10.1016/J.FOODPOL.2010.10.010

Goddek S (2017) Opportunities and challenges of multi-loop aquaponic systems. Wageningen University, Wageningen. https://doi.org/10.18174/412236

Goddek S, Keesman KJ (2018) The necessity of desalination technology for designing and sizing multi-loop aquaponics systems. Desalination 428:76–85. https://doi.org/10.1016/j.desal.2017.11.024

Goddek S, Vermeulen T (2018) Comparison of *Lactuca sativa* growth performance in conventional and RAS-based hydroponic systems. Aquac Int 26:1377. https://doi.org/10.1007/s10499-018-0293-8

Goddek S, Delaide B, Mankasingh U, Ragnarsdottir K, Jijakli H, Thorarinsdottir R (2015) Challenges of sustainable and commercial aquaponics. Sustainability 7:4199–4224. https://doi.org/10.3390/su7044199

Goddek S, Schmautz Z, Scott B, Delaide B, Keesman K, Wuertz S, Junge R (2016) The effect of anaerobic and aerobic fish sludge supernatant on hydroponic lettuce. Agronomy 6:37. https://doi.org/10.3390/agronomy6020037

Goddek S, Delaide BPL, Joyce A, Wuertz S, Jijakli MH, Gross A, Eding EH, Bläser I, Reuter M, Keizer LCP, Morgenstern R, Körner O, Verreth J, Keesman KJ (2018) Nutrient mineralization and organic matter reduction performance of RAS-based sludge in sequential UASB-EGSB reactors. Aquac Eng 83:10–19. https://doi.org/10.1016/J.AQUAENG.2018.07.003

Goddek S, Körner O (2019) A fully integrated simulation model of multi-loop aquaponics: a case study for system sizing in different environments. Agric Syst 171:143–154. https://doi.org/10.1016/j.agsy.2019.01.010

Godfray HCJ, Beddington JR, Crute IR, Haddad L, Lawrence D, Muir JF, Pretty J, Robinson S, Thomas SM, Toulmin C (2010) Food security: the challenge of feeding 9 billion people. Science 327:812–818. https://doi.org/10.1126/science.1185383

Graber A, Junge R (2009) Aquaponic systems: nutrient recycling from fish wastewater by vegetable production. Desalination 246:147–156

Grewal SS, Grewal PS (2012) Can cities become self-reliant in food? Cities 29:1–11. https://doi.org/10.1016/J.CITIES.2011.06.003

Hui SCM (2011) Green roof urban farming for buildings in high-density urban cities. In: World green roof conference. pp 1–9

Joly A, Junge R, Bardocz T (2015) Aquaponics business in Europe: some legal obstacles and solutions. Ecocycles 1:3–5. https://doi.org/10.19040/ecocycles.v1i2.30

Junge R, König B, Villarroel M, Komives T, Jijakli M (2017) Strategic points in aquaponics. Water 9:182. https://doi.org/10.3390/w9030182

Kahiluoto H, Kuisma M, Kuokkanen A, Mikkilä M, Linnanen L (2014) Taking planetary nutrient boundaries seriously: can we feed the people? Glob Food Sec 3:16–21. https://doi.org/10.1016/J.GFS.2013.11.002

Kloas W, Groß R, Baganz D, Graupner J, Monsees H, Schmidt U, Staaks G, Suhl J, Tschirner M, Wittstock B, Wuertz S, Zikova A, Rennert B (2015) A new concept for aquaponic systems to improve sustainability, increase productivity, and reduce environmental impacts. Aquac Environ Interact 7:179–192. https://doi.org/doi. https://doi.org/10.3354/aei00146

Kotzen B, Appelbaum S (2010) An investigation of aquaponics using brackish water resources in the Negev desert. J Appl Aquac 22:297–320. https://doi.org/10.1080/10454438.2010.527571

Manelli A (2016) New paradigms for a sustainable Well-being. Agric Agric Sci Procedia 8:617–627. https://doi.org/10.1016/J.AASPRO.2016.02.084

McKinnon A (2007) CO_2 emissions from freight transport: an analysis of UK data. Prepared for the climate change working group of the commission for integrated transport. London, England

Miličić V, Thorarinsdottir R, Santos M, Hančič M (2017) Commercial aquaponics approaching the European market: to consumers' perceptions of aquaponics products in Europe. Water 9:80. https://doi.org/10.3390/w9020080

Mogk J, Kwiatkowski S, Weindorf M (2010) Promoting urban agriculture as an alternative land use for vacant properties in the city of Detroit: benefits, problems, and proposals for a regulatory framework for successful land use integration. Law Fac Res Publ 56:1521

Monsees H, Keitel J, Kloas W, Wuertz S (2015) Potential reuse of aquacultural waste for nutrient solutions in aquaponics. In: Proc of Aquaculture Europe. Rotterdam, The Netherlands

Mundler P, Criner G (2016) Food systems: food miles. In: Encyclopedia of food and health. Elsevier, Amsterdam, pp 77–82. https://doi.org/10.1016/B978-0-12-384947-2.00325-1

Rickman JC, Barrett DM, Bruhn CM (2007) Review nutritional comparison of fresh, frozen and canned fruits and vegetables. Part 1. Vitamins C and B and phenolic compounds. J Sci Food Agric J Sci Food Agric 87:930–944. https://doi.org/10.1002/jsfa.2825

Rockström J, Steffen W, Noone K, Persson Å, Chapin FS, Lambin EF, Lenton TM, Scheffer M, Folke C, Schellnhuber HJ, Nykvist B, de Wit CA, Hughes T, van der Leeuw S, Rodhe H, Sörlin S, Snyder PK, Costanza R, Svedin U, Falkenmark M, Karlberg L, Corell RW, Fabry VJ,

Hansen J, Walker B, Liverman D, Richardson K, Crutzen P, Foley JA (2009) A safe operating space for humanity. Nature 461:472–475. https://doi.org/10.1038/461472a

Schilling J, Logan J (2008) Greening the rust belt: a green infrastructure model for right sizing America's shrinking cities. J Am Plan Assoc 74:451–466. https://doi.org/10.1080/01944360802354956

Sirakov I, Lutz M, Graber A, Mathis A, Staykov Y, Smits T, Junge R (2016) Potential for combined biocontrol activity against fungal fish and plant pathogens by bacterial isolates from a model aquaponic system. Water 8:518. https://doi.org/10.3390/w8110518

Steffen W, Richardson K, Rockström J, Cornell SE, Fetzer I, Bennett EM, Biggs R, Carpenter SR, de Vries W, de Wit CA, Folke C, Gerten D, Heinke J, Mace GM, Persson LM, Ramanathan V, Reyers B, Sörlin S (2015) Planetary boundaries: guiding human development on a changing planet. Science 347(80):736

Suweis S, Carr JA, Maritan A, Rinaldo A, D'Odorico P (2015) Resilience and reactivity of global food security. Proc Natl Acad Sci U S A 112:6902–6907. https://doi.org/10.1073/pnas.1507366112

UN (2014) World's population increasingly urban with more than half living in urban areas | UN DESA | United Nations Department of Economic and Social Affairs [WWW Document]. URL http://www.un.org/en/development/desa/news/population/world-urbanization-prospects-2014.html. Accessed 19 Oct 2018

UN (2017) The sustainable development goals report. UN, New York

van Gorcum B, Goddek S, Keesman KJ (2019) Gaining market insights for aquaponically produced vegetables in Kenya. Aquac Int:1–7. https://link.springer.com/article/10.1007/s10499-019-00379-1

Van Vuuren DP, Bouwman AF, Beusen AHW (2010) Phosphorus demand for the 1970–2100 period: a scenario analysis of resource depletion. Glob Environ Chang 20:428–439. https://doi.org/10.1016/J.GLOENVCHA.2010.04.004

van Woensel L, Archer G, Panades-Estruch L, Vrscaj D (2015) Ten technologies which could change our lives – potential impacts and policy implications. European Commission, Brussels

Vermeulen T, Kamstra A (2013) The need for systems design for robust aquaponic systems in the urban environment

Weber CL, Matthews HS (2008) Food-miles and the relative climate impacts of food choices in the United States. Environ Sci Technol 42:3508–3513. https://doi.org/10.1021/es702969f

World Bank (2018) Agricultural land (% of land area) [WWW Document]. URL https://data.worldbank.org/indicator/ag.lnd.agri.zs. Accessed19 Oct 2018

2

Nutrient Cycling in Aquaponics Systems

Alyssa Joyce, Simon Goddek, Benz Kotzen, and Sven Wuertz

Abstract Hydroponics initially developed in arid regions in response to freshwater shortages, while in areas with poor soil, it was viewed as an opportunity to increase productivity with fewer fertilizer inputs. In the 1950s, recirculating aquaculture also emerged in response to similar water limitations in arid regions in order to make better use of available water resources and better contain wastes. However, disposal of sludge from such systems remained problematic, thus leading to the advent of aquaponics, wherein the recycling of nutrients produced by fish as fertilizer for plants proved to be an innovative solution to waste discharge that also had economic advantages by producing a second marketable product. Aquaponics was also shown to be an adaptable and cost-effective technology given that farms could be situated in areas that are otherwise unsuitable for agriculture, for instance, on rooftops and on unused, derelict factory sites. A wide range of cost savings could be achieved through strategic placement of aquaponics sites to reduce land acquisition costs, and by also allowing farming closer to suburban and urban areas, thus reducing transportation costs to markets and hence also the fossil fuel and CO_2 footprints of production.

Keywords Aquaponics · Sustainable agriculture · Eutrophication · Soil degradation · Nutrient cycling

---

A. Joyce (✉)
Department of Marine Science, University of Gothenburg, Gothenburg, Sweden
e-mail: alyssa.joyce@gu.se

S. Goddek
Mathematical and Statistical Methods (Biometris), Wageningen University, Wageningen, The Netherlands
e-mail: simon.goddek@wur.nl; simon@goddek.nl

B. Kotzen
School of Design, University of Greenwich, London, UK
e-mail: b.kotzen@greenwich.ac.uk

S. Wuertz
Department Ecophysiology and Aquaculture, Leibniz-Institute of Freshwater Biology and Inland Fisheries, Berlin, Germany
e-mail: wuertz@igb-berlin.de

2.1 Introduction

The term 'tipping point' is currently being used to describe natural systems that are on the brink of significant and potentially catastrophic change (Barnosky et al. 2012). Agricultural food production systems are considered one of the key ecological services that are approaching a tipping point, as climate change increasingly generates new pest and disease risks, extreme weather phenomena and higher global temperatures. Poor land management and soil conservation practices, depletion of soil nutrients and risk of pandemics also threaten world food supplies.

Available arable land for agricultural expansion is limited, and increased agricultural productivity in the past few decades has primarily resulted from increased cropping intensity and better crop yields as opposed to expansion of the agricultural landmass (e.g. 90% of gains in crop production have been a result of increased productivity, but only 10% due to land expansion) (Alexandratos and Bruinsma 2012; Schmidhuber 2010). Global population is estimated to reach 8.3–10.9 billion people by 2050 (Bringezu et al. 2014), and this growing world population, with a corresponding increase in total as well as per capita consumption, poses a wide range of new societal challenges. The United Nations Convention to Combat Desertification (UNCCD) *Global Land Outlook Working Paper* 2017 report notes worrying trends affecting food production (Thomas et al. 2017) including land degradation, loss of biodiversity and ecosystems, and decreased resilience in response to environmental stresses, as well as a widening gulf between food production and demand. The uneven distribution of food supplies results in inadequate quantities of food, or lack of food of sufficient nutritional quality for part of the global population, while in other parts of the world overconsumption and diseases related to obesity have become increasingly common. This unbalanced juxtaposition of hunger and malnutrition in some parts of the world, with food waste and overconsumption in others, reflects complex interrelated factors that include political will, resource scarcity, land affordability, costs of energy and fertilizer, transportation infrastructure and a host of other socioeconomic factors affecting food production and distribution.

Recent re-examinations of approaches to food security have determined that a 'water-energy-food nexus' approach is required to effectively understand, analyse and manage interactions among global resource systems (Scott et al. 2015). The nexus approach acknowledges the interrelatedness of the resource base – land, water, energy, capital and labour – with its drivers, and encourages inter-sectoral consultations and collaborations in order to balance different resource user goals and interests. It aims to maximize overall benefits while maintaining ecosystem integrity in order to achieve food security. Sustainable food production thus requires reduced utilization of resources, in particular, water, land and fossil fuels that are limited, costly and often poorly distributed in relation to population growth, as well as recycling of existing resources such as water and nutrients within production systems to minimize waste.

In this chapter, we discuss a range of current challenges in relation to food security, focusing on resource limitations and ways that new technologies and interdisciplinary approaches such as aquaponics can help address the water-food-energy nexus in relation to the UN's goals for sustainable development. We concentrate on the need for increased nutrient recycling, reductions in water consumption and non-renewable energy, as well as increased food production on land that is marginal or unsuitable for agriculture.

2.2 Food Supply and Demand

2.2.1 Predictions

Over the last 50 years, total food supply has increased almost threefold, whereas the world's population has only increased twofold, a shift that has been accompanied by significant changes in diet related to economic prosperity (Keating et al. 2014). Over the last 25 years, the world's population increased by 90% and is expected to reach the 7.6 billion mark in the first half of 2018 (Worldometers). Estimates of increased world food demand in 2050 relative to 2010 vary between 45% and 71% depending on assumptions around biofuels and waste, but clearly there is a production gap that needs to be filled. In order to avoid a reversal in recent downward trends undernourishment, there must be reductions in food demand and/or fewer losses in food production capacity (Keating et al. 2014). An increasingly important reason for rising food demand is per capita consumption, as a result of rising per capita income, which is marked by shifts towards high protein foods, particularly meat (Ehrlich and Harte 2015b). This trend creates further pressures on the food supply chain, since animal-based production systems generally require disproportionately more resources, both in water consumption and feed inputs (Rask and Rask 2011; Ridoutt et al. 2012; Xue and Landis 2010). Even though the rate of increasing food demand has declined in recent decades, if current trajectories in population growth and dietary shifts are realistic, global demand for agricultural products will grow at 1.1–1.5% per year until 2050 (Alexandratos and Bruinsma 2012).

Population growth in urban areas has put pressure on land that has been traditionally used for soil-based crops: demands for housing and amenities continue to encroach on prime agricultural land and raise its value well beyond what farmers could make from cultivation. Close to 54% of the world's population now lives in urban areas (Esch et al. 2017), and the trend towards urbanization shows no signs of abating. Production systems that can reliably supply fresh foods close to urban centres are in demand and will increase as urbanization increases. For instance, the rise of vertical farming in urban centres such as Singapore, where land is at a premium, provides a strong hint that concentrated, highly productive farming systems will be an integral part of urban development in the future. Technological advances are increasingly making indoor farming systems economical, for instance the development of LED horticultural lights that are extremely long lasting and energy efficient has increased competitiveness of indoor farming as well as production in high latitudes.

Analysis of agrobiodiversity consistently shows that high- and middle-income countries obtain diverse foods through national or international trade, but this also implies that production and food diversity are uncoupled and thus more vulnerable to interruptions in supply lines than in low-income countries where the majority of food is produced nationally or regionally (Herrero et al. 2017). Also, as farm sizes increase, crop diversity, especially for crops belonging to highly nutritious food groups (vegetables, fruits, meat), tends to decrease in favour of cereals and legumes, which again risks limiting local and regional availability of a range of different food groups (Herrero et al. 2017).

2.3 Arable Land and Nutrients

2.3.1 Predictions

Even as more food needs to be produced, usable land for agricultural practices is inherently limited to roughly 20–30% of the world's land surface. The availability of agricultural land is decreasing, and there is a shortage of suitable land where it is most needed, i.e. particularly near population centres. Soil degradation is a major contributor to this decline and can generally be categorized in two ways: displacement (wind and water erosion) and internal soil chemical and physical deterioration (loss of nutrients and/or organic matter, salinization, acidification, pollution, compaction and waterlogging). Estimating total natural and human-induced soil degradation worldwide is fraught with difficulty given the variability in definitions, severity, timing, soil categorization, etc. However, it is generally agreed that its consequences have resulted in the loss of net primary production over large areas (Esch et al. 2017), thus restricting increases in arable and permanently cropped land to 13% in the four decades from the early 1960s to late 1990s (Bruinsma 2003). More importantly in relation to population growth during that time period, arable land per capita declined by about 40% (Conforti 2011). The term 'arable land' implies availability of adequate nutrients to support crop production. To counteract nutrient depletion, worldwide fertilizer consumption has risen from 90 kg/ha in 2002 to 135 kg in 2013 (Pocketbook 2015). Yet the increased use of fertilizers often results in excesses of nitrate and phosphates ending up in aquatic ecosystems (Bennett et al. 2001), causing algal blooms and eutrophication when decaying algal biomass consumes oxygen and limits the biodiversity of aquatic life. Large-scale nitrate and phosphate-induced environmental changes are particularly evident in watersheds and coastal zones.

Nitrogen, potassium and phosphorus are the three major nutrients essential for plant growth. Even though demand for phosphorus fertilizers continues to grow exponentially, rock phosphate reserves are limited and estimates suggest they will be depleted within 50–100 years (Cordell et al. 2011; Steen 1998; Van Vuuren et al. 2010). Additionally, anthropogenic nitrogen input is expected to drive terrestrial ecosystems towards greater phosphorous limitations, although a better understanding of the processes is critical (Deng et al. 2017; Goll et al. 2012; Zhu et al. 2016). Currently, there are no substitutes for phosphorus in agriculture, thus putting constraints on future agricultural productivity that relies on key fertilizer input of mined phosphate (Sverdrup and Ragnarsdottir 2011). The 'P-paradox', in other words, an excess of P impairing water quality, alongside its shortage as a depleting non-renewable resource, means that there must be substantial increases in recycling and efficiency of its use (Leinweber et al. 2018).

Modern intensive agricultural practices, such as the frequency and timing of tillage or no-till, application of herbicides and pesticides, and infrequent addition of organic matter containing micronutrients can alter soil structure and its microbial biodiversity such that the addition of fertilizers no longer increases productivity per hectare. Given that changes in land usage have resulted in losses of soil organic carbon estimated to be around 8%, and projected losses between 2010 and 2050 are 3.5 times that figure, it is assumed that soil water-holding capacity and nutrient losses will continue, especially in view of global warming (Esch et al. 2017). Obviously there are trade-offs between satisfying human needs and not compromising the ability of the biosphere to support life (Foley et al. 2005). However, it is clear when modelling planetary boundaries in relation to current land use practices that it is necessary to improve N and P cycling, principally by reducing both nitrogen and phosphorus emissions and runoff from agricultural land, but also by better capture and reuse (Conijn et al. 2018).

2.3.2 Aquaponics and Nutrients

One of the principal benefits of aquaponics is that it allows for the recycling of nutrient resources. Nutrient input into the fish component derives from feed, the composition of which depends on the target species, but feed in aquaculture typically constitutes a significant portion of input costs and can be more than half the total annual cost of production. In certain aquaponics designs, bacterial biomass can also be harnessed as feed, for instance, where biofloc production makes aquaponic systems increasingly self-contained (Pinho et al. 2017).

Wastewater from open-cage pens or raceways is often discharged into waterbodies, where it results in nutrient pollution and subsequent eutrophication. By contrast, aquaponic systems take the dissolved nutrients from uneaten fish feed and faeces, and utilizing microbes that can break down organic matter, convert the nitrogen and phosphorous into bioavailable forms for use by plants in the hydroponics unit. In order to achieve economically acceptable plant production levels, the presence of appropriate microbial assemblages reduces the need to add much of the supplemental nutrients that are routinely used in stand-alone hydroponic units. Thus aquaponics is a near-zero discharge system that offers not only economic benefit from both fish and plant production streams, but also significant reductions in both environmentally noxious discharges from aquaculture sites. It also eliminates the problem of N- and P-rich runoff from fertilizers used in soil-based agriculture. In decoupled aquaponic systems, aerobic or anaerobic bioreactors can also used to treat sludge and recover significant macro- and micronutrients in bioavailable forms for subsequent use in hydroponic production (Goddek et al. 2018) (see Chap. 8). Exciting new developments such as these, many of which are now being realized for commercial prodution, continue to refine the circular economy concept by increasingly allowing for nutrient recovery.

2.4 Pest, Weed and Disease Control

2.4.1 Predictions

It is generally recognized that control of diseases, pests and weeds is a critical component of curbing production losses that threaten food security (Keating et al. 2014). In fact, increasing the use of antibiotics, insecticides, herbicides and fungicides to cut losses and enhance productivity has allowed dramatic increases in agricultural output in the latter half of the twentieth century. However, these practices are also linked to a host of problems: pollution from persistent organic compounds in soils and irrigation water, changes in rhizobacterial and mycorrhizal activity in soils, contamination of crops and livestock, development of resistant strains, detrimental effects on pollinators and a wide range of human health risks (Bringezu et al. 2014; Ehrlich and Harte 2015a; Esch et al. 2017; FAO 2015b). Tackling pest, weed and disease control in ways that reduce the use of these substances is mentioned in virtually every call to provide food security for a growing world population.

2.4.2 Control of Pests, Weeds and Diseases

As a closed system with biosecurity measures, aquaponic systems require far fewer chemical pesticide applications in the plant component. If seed and transplant stocks are carefully handled and monitored, weed, fungal and bacterial/algal contaminants can be controlled in hydroponic units with targeted measures rather than the widespread preventive application of herbicides and fungicides prevalent in soil-based agriculture. As technology continues to advance, developments such as positive pressure greenhouses can further reduce pest problems (Mears and Both 2001). Design features to reduce pest risks can cut costs in terms of chemicals, labour, application time and equipment, especially since the land footprint of industrial-scale aquaponics systems is small, and systems are compact and tightly contained, as compared to the equivalent open production area of vegetable and fruit crops of conventional soil-based farms.

The use of RAS in aquaponic systems also prevents disease transmissions between farmed stocks and wild populations, which is a pressing concern in flow-through and open-net pen aquaculture (Read et al. 2001; Samuel-Fitwi et al. 2012). Routine antibiotic use is generally not required in the RAS component, since it is a closed system with few available vectors for disease introduction. Furthermore, the use of antimicrobials and antiparasitics is generally discouraged, as it can be detrimental to the microbiota that are crucial for converting organic and inorganic wastes into usable compounds for plant growth in the hydroponic unit (Junge et al. 2017). If disease does emerge, containment of both fish and plants from the surrounding environment makes decontamination and eradication more manageable. Although closed systems clearly do not completely alleviate all disease and pest problems (Goddek et al. 2015), proper biocontrol measures that are already practised in stand-alone RAS and hydroponics result in significant reductions of risk. These issues are discussed in further detail in subsequent chapters (for fish, see Chap. 6; for plants, further details in Chap. 14).

2.5 Water Resources

2.5.1 Predictions

In addition to requiring fertilizer applications, modern intensive agricultural practices also place high demands on water resources. Among biochemical flows (Fig. 2.1), water scarcity is now believed to be one of the most important factors constraining food production (Hoekstra et al. 2012; Porkka et al. 2016). Projected global population increases and shifts in terrestrial water availability due to climate change, demand more efficient use of water in agriculture. As noted previously, by 2050, aggregate agriculture production will need to produce 60% more food globally (Alexandratos and Bruinsma 2012), with an estimated 100% more in developing countries, based on population growth and rising expectations for standards of living (Alexandratos and Bruinsma 2012; WHO 2015). Famine in some regions of the world, as well as malnutrition and hidden hunger, indicates that the balance between food demand and availability has already reached critical levels, and that food and water security are directly linked (McNeill et al. 2017). Climate change predictions suggest reduced freshwater availability, and a corresponding decrease in agricultural yields by the end of the twenty-first century (Misra 2014).

The agriculture sector currently accounts for roughly 70% of the freshwater use worldwide, and the withdrawal rate even exceeds 90% in most of the world's least-developed countries. Water scarcity will increase in the next 25 years due to expected population growth (Connor et al. 2017; Esch et al. 2017), with the latest

Fig. 2.1 Water footprint (L per kg). Fish in RAS systems use the least water of any food production system

modelling forecasting declining water availability in the near future for nearly all countries (Distefano and Kelly 2017). The UN predicts that the pursuit of business-as-usual practices will result in a global water deficit of 40% by 2030 (Water 2015). In this respect, as groundwater supplies for irrigation are depleted or contaminated, and arid regions experience more drought and water shortages due to climate change, water for agricultural production will become increasingly valuable (Ehrlich and Harte 2015a). Increasing scarcity of water resources compromises not only water security for human consumption but also global food production (McNeill et al. 2017). Given that water scarcity is expected even in areas that currently have relatively sufficient water resources, it is important to develop agricultural techniques with low water input requirements, and to improve ecological management of wastewater through better reuse (FAO 2015a).

The UN World Water Development Report for 2017 (Connor et al. 2017) focuses on wastewater as an untapped source of energy, nutrients and other useful by-products, with implications not only for human and environmental health but also for food and energy security as well as climate change mitigation. This report calls for appropriate and affordable technologies, along with legal and regulatory frameworks, financing mechanisms and increased social acceptability of wastewater treatment, with the goal of achieving water reuse within a circular economy. The report also points to a 2016 World Economic Forum report that lists the water crisis as the global risk of highest concern in the next 10 years.

The concept of a water footprint as a measure of humans' use of freshwater resources has been put forwards in order to inform policy development on water use. A water footprint has three components: (1) blue water, which comprises the surface and groundwater consumed while making products or lost through evaporation, (2) green water that is rainwater used especially in crop production and (3) grey water, which is water that is polluted but still within existing water quality standards (Hoekstra and Mekonnen 2012). These authors mapped water footprints of countries worldwide and found that agricultural production accounts for 92% of global freshwater use, and industrial production uses 4.4% of the total, while domestic water only 3.6%. This raises concerns about water availability and has resulted in public education efforts aimed at raising awareness about the amounts of water required to produce various types of food, as well as national vulnerabilities, especially in water-scarce countries in North Africa and the Middle East.

2.5.2 Aquaponics and Water Conservation

The economic concept of comparative productivity measures the relative amount of a resource needed to produce a unit of goods or services. Efficiency is generally construed to be higher when the requirement for resource input is lower per unit of goods and services. However, when water-use efficiency is examined in an environmental context, water quality also needs to be taken into account, because maintaining or enhancing water quality also enhances productivity (Hamdy 2007).

The growing problem of water scarcity demands improvements in water-use efficiencies especially in arid and semiarid regions, where availability of water for agriculture, and water quality of discharge, are critical factors in food production. In these regions, recirculation of water in aquaponic units can achieve remarkable water re-use efficiency of 95–99% (Dalsgaard et al. 2013). Water demand is also less than 100 L/kg of fish harvested, and water quality is maintained within the system for production of crops (Goddek et al. 2015). Obviously, such systems must be constructed and operated to minimize water losses; they must also optimize their ratios of fish water to plants, as this ratio is very important in maximizing water re-use efficiency and ensuring maximal nutrient recycling. Modelling algorithms and technical solutions are being developed to integrate improvements in individual units, and to better understand how to effectively and efficiently manage water (Vilbergsson et al. 2016). Further information is provided in Chaps. 9 and 11.

In light of soil, water and nutrient requirements, the water footprint of aquaponic systems is considerably better than traditional agriculture, where water quality and demand, along with availability of arable land, costs of fertilizers and irrigation are all constraints to expansion (Fig. 2.1).

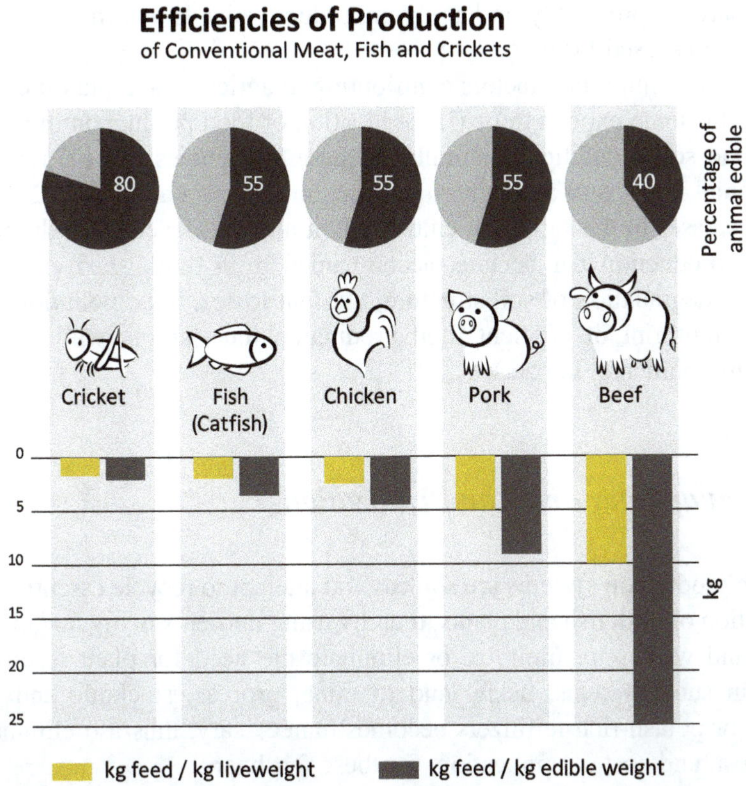

Fig. 2.2 Feed conversion ratios (FCRs) based as kg of feed per live weight and kg of feed for edible portion. Only insects, which are eaten whole in some parts of the world, have a better FCR than fish

2.6 Land Utilization

2.6.1 Predictions

Globally, land-based crops and pasture occupy approximately 33% of total available land, and expansion for agricultural uses between 2000 and 2050 is estimated to increase by 7–31% (350–1500 Mha, depending on source and underlying assumptions), most often at the expense of forests and wetlands (Bringezu et al. 2014). While there is currently still land classed as 'good' or 'marginal' that is available for rain-fed agriculture, significant portions of it are far from markets, lack infrastructure or have endemic diseases, unsuitable terrain or other conditions that limit development potential. In other cases, remaining lands are already protected, forested or developed for other uses (Alexandratos and Bruinsma 2012). By contrast, dryland ecosystems, defined in the UN's Commission on Sustainable Development as arid, semiarid and dry subhumid areas that typically have low productivity, are threatened by desertification and are therefore unsuitable for agricultural expansion but nevertheless have many millions of people living in close proximity (Economic 2007). These facts point to the need for more sustainable intensification of food production closer to markets, preferably on largely unproductive lands that may never become suitable for soil-based farming.

The two most important factors contributing to agricultural input efficiencies are considered by some experts to be (i) the location of food production in areas where climatic (and soil) conditions naturally increase efficiencies and (ii) reductions in environmental impacts of agricultural production (Michael and David 2017). There must be increases in the supply of cultivated biomass achieved through the intensification of production per hectare, accompanied by a diminished environmental burden (e.g. degradation of soil structure, nutrient losses, toxic pollution). In other words, the footprint of efficient food production must shrink while minimizing negative environmental impacts.

2.6.2 *Aquaponics and Land Utilization*

Aquaponic production systems are soilless and attempt to recycle essential nutrients for cultivation of both fish and plants, thereby using nutrients in organic matter from fish feed and wastes to minimize or eliminate the need for plant fertilizers. For instance, in such systems, using land to mine, process, stockpile and transport phosphate or potash-rich fertilizers becomes unnecessary, thus aso eliminating the inherent cost, and cost of application, for these fertilizers.

Aquaponics production contributes not only to water usage efficiency (Sect. 2.5.2) but also to agricultural input efficiency by reducing the land footprint needed for production. Facilities for instance, can be situated on nonarable land and in suburban or urban areas closer to markets, thus reducing the carbon footprint associated with rural farms and transportation of products to city markets. With a smaller footprint, production capacity can be located in otherwise unproductive areas such as on rooftops or old factory sites, which can also reduce land acquisition costs if those areas are deemed unsuitable for housing or retail businesses. A smaller footprint for production of high-quality protein and vegetables in aquaponics can also take pressure away from clearing ecologically valuable natural and semi-natural areas for conventional agriculture.

2.7 Energy Resources

2.7.1 *Predictions*

As mechanization spreads globally, open-field intensive agriculture increasingly relies heavily on fossil fuels to power farm machinery and for transportation of fertilizers as well as farm products, as well as to run the equipment for processing, packaging and storage. In 2010, the OECD International Energy Agency predicted that global energy consumption would grow by up to 50% by 2035; the FAO has also estimated that 30% of global energy consumption is devoted to food production and its supply chain (FAO 2011). Greenhouse gas (GHG) emissions associated with fossil fuels (approximately 14% in lifecycle analysis) added to those from fertilizer manufacturing (16%) and nitrous oxide from average soils (44%) (Camargo et al. 2013), all contribute substantially to the environmental impacts of farming. A trend in the twenty-first century to produce crop-based biofuels (e.g. corn for ethanol) to replace fossil fuels has increased pressure on the clearing of rainforests, peatlands, savannas and grasslands for agricultural production. However, studies point to creation of a 'carbon debt' from such practices, since the overall release of CO_2 exceeds the reductions in GHGs they provide by displacing fossil fuels (Fargione et al. 2008). Arguably a similar carbon debt exists when clearing land to raise food crops via conventional agriculture that relies on fossil fuels.

In a comparative analysis of agricultural production systems, trawling fisheries and recirculating aquaculture systems (RAS) were found to emit GHGs 2–2.5 times that of non-trawling fisheries and non-RAS (pen, raceway) aquaculture. In RAS, these energy requirements relate primarily to the functioning of pumps and filters (Michael and David 2017). Similarly, greenhouse production systems can emit up to three times more GHGs than open-field crop production if energy is required to maintain heat and light within optimal ranges (ibid.). However, these GHG figures do not take into account other environmental impacts of non-RAS systems, such as eutrophication or potential pathogen transfers to wild stocks. Nor do they consider GHG from the production, transportation and application of herbicides and pesticides used in open-field cultivation, nor methane and nitrous oxide from associated livestock production, both of which have a 100-year greenhouse warming potential (GWP) 25 and 298 times that of CO_2, respectively (Camargo et al. 2013; Eggleston et al. 2006).

These sobering estimates of present and future energy consumption and GHG emissions associated with food production have prompted new modelling and approaches, for example, the UN's water-food-energy nexus approach mentioned in Sect. 2.1. The UN's Sustainable Development Goals have pinpointed the vulnerability of food production to fluctuations in energy prices as a key driver of food insecurity. This has prompted efforts to make agrifood systems 'energy smart' with an emphasis on improving energy efficiencies, increasing use of renewable energy sources and encouraging integration of food and energy production (FAO 2011).

2.7.2 Aquaponics and Energy Conservation

Technological advances in aquaponic system operations are moving towards being increasingly 'energy smart' and reducing the carbon debt from pumps, filters and heating/cooling devices by using electricity generated from renewable sources. Even in temperate latitudes, many new designs allow the energy involved in heating and cooling of fish tanks and greenhouses to be fully reintegrated, such that these systems do not require inputs beyond solar arrays or the electricity/heat generated from bacterial biogas production of aquaculture-derived sludge (Ezebuiro and Körner 2017; Goddek and Keesman 2018; Kloas et al. 2015; Yogev et al. 2016). In addition, aquaponic systems can use microbial denitrification to convert nitrous oxide to nitrogen gas if enough carbon sources from wastes are available, such that heterotrophic and facultative anaerobic bacteria can convert excess nitrates to nitrogen gas (Van Rijn et al. 2006). As noted in Sect. 2.7.1, nitrous oxide is a potent GHG and microbes already present in closed aquaponics systems can facilitate its conversion into nitrogen gas.

2.8 Summary

As the human population continues to increase, there is increasing demand for high-quality protein worldwide. Compared to meat sources, fish are widely recognized as being a particularly healthy source of protein. In relation to the world food supply, aquaculture now provides more fish protein than capture fisheries (FAO 2016). Globally, human per capita fish consumption continues to rise at an annual average rate of 3.2% (1961–2013), which is double the rate of population growth. In the period from 1974 to 2013, biologically unsustainable 'overfishing' has increased by 22%. During the same period, the catch from what are deemed to be 'fully exploited' fisheries has decreased by 26%. Aquaculture therefore provides the only possible solution for meeting increased market demand. It is now the fastest growing food sector and therefore an important component of food security (ibid.)

With the global population estimated to reach 8.3–10.9 billion people by 2050 (Bringezu et al. 2014), sustainable development of the aquaculture and agricultural sectors requires optimization in terms of production efficiency, but also reductions in utilization of limited resources, in particular, water, land and fertilizers. The benefits of aquaponics relate not just to the efficient uses of land, water and nutrient resources but also allow for increased integration of smart energy opportunities such as biogas and solar power. In this regard, aquaponics is a promising technology for producing both high-quality fish protein and vegetables in ways that can use substantially less land, less energy and less water while also minimizing chemical and fertilizer inputs that are used in conventional food production.

References

Alexandratos N, Bruinsma J (2012) World agriculture towards 2030/50: the 2012 revision, ESA Work. Paper 12–03. UN Food Agriculture Organization (FAO), Rome

Barnosky AD, Hadly EA, Bascompte J, Berlow EL, Brown JH, Fortelius M, Getz WM, Harte J, Hastings A, Marquet PA (2012) Approaching a state shift in Earth/'s biosphere. Nature 486:52–58

Bennett EM, Carpenter SR, Caraco NF (2001) Human impact on erodable phosphorus and eutrophication: a global perspective: increasing accumulation of phosphorus in soil threatens rivers, lakes, and coastal oceans with eutrophication. AIBS Bull 51:227–234

Bringezu S, Schütz H, Pengue W, OBrien M, Garcia F, Sims R (2014) Assessing global land use: balancing consumption with sustainable supply. UNEP/International Resource Panel, Nairobi/Paris

Bruinsma J (2003) World agriculture: towards 2015/2030: an FAO perspective. Earthscan, London

Camargo GG, Ryan MR, Richard TL (2013) Energy use and greenhouse gas emissions from crop production using the farm energy analysis tool. Bioscience 63:263–273

Conforti P (2011) Looking ahead in world food and agriculture: perspectives to 2050. Food and Agriculture Organization of the United Nations (FAO), Rome

Conijn J, Bindraban P, Schröder J, Jongschaap R (2018) Can our global food system meet food demand within planetary boundaries? Agric Ecosyst Environ 251:244–256

Connor R, Renata A, Ortigara C, Koncagül E, Uhlenbrook S, Lamizana-Diallo BM, Zadeh SM, Qadir M, Kjellén M, Sjödin J (2017) The United Nations world water development report 2017. In: Wastewater: the untapped resource, The United Nations world water development report. UNESCO, Paris

Cordell D, Rosemarin A, Schröder J, Smit A (2011) Towards global phosphorus security: a systems framework for phosphorus recovery and reuse options. Chemosphere 84:747–758

Dalsgaard J, Lund I, Thorarinsdottir R, Drengstig A, Arvonen K, Pedersen PB (2013) Farming different species in RAS in Nordic countries: current status and future perspectives. Aquac Eng 53:2–13

Deng Q, Hui D, Dennis S, Reddy KC (2017) Responses of terrestrial ecosystem phosphorus cycling to nitrogen addition: a meta-analysis. Glob Ecol Biogeogr 26:713–728

Distefano T, Kelly S (2017) Are we in deep water? Water scarcity and its limits to economic growth. Ecol Econ 142:130–147

Economic UNDo (2007) Indicators of sustainable development: guidelines and methodologies. United Nations Publications, New York

Eggleston H, Buendia L, Miwa K, Ngara T, Tanabe K (2006) IPCC guidelines for national greenhouse gas inventories. Inst Glob Environ Strateg, Hayama, Japan 2:48–56

Ehrlich PR, Harte J (2015a) Food security requires a new revolution. Int J Environ Stud 72:908–920

Ehrlich PR, Harte J (2015b) Opinion: to feed the world in 2050 will require a global revolution. Proc Natl Acad Sci 112:14743–14744

Esch Svd, Brink Bt, Stehfest E, Bakkenes M, Sewell A, Bouwman A, Meijer J, Westhoek H, Berg Mvd, Born GJvd (2017) Exploring future changes in land use and land condition and the impacts on food, water, climate change and biodiversity: scenarios for the UNCCD Global Land Outlook. PBL Netherlands Environmental Assessment Agency, The Hague

Ezebuiro NC, Körner I (2017) Characterisation of anaerobic digestion substrates regarding trace elements and determination of the influence of trace elements on the hydrolysis and acidification phases during the methanisation of a maize silage-based feedstock. J Environ Chem Eng 5:341–351

FAO (2011) Energy-smart food for people and climate. Food and Agriculture Organization of the United Nations, Rome

FAO (2015a) Environmental and social management guidelines. Food and Agriculture Organization of the United Nations, Rome

FAO (2015b) Statistical pocketbook 2015. Food and Agriculture Organization of the United Nations, Rome

FAO (2016) The state of world fisheries and aquaculture 2016. Contributing to food security and nutrition for all. Food and Agriculture Organization of the United Nations, Rome, p 200

Fargione J, Hill J, Tilman D, Polasky S, Hawthorne P (2008) Land clearing and the biofuel carbon debt. Science 319:1235–1238

Foley JA, DeFries R, Asner GP, Barford C, Bonan G, Carpenter SR, Chapin FS, Coe MT, Daily GC, Gibbs HK (2005) Global consequences of land use. Science 309:570–574

Goddek S, Keesman KJ (2018) The necessity of desalination technology for designing and sizing multi-loop aquaponics systems. Desalination 428:76–85

Goddek S, Delaide B, Mankasingh U, Ragnarsdottir KV, Jijakli H, Thorarinsdottir R (2015) Challenges of sustainable and commercial aquaponics. Sustainability 7:4199–4224

Goddek S, Delaide BPL, Joyce A, Wuertz S, Jijakli MH, Gross A, Eding EH, Bläser I, Reuter M, Keizer LCP, Morgenstern R, Körner O, Verreth J, Keesman KJ (2018) Nutrient mineralization and organic matter reduction performance of RAS-based sludge in sequential UASB-EGSB reactors. Aquac Eng 83:10–19. https://doi.org/10.1016/J.AQUAENG.2018.07.003

Goll DS, Brovkin V, Parida B, Reick CH, Kattge J, Reich PB, Van Bodegom P, Niinemets Ü (2012) Nutrient limitation reduces land carbon uptake in simulations with a model of combined carbon, nitrogen and phosphorus cycling. Biogeosciences 9:3547–3569

Hamdy A (2007) Water use efficiency in irrigated agriculture: an analytical review. Water use efficiency and water productivity: WASAMED project, pp 9–19

Herrero M, Thornton PK, Power B, Bogard JR, Remans R, Fritz S, Gerber JS, Nelson G, See L, Waha K (2017) Farming and the geography of nutrient production for human use: a transdisciplinary analysis. Lancet Planetary Health 1:e33–e42

Hoekstra AY, Mekonnen MM (2012) The water footprint of humanity. Proc Natl Acad Sci 109:3232–3237

Hoekstra AY, Mekonnen MM, Chapagain AK, Mathews RE, Richter BD (2012) Global monthly water scarcity: blue water footprints versus blue water availability. PLoS One 7:e32688

Junge R, König B, Villarroel M, Komives T, Jijakli MH (2017) Strategic points in aquaponics. Water 9:182

Keating BA, Herrero M, Carberry PS, Gardner J, Cole MB (2014) Food wedges: framing the global food demand and supply challenge towards 2050. Glob Food Sec 3:125–132

Kloas W, Groß R, Baganz D, Graupner J, Monsees H, Schmidt U, Staaks G, Suhl J, Tschirner M, Wittstock B, Wuertz S, Zikova A, Rennert B (2015) A new concept for aquaponic systems to improve sustainability, increase productivity, and reduce environmental impacts. Aquac Environ Interact 7:179–192

Leinweber P, Bathmann U, Buczko U, Douhaire C, Eichler-Löbermann B, Frossard E, Ekardt F, Jarvie H, Krämer I, Kabbe C (2018) Handling the phosphorus paradox in agriculture and natural ecosystems: scarcity, necessity, and burden of P. Ambio 47:3–19

McNeill K, Macdonald K, Singh A, Binns AD (2017) Food and water security: analysis of integrated modeling platforms. Agric Water Manag 194:100–112

Mears D, Both A (2001) A positive pressure ventilation system with insect screening for tropical and subtropical greenhouse facilities. Int Symp Des Environ Control Trop Subtrop Greenh 578:125–132

Michael C, David T (2017) Comparative analysis of environmental impacts of agricultural production systems, agricultural input efficiency, and food choice. Environ Res Lett 12:064016

Misra AK (2014) Climate change and challenges of water and food security. Int J Sustain Built Environ 3:153–165

Pinho SM, Molinari D, de Mello GL, Fitzsimmons KM, Emerenciano MGC (2017) Effluent from a biofloc technology (BFT) tilapia culture on the aquaponics production of different lettuce varieties. Ecol Eng 103:146–153

Pocketbook FS (2015) World food and agriculture (2015). Food and Agriculture Organization of the United Nations, Rome

Porkka M, Gerten D, Schaphoff S, Siebert S, Kummu M (2016) Causes and trends of water scarcity in food production. Environ Res Lett 11:015001

Rask KJ, Rask N (2011) Economic development and food production–consumption balance: a growing global challenge. Food Policy 36:186–196

Read P, Fernandes T, Miller K (2001) The derivation of scientific guidelines for best environmental practice for the monitoring and regulation of marine aquaculture in Europe. J Appl Ichthyol 17:146–152

Ridoutt BG, Sanguansri P, Nolan M, Marks N (2012) Meat consumption and water scarcity: beware of generalizations. J Clean Prod 28:127e133

Samuel-Fitwi B, Wuertz S, Schroeder JP, Schulz C (2012) Sustainability assessment tools to support aquaculture development. J Clean Prod 32:183–192

Schmidhuber J (2010) FAO's long-term outlook for global agriculture–challenges, trends and drivers. International Food & Agriculture Trade Policy Council

Scott CA, Kurian M, Wescoat JL Jr (2015) The water-energy-food nexus: enhancing adaptive capacity to complex global challenges, Governing the nexus. Springer, Cham, pp 15–38

Steen I (1998) Management of a non-renewable resource. Phosphorus Potassium 217:25–31

Sverdrup HU, Ragnarsdottir KV (2011) Challenging the planetary boundaries II: assessing the sustainable global population and phosphate supply, using a systems dynamics assessment model. Appl Geochem 26:S307–S310

Thomas, R., Reed, M., Clifton, K., Appadurai, A., Mills, A., Zucca, C., Kodsi, E., Sircely, J., Haddad, F., vonHagen, C., 2017. Scaling up sustainable land management and restoration of degraded land

Van Rijn J, Tal Y, Schreier HJ (2006) Denitrification in recirculating systems: theory and applications. Aquac Eng 34:364–376

Van Vuuren DP, Bouwman AF, Beusen AH (2010) Phosphorus demand for the 1970–2100 period: a scenario analysis of resource depletion. Glob Environ Chang 20:428–439

Vilbergsson B, Oddsson GV, Unnthorsson R (2016) Taxonomy of means and ends in aquaculture production–part 2: the technical solutions of controlling solids, dissolved gasses and pH. Water 8:387

Water U (2015) Water for a sustainable world, The United Nations world water development report. United Nations Educational, Scientific and Cultural Organization, Paris

WHO (2015) Progress on sanitation and drinking water: 2015 update and MDG assessment. World Health Organization, Geneva

Xue X, Landis AE (2010) Eutrophication potential of food consumption patterns. Environ Sci Technol 44:6450–6456

Yogev U, Barnes A, Gross A (2016) Nutrients and energy balance analysis for a conceptual model of a three loops off grid, aquaponics. Water 8:589

Zhu Q, Riley W, Tang J, Koven C (2016) Multiple soil nutrient competition between plants, microbes, and mineral surfaces: model development, parameterization, and example applications in several tropical forests. Biogeosciences 13:341

3

Recirculating Aquaculture Systems: Past and Future

Carlos A. Espinal and Daniel Matulić

Abstract Recirculating aquaculture technology, which includes aquaponics, has been under development for the past 40 years from a combination of technologies derived from the wastewater treatment and aquaculture sectors. Until recently, recirculating aquaculture systems (RAS) farms have been relatively small compared with other types of modern aquaculture production. The last two decades have seen a significant increase in the development of this technology, with increased market acceptance and scale. This chapter provides a brief overview of the history, water quality control processes, new developments and ongoing challenges of RAS.

Keywords Recirculating aquaculture systems (RAS) · Wastewater treatment · Biofilter · Denitrification · Membrane technology

3.1 Introduction

Recirculating aquaculture systems (RAS) describe intensive fish production systems which use a series of water treatment steps to depurate the fish-rearing water and facilitate its reuse. RAS will generally include (1) devices to remove solid particles from the water which are composed of fish faeces, uneaten feed and bacterial flocs (Chen et al. 1994; Couturier et al. 2009), (2) nitrifying biofilters to oxidize ammonia excreted by fish to nitrate (Gutierrez-Wing and Malone 2006) and (3) a number of gas exchange devices to remove dissolved carbon dioxide expelled by the fish as well as/or adding oxygen required by the fish and nitrifying bacteria (Colt and Watten 1988; Moran 2010; Summerfelt 2003; Wagner et al. 1995). In addition,

C. A. Espinal (✉)
Landing Aquaculture, Oirschot, The Netherlands
e-mail: carlos@landingaquaculture.com

D. Matulić
Department of Fisheries, Beekeeping, Game management and Special Zoology, Faculty of Agriculture, University of Zagreb, Zagreb, Croatia
e-mail: dmatulic@agr.hr

RAS may also use UV irradiation for water disinfection (Sharrer et al. 2005; Summerfelt et al. 2009), ozonation and protein skimming for fine solids and microbial control (Attramadal et al. 2012a; Gonçalves and Gagnon 2011; Summerfelt and Hochheimer 1997) and denitrification systems to remove nitrate (van Rijn et al. 2006).

Modern recirculating aquaculture technology has been developing for more than 40 years, but novel technologies increasingly offer ways to change the paradigms of traditional RAS including improvements on classic processes such as solids capture, biofiltration and gas exchange. RAS has also experienced important developments in terms of scale, production capacities and market acceptance, with systems becoming progressively larger and more robust.

This chapter discusses how RAS technology has developed over the past two decades from a period of technological consolidation to a new era of industrial implementation.

3.1.1 History of RAS

The earliest scientific research on RAS conducted in Japan in the 1950s focused on biofilter design for carp production driven by the need to use locally limited water resources more productively (Murray et al. 2014). In Europe and the United States, scientists similarly attempted to adapt technologies developed for domestic wastewater treatment in order to better reuse water within recirculating systems (e.g. activated sludge processes for sewage treatment, trickling, submerged and down-flow biofilters and several mechanical filtration systems). These early efforts included primarily work on marine systems for fish and crustacean production, but were soon adopted in arid regions where the agriculture sector is restricted by water supply. In aquaculture, different solutions have been designed to maximize water use including highly intensive recirculating systems that incorporate water filtration systems such as drum filters, biological filters, protein skimmers and oxygen injection systems (Hulata and Simon 2011). Despite a strong conviction by pioneers in the industry about the commercial viability of their work, most of the early studies focused exclusively on the oxidation of toxic inorganic nitrogen wastes derived from protein metabolism. The trust in technology was reinforced by the successful operation of public as well as domestic aquaria, which generally feature over-sized treatment units to ensure crystal-clear water. Additionally, extremely low stocking densities and associated feed inputs meant that such over-engineering still made a relatively small contribution to capital and operational costs of the system compared to intensive RAS. Consequently, the changes in process dynamics associated with scale-change were unaccounted for, resulting in the under-sizing of RAS treatment units in order to minimize capital costs. As a consequence, safety margins were far too narrow or non-existent (Murray et al. 2014). Because many of the pioneering scientists had biological rather than engineering backgrounds, technical improvements were also constrained by miscommunications between scientists, designers, construction personnel and operators. The development of a standardized terminology, units of measurement and reporting formats in 1980 (EIFAC/ICES 1980), helped address the situation, though regional differences still persisted. It was not until the mid-1980s that cyclic water quality parameters became well recognized as being important in pond production, e.g. periodically measuring the concentrations of pH, oxygen, TAN (total ammonia nitrogen), NO2 (nitrate), BOD (biochemical oxygen demand) and COD (chemical oxygen demand).

In the latter part of the last century, numerous articles were published on the early development of RAS. Rosenthal (1980) elaborated on the state of recirculation systems in Western Europe, while Bovendeur et al. (1987) developed a water recirculation system for the culture of African catfish in relation to waste production and waste removal kinetics (a design was presented for a water treatment system consisting of a primary clarifier and an aerobic fixed-film reactor that demonstrated satisfactory results for high-density culture of African catfish). This work was part of the rapid development in fish culture systems up to the mid-1990s in Northern and Western Europe (Rosenthal and Black 1993), as well as in North America (Colt 1991). New classifications, such as the classification according to how water flows through an aquaculture system, provided key insights with respect to the water quality processes that are important for fish production (Krom and van Rijn 1989). In subsequent work by van Rijn (1996), concepts were introduced focused on the biological processes underlying the treatment systems. The conclusions from this work were that incorporating methods for reducing the accumulation of sludge and nitrate resulted in more stable water quality conditions within the culture units. During this period, RAS production increased significantly in volume and species diversity (Rosenthal 1980; Verreth and Eding 1993; Martins et al. 2005). Today, more than 10 species are produced in RAS (African catfish, eel and trout as major freshwater species and turbot, seabass and sole as major marine species) (Martins et al 2010b), with RAS also becoming a crucial element in the production of larvae and juveniles of diverse species.

While maximum sustainable yields of many aquatic wild stock species have been or will soon be reached, and many species are already overfished, RAS is considered a key technology that will help the aquaculture sector meet the needs for aquatic species over the coming decades (Ebeling and Timmons 2012).

3.1.2 A Short History of Aquaponics in the Context of RAS

Aquaponics is a term that has been 'coined' in the 1970s, but in practice has ancient roots – although there are still discussions about its first occurrence. The Aztec cultivated agricultural islands known as *chinampas* (the earliest 1150–1350CE), in a system considered by some to be the first form of aquaponics for agricultural use (Fig. 3.1). In such systems, plants were raised on stable, or sometime movable and floating islands placed in lake shallows wherein nutrient rich mud could be dredged

Fig. 3.1 Chinampas (floating gardens) in Central America – artificial island construction as antecedent of aquaponic technology. (From Marzolino/Shutterstock.com)

from the chinampa canals and placed on the islands to support plant growth (Crossley 2004).

An even earlier example of aquaponics started on the other side of the world in south China and is believed to have spread within South East Asia where Chinese settlers from Yunnan settled around 5 CE. Farmers cultivated and farmed rice in paddy fields in combination with fish (FAO 2001). These polycultural farming systems existed in many Far Eastern countries to raise fish such as oriental loach (*Misgurnus anguillicaudatus*) (Tomita-Yokotani et al. 2009), swamp eel (fam. Synbranchidae), common carp (*Cyprinus carpio*) and crucian carp (*Carassius carassius*) (FAO 2004). In essence, however, these were not aquaponic systems but can be best described as early examples of integrated aquaculture systems (Gomez 2011). In the twentieth century, the first attempts to create practical, efficient and integrated fish production systems alongside vegetables were made in the 1970s with the work of Lewis and Naegel (Lewis and Wehr 1976; Naegel 1977; Lewis et al. 1978). Further early systems were designed by Waten and Busch in 1984 and Rakocy in 1989 (Palm et al. 2018).

3.2 Review of Water Quality Control in RAS

RAS are complex aquatic production systems that involve a range of physical, chemical and biological interactions (Timmons and Ebeling 2010). Understanding these interactions and the relationships between the fish in the system and the equipment used is crucial to predict any changes in water quality and system performance. There are more than 40 water quality parameters than can be used to determine water quality in aquaculture (Timmons and Ebeling 2010). Of these, only a few (as described in Sects. 3.2.1, 3.2.2, 3.2.3, 3.2.4, 3.2.5, 3.2.6 and 3.2.7) are traditionally controlled in the main recirculation processes, given that these processes can rapidly affect fish survival and are prone to change with the addition of feed to the system. Many other water quality parameters are not normally monitored or controlled because (1) water quality analytics may be expensive, (2) the pollutant to be analysed can be diluted with daily water exchange, (3) potential water sources containing them are ruled out for use or (4) because their potential negative effects have not been observed in practice. Therefore, the following water quality parameters are normally monitored in RAS.

3.2.1 Dissolved Oxygen (DO)

Dissolved oxygen (DO) is generally the most important water quality parameter in intensive aquatic systems, as low DO levels may quickly result in high stress in fish, nitrifying biofilter malfunction and indeed significant fish losses. Commonly, stocking densities, feed addition, temperature and the tolerance of the fish species to hypoxia will determine the oxygen requirements of a system. As oxygen can be transferred to water in concentrations higher than its saturation concentration under atmospheric conditions (this is called supersaturation), a range of devices and designs exist to ensure that the fish are provided with sufficient oxygen.

In RAS, DO can be controlled via aeration, addition of pure oxygen, or a combination of these. Since aeration is only capable of raising the DO concentrations to the atmospheric saturation point, the technique is generally reserved for lightly loaded systems or systems with tolerant species such as tilapia or catfish. However, aerators are also an important component of commercial RAS where the use of expensive technical oxygen is reduced by aerating water with a low dissolved oxygen content back to the saturation point before supersaturating the water with technical oxygen.

There are several types of aerators and oxygenators that can be used in RAS and these fall within two broad categories: gas-to-liquid and liquid-to-gas systems (Lekang 2013). Gas-to-liquid aerators mostly comprise diffused aeration systems where gas (air or oxygen) is transferred to the water, creating bubbles which exchange gases with the liquid medium (Fig. 3.2). Other gas-to-liquid systems include passing gases through diffusers, perforated pipes or perforated plates to

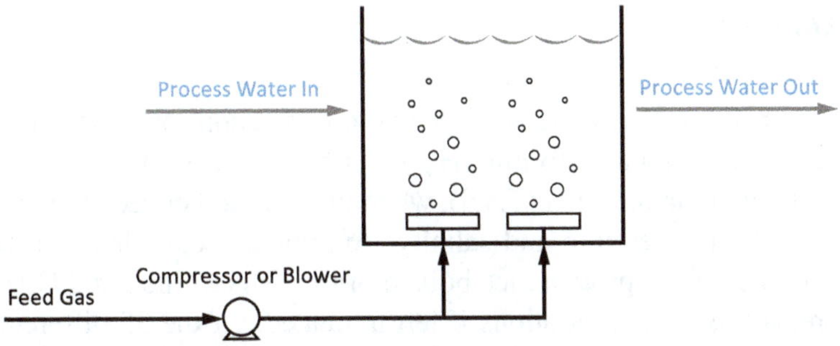

Fig. 3.2 Diagrams of two gas-to-liquid transfer examples: diffused aeration and Venturi injectors/aspirators

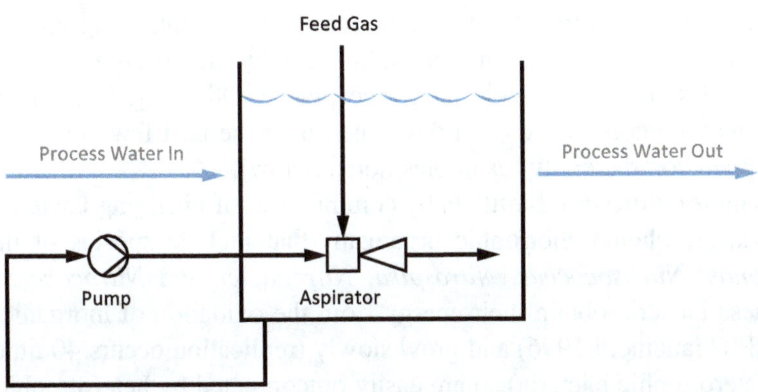

Fig. 3.3 Diagrams of two liquid-to-gas transfer examples: the packed column aerator and surface splashers in an enclosed tank. The packed column aerator allows water to trickle down an enclosed vessel, usually packed with structured media, where air is forced through using a fan or blower. Surface splashers found in pond aquaculture can also be used in enclosed atmospheres enriched with gases – normally oxygen – for gas transfer

create bubbles using Venturi injectors which create masses of small bubbles or devices which trap gas bubbles in the water stream such as the Speece Cone and the U-tube oxygenator.

Liquid-to-gas aerators are based on diffusing the water into small droplets to increase the surface area available for contact with the air, or creating an atmosphere enriched with a mixture of gases (Fig. 3.3). The packed column aerator (Colt and Bouck 1984) and the low-head oxygenators (LHOs) (Wagner et al. 1995) are examples of liquid-to-gas systems used in recirculating aquaculture. However, other liquid-to-gas systems popular in ponds and outdoor farms such as paddlewheel aerators (Fast et al. 1999) are also used in RAS.

Considerable literature is available on gas exchange theory and the fundamentals of gas transfer in water, and the reader is encouraged not only to consult aquaculture and aquaculture engineering texts, but also to refer to process engineering and wastewater treatment materials for a better understanding of these processes.

3.2.2 *Ammonia*

In an aqueous medium, ammonia exists in two forms: a non-ionized form (NH_3) that is toxic to fish and an ionized form (NH_4^+) that has low toxicity to fish. These two form the total ammonia nitrogen (TAN), wherein the ratio between the two forms is controlled by pH, temperature and salinity. Ammonia accumulates in the rearing water as a product of the protein metabolism of the fish (Altinok and Grizzle 2004) and can achieve toxic concentrations if left untreated. Of the 35 different types of freshwater fish that have been studied, the average acute toxicity value for ammonia is 2.79 mg NH3/l (Randall and Tsui 2002).

Ammonia has been traditionally treated in recirculation systems with nitrifying biofilters, devices that are designed to promote microbial communities that can oxidize ammonia into nitrate (NO_3). Although the use of nitrifying biofilters is not new, contemporary RAS has seen a streamlining of biofilter designs, with just a few, well-studied designs having widespread acceptance. Other highly innovative techniques to treat ammonia have been developed over the past few years, but are not widely applied commercially (examples noted below).

Ammonia is oxidized in biofilters by communities of nitrifying bacteria. Nitrifying bacteria are chemolithotrophic organisms that include species of the genera *Nitrosomonas*, *Nitrosococcus*, *Nitrospira*, *Nitrobacter* and *Nitrococcus* (Prosser 1989). These bacteria obtain their energy from the oxidation of inorganic nitrogen compounds (Mancinelli 1996) and grow slowly (replication occurs 40 times slower than for heterotrophic bacteria) so are easily outcompeted by heterotrophic bacteria if organic carbon, mostly present in biosolids suspended in the culture water, are allowed to accumulate (Grady and Lim 1980). During RAS operation, good system management greatly relies on minimizing suspended solids through adequate solids removal techniques (Fig. 3.4).

Nitrifying biofilters or biofilter reactors have been roughly classified into two main categories: suspended growth and attached growth systems (Malone and Pfeiffer 2006). In suspended growth systems, the nitrifying bacterial communities grow freely in the water, forming bacterial flocs which also harbour rich ecosystems where protozoa, ciliates, nematodes and algae are present (Manan et al. 2017). With appropriate mixing and aeration, algae, bacteria, zooplankton, feed particles and faecal matter remain suspended in the water column and naturally flocculate together, forming the particles that give biofloc culture systems their name (Browdy et al. 2012). The main disadvantage of suspended growth systems is their tendency to lose their bacterial biomass as process water flows out of the reactor, thus requiring a means to capture and return it to the system. In attached growth systems, solid forms (sand grains, stones, plastic elements) are used as substrates to retain the bacteria inside the reactor and thus, do not need a post-treatment solids capture step. Generally, attached growth systems provide more surface area for bacterial attachment than suspended growth systems, and do not produce significant solids in their outflow, which is one of the main reasons why attached growth biofilters have been so commonly used in RAS.

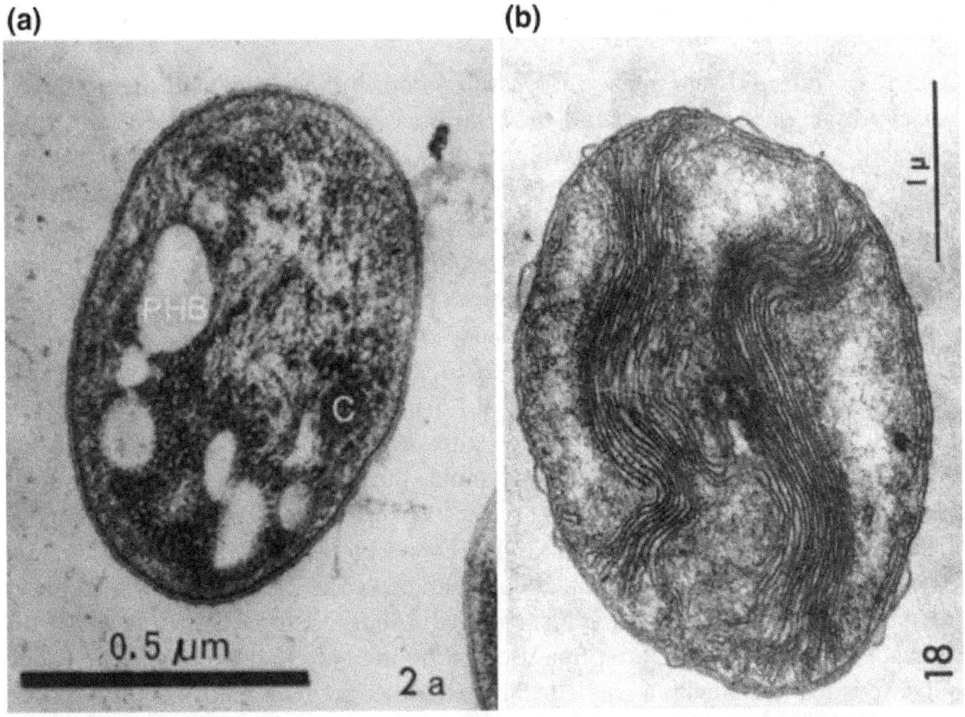

Fig. 3.4 Nitrifying bacteria *Nitrosomonas* (left), and *Nitrobacter* (right). (Left photo: Bock et al. 1983. Right photo: Murray and Watson 1965)

Efforts have been made to classify biofilters and to document their performance in order to help farmers and designers specify systems with a better degree of reliability (Drennan et al. 2006; Gutierrez-Wing and Malone 2006). In recent years, the aquaculture industry has opted for biofilter designs which have been widely studied and thus can offer predictable performance. The moving bed bioreactor (Rusten et al. 2006), the fluidized sand filter bioreactor (Summerfelt 2006) and the fixed-bed bioreactor (Emparanza 2009; Zhu and Chen 2002) are examples of attached growth biofilter designs which have become standard in modern commercial RAS. Trickling filters (Díaz et al. 2012), another popular design, have seen their popularity reduced due to their relatively high pumping requirements and relatively large sizes.

3.2.3 *Biosolids*

Biosolids in RAS originate from fish feed, faeces and biofilms (Timmons and Ebeling 2010) and are one of the most critical and difficult water quality parameters to control. As biosolids serve as a substrate for heterotrophic bacterial growth, an increase in their concentration may eventually result in increased oxygen consumption, poor biofilter performance (Michaud et al. 2006), increased water turbidity and even mechanical blockage of parts of the system (Becke et al. 2016; Chen et al. 1994; Couturier et al. 2009).

In RAS, biosolids are generally classified both by their size and their removal capacity by certain techniques. Of the total fraction of solids produced in a RAS, settleable solids are those generally bigger than 100 μm and that can be removed by gravity separation. Suspended solids, with sizes ranging from 100 μm to 30 μm, are those which do not settle out of suspension, but that can be removed by mechanical (i.e. sieving) means. Fine solids, with sizes of less than 30 μm, are generally those that cannot be removed by sieving, and must be controlled by other means such as physico-chemical processes, membrane filtration processes, dilution or bioclarification (Chen et al. 1994; Lee 2014; Summerfelt and Hochheimer 1997; Timmons and Ebeling 2010; Wold et al. 2014). The techniques for controlling settleable and suspended solids are well known and developed, and an extensive literature exists on the subject. For example, the use of dual-drain tanks, swirl separators, radial flow separators and settling basins is a popular means to control settleable solids (Couturier et al. 2009; Davidson and Summerfelt 2004; De Carvalho et al. 2013; Ebeling et al. 2006; Veerapen et al. 2005). Microscreen filters are the most popular method for suspended solids control (Dolan et al. 2013; Fernandes et al. 2015) and are often used in the industry to control both settleable and suspended solids with a single technique. Other popular solids capture devices are depth filters such as the bead filters (Cripps and Bergheim 2000) and rapid sand filters, which are also popular in swimming pool applications. Moreover, design guidelines to prevent the accumulation of solids in tanks, pipework, sumps and other system components are also available in the literature (Davidson and Summerfelt 2004; Lekang 2013; Wong and Piedrahita 2000). Lastly, fine solids in RAS are commonly treated by ozonation, bioclarification, foam fractionation or a combination of these techniques. The last few years in RAS development have focused on a greater understanding of how to control the fine solids fraction and to understand its effect on fish welfare and system performance.

3.2.4 Carbon Dioxide (CO_2)

In RAS, the control of dissolved gases does not stop with supplying oxygen to the fish. Other gases dissolved in the rearing water may affect fish welfare if not controlled. High dissolved carbon dioxide (CO_2) concentrations in the water inhibit the diffusion of CO_2 from the blood of fish. In fish, increased CO_2 in blood reduces the blood's pH and in turn, the affinity of haemoglobin for oxygen (Noga 2010). High CO_2 concentrations have also been associated with nephrocalcinosis, systemic granulomas and chalky deposits in organs in salmonids (Noga 2010). CO_2 in RAS originates as a product of heterotrophic respiration by fish and bacteria. As a highly soluble gas, carbon dioxide does not reach atmospheric equilibrium as easily as oxygen or nitrogen and thus, it must be put in contact with high volumes of air with a low concentration of CO_2 to ensure transfer out of water (Summerfelt 2003). As a general rule, RAS which are supplied with pure oxygen will require some form of carbon dioxide stripping, while RAS which are supplied with aeration for oxygen supplementation will not require active CO_2 stripping (Eshchar et al. 2003; Loyless and Malone 1998).

In theory, any gas transfer/aeration device open to the atmosphere will offer some form of CO_2 stripping. However, specialized carbon dioxide stripping devices require that large volumes of air are put in contact with the process water. CO_2 stripper designs have mostly focused on cascade-type devices such as cascade aerators, trickling biofilters and, more importantly, the packed column aerator (Colt and Bouck 1984; Moran 2010; Summerfelt 2003), which has become a standard piece of equipment in commercial RAS operating with pure oxygen. Although the development of packed column aeration technology has advanced over past years, most of the research done on this device has been focused on understanding its performance under different conditions (i.e. freshwater vs seawater) and design variations such as heights, packing types and ventilation rates. The effect of the hydraulic loading rate (unit flow per unit area of degasser) is known to have an effect on the efficiency of a degasser, but further research is needed to have a better understanding of this design parameter.

3.2.5 Total Gas Pressure (TGP)

Total gas pressure (TGP) is defined as the sum of the partial pressures of all the gases dissolved in an aqueous solution. The less soluble a gas is, the more 'room' it occupies in the aqueous solution and thus, the more pressure it exerts in it. Of the main atmospheric gases (nitrogen, oxygen and carbon dioxide) nitrogen is the least soluble (e.g. 2.3 times less soluble than oxygen and more than 90 times less soluble than carbon dioxide). Thus, nitrogen contributes to total gas pressure more than any other gas, but is not consumed by fish or heterotrophic bacteria, so it will accumulate in the water unless stripped. It is also important to note that oxygen will also contribute to high TGP if the gas transfer process does not allow excess gases to be displaced out of the solution. A classic example of this are ponds with photoautotrophic activity in them. Photoautotrophs (usually plant organisms that carry out photosynthesis) release oxygen into the water while a quiet water surface may not provide enough gas exchange for excess gas to escape to the atmosphere and thus, supersaturation may occur.

Fish require total gas pressures equal to atmospheric pressure. If fish breathe water with a high total gas pressure, excess gas (generally nitrogen) exits the bloodstream and forms bubbles, with often serious health effects for the fish (Noga 2010). In aquaculture this is known as gas bubble disease.

Avoiding high TGP requires careful examination of all areas in the RAS where gas transfer may occur. High-pressure oxygen injection without off-gassing (allowing excess nitrogen to be displaced out of the water) may also contribute to high TGP. In systems with fish which are very sensitive to TGP, the use of vacuum degassers is an option (Colt and Bouck 1984). However, maintaining a RAS free from areas of uncontrolled gas pressurization, using carbon dioxide strippers (which will also strip nitrogen) and dosing technical oxygen with care, is enough to keep TGP at safe levels in commercial RAS.

3.2.6 Nitrate

Nitrate (NO_3) is the end product of nitrification and commonly the last parameter to be controlled in RAS, due to its relatively low toxicity (Davidson et al. 2014; Schroeder et al. 2011; van Rijn 2013). This is mostly attributed to its low permeability at the fish gill membrane (Camargo and Alonso 2006). The toxic action of nitrate is similar to that of nitrite, affecting the capacity of oxygen-carrying molecules. The control of nitrate concentrations in RAS has traditionally been achieved by dilution, by effectively controlling the hydraulic retention time or daily exchange rate. However, the biological control of nitrate using denitrification reactors is a growing area of research and development in RAS.

Tolerance to nitrate may vary by aquatic species and life stage, with salinity having an ameliorating effect over its toxicity. It is important for RAS operators to understand the chronic effects of nitrate exposure rather than the acute effects, as acute concentrations will probably not be reached during normal RAS operation.

3.2.7 Alkalinity

Alkalinity is, in broad terms, defined as the pH buffering capacity of water (Timmons and Ebeling 2010). Alkalinity control in RAS is important as nitrification is an acid-forming process which destroys it. In addition, nitrifying bacteria require a constant supply of alkalinity. Low alkalinity in RAS will result in pH swings and nitrifying biofilter malfunction (Summerfelt et al. 2015; Colt 2006). Alkalinity addition in RAS will be determined by nitrification activity in the systems, which is in turn related to feed addition, by the alkalinity content of the make-up (daily exchange) water and by the presence of denitrifying activity, which restores alkalinity (van Rijn et al. 2006).

3.3 Developments in RAS

The last few years have seen an increase in the number and sizes of recirculating aquaculture farms, especially in Europe. With the increase in acceptance of the technology, improvements over traditional engineering approaches, innovations and new technical challenges keep emerging. The following section describes the key design and engineering trends and new challenges that recirculating aquaculture technology is facing.

3.3.1 Main Flow Oxygenation

The control of dissolved oxygen in modern RAS aims to increase the efficiency of oxygen transfer and decrease the energy requirements of this process. Increasing the oxygen transfer efficiency can be achieved by devising systems which retain oxygen gas in contact with water for longer, while a decrease in energy requirements may be achieved by the use of low-head oxygen transfer systems or using systems which do not use electricity at all, such as liquid oxygen systems connected to oxygen diffusers operating only by pressure. A defining factor of low-head oxygenators is the relatively low dissolved concentration that can be achieved compared to high-pressure systems. To overcome this limitation, low-head oxygenation devices are strategically placed to treat the full recirculating flow instead of using a smaller bypass of highly supersaturated water, thus ensuring sufficient mass transport of oxygen. Using oxygenation devices installed in the main recirculating flow generates savings in electricity consumption because the use of energy-intensive high-pressure systems that are necessary to achieve high DO concentrations in small flows is avoided. Low-head oxygenation systems may also reduce the amount of pumping systems needed, as high-pressure oxygenation systems are commonly placed on a bypass in the pipelines going to the fish tanks. In contrast, low-head oxygenation devices tend to be comparatively larger because of their need to handle larger flows and thus, their initial cost may be higher. Examples of devices that can treat the totality of the flow include the low-head oxygenator (LHO) (Wagner et al. 1995), operated by gravity as water is firstly pumped into a biofilter and a packed column (Summerfelt et al. 2004), low-head oxygen cones, variants of the Speece Cone (Ashley et al. 2008; Timmons and Losordo 1994) operated at low pressure, the deep shaft cones (Kruger Kaldnes, Norway), also a variant of the Speece cone designed to reach higher operating pressures by means of increased hydrostatic pressure resulting from placing the devices lower than the fish tanks and pump sumps, the U-tube oxygenator and its design variants such as the Farrell tube or the patented oxygen dissolver system (AquaMAOF, Israel) and the use of diffused oxygenation in deep fish tanks (Fig. 3.5).

3.3.2 Nitrifying Biofiltration Alternatives

Although nitrifying biofilters continue to be the main commercially accepted method of ammonia removal in commercial RAS, new nitrogen removal technologies have been developed over recent years. Some of these technologies consider alternative biological pathways to remove ammonia from the culture water, while others aim to

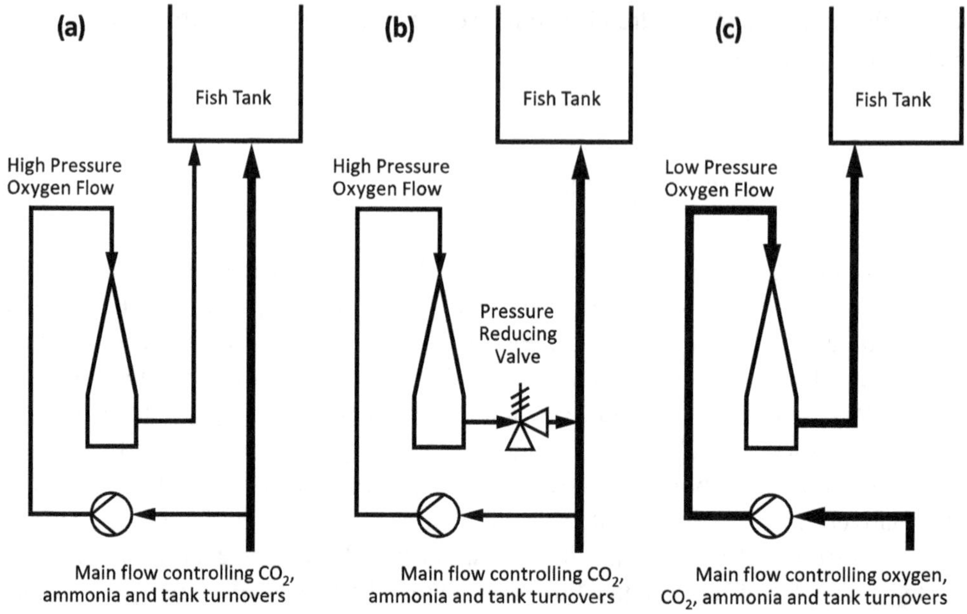

Fig. 3.5 Gas transfer alternatives for recirculating water returning into fish tanks. If the gas contacting vessel allows for pressurization, oxygen can be transferred in high concentrations in relatively small, high-pressure streams (**a**, **b**). However, oxygen at lower concentrations can be transferred into the main recirculation loop, but for this, the oxygen transfer device must be much larger to handle full flow of the system (**c**)

replace or work in parallel with nitrifying biofilters in order to reduce inherent limitations. These include large reactor sizes, susceptibility to crashing, long start-up times and poorer performance in both cold water and marine systems.

Anammox-based Processes
An alternative biological ammonia removal pathway considered for RAS is the anammox process (Tal et al. 2006), which occurs under anaerobic conditions. Anaerobic ammonia oxidation is a process which eliminates nitrogen by combining ammonia and nitrite to produce nitrogen gas (van Rijn et al. 2006). The anammox process is of interest to RAS because it allows for complete autotrophic nitrogen removal, in contrast to traditional combinations of nitrifying biofilters with heterotrophic denitrification systems requiring organic carbon addition (van Rijn et al. 2006). Moreover, in the anammox pathway, only half of the ammonia released by the fish is aerobically oxidized to nitrite (requiring oxygen), while the other half is anaerobically converted to nitrogen gas along with the nitrite produced. This may provide savings in oxygen and energy use in RAS (van Rijn et al. 2006).

Anammox reactor prototypes have been demonstrated successfully (Tal et al. 2006, 2009), while anammox activity has been suspected to occur in marine denitrification systems (Klas et al. 2006). The European FP7 project DEAMNRECIRC was also successful in creating anammox reactor prototypes for cold water and seawater aquaculture applications. However, commercial applications of the technology have as yet not been identified by the authors.

Chemical Removal of Ammonia
Ammonia removal systems based on ion exchange and electrochemical oxidation processes are being proposed as alternatives to nitrifying biofilters. Ion exchange processes rely on using adsorptive materials such as zeolites or ion-selective resins to extract dissolved ammonia from the water (Lekang 2013), while electrochemical oxidation processes convert ammonia to nitrogen gas through a number of complex oxidation reactions (Lahav et al. 2015). By comparison, ion exchange processes are suitable for waters with low concentrations of ions (i.e. freshwater), while electrochemical oxidation processes take advantage of the chloride ions present in the water to produce active chlorine species which readily react with ammonia (Lahav et al. 2015) and are thus suitable for waters with higher concentrations of chloride ions (i.e. brackish and marine waters).

Although ion exchange processes are not new, their application into RAS has been limited by their capacity to maintain performance over time: the filtering material eventually becomes 'saturated', losing its adsorptive capacity and, thus, must be regenerated. Gendel and Lahav (2013), proposed a novel approach to an ion exchange-based ammonia process in tandem with an innovative adsorbent regeneration process using electrochemical oxidation. Electrochemical oxidation of ammonia is a process which has received greater attention in recent years, and several concepts have been investigated and have been launched commercially, for example, EloxiRAS in Spain.

Factors limiting the application of these technologies into commercial RAS include, in the case of ion exchange processes, poor economic performance, difficulty to regenerate large amounts of adsorbent materials on demand (Lekang 2013), system complexity requiring the addition of chemical reagents, high electricity consumption and a high degree of suspended solids removal (Lahav et al. 2015), which is often impractical in large-scale RAS. In the case of ammonia electrooxidation processes, the production of toxic reactive species requiring active removal is their most important limitation, although their high solids control requirement, often possible only with pressurized mechanical filters, is also a challenge in RAS operating with large flows and low pressure.

3.3.3 *Fine Solids Control*

Fine solids are the dominant solids fraction in RAS with particles <30 μm forming more than 90% of the total suspended solids in the culture water. Recent investigations have found that more than 94% of the solids present in the culture water of a RAS are <20 μm in size or 'fine' (Fernandes et al. 2015). The accumulation of fine solids mainly occur as larger solids bypass the mechanical filters (which are not 100% efficient) and are eventually broken down by pumps, friction with surfaces and bacterial activity. Once solids sizes are reduced, traditional mechanical filtration techniques are rendered useless.

In recent years, the production, control, fish welfare effects and system performance effects of fine solids continue to be explored. The effects of fine solids on fish welfare were initially investigated through fisheries research (Chen et al. 1994). However, the direct effects of fine solids in RAS on fish welfare have not been thoroughly investigated until recently. Surprisingly, separate work on rainbow trout by Becke et al. (2016) and Fernandes et al. (2015) showed no negative welfare effects in systems with suspended solids concentrations of up to 30 mg/l in exposure trials lasting 4 and 6 weeks, respectively. Despite these findings, the indirect effects of fine solids accumulation in RAS are known (Pedersen et al. 2017) and are reported to be mostly linked to the proliferation of opportunistic microorganisms (Vadstein et al. 2004;Attramadal et al. 2014; Pedersen et al. 2017) since fine solids provide a high-surface area substrate for bacteria to colonize. Another important negative effect of fine solids accumulation is the increase in turbidity, which makes visual inspection of fish difficult and may hamper photoperiod control strategies which require light penetration in the water column to occur. Fine solids control strategies used in modern RAS include ozonation, protein skimming, floatation, cartridge filtration and membrane filtration (Couturier et al. 2009; Cripps and Bergheim 2000; Summerfelt and Hochheimer 1997; Wold et al. 2014). Protein skimmers, also known as foam fractionators, are also relatively popular fine solids control devices, especially in marine systems (Badiola et al. 2012).

3.3.4 Ozonation

Knowledge of ozone (O_3) application in RAS has existed since the 1970s and 1980s (Summerfelt and Hochheimer 1997). However, its application has not been as widespread as other processes such as nitrifying biofilters or mechanical filters (Badiola et al. 2012). Aside from fine solids treatment, ozone, as a powerful oxidizer, can be used in RAS to eliminate microorganisms, nitrite and humic substances (Gonçalves and Gagnon 2011). Recent years have seen an increase in knowledge about the potentials and limitations of ozone applied in both freshwater and marine RAS. Importantly, the ozone doses that can be safely achieved to improve water quality in both freshwater and seawater systems have been confirmed in several publications (Li et al. 2015; Park et al. 2013, 2015; Schroeder et al. 2011; Summerfelt 2003; Timmons and Ebeling 2010), with the conclusion that ozone doses over recommended limits (1) do not improve water quality further and (2) may cause negative welfare effects, especially in seawater systems where excessive ozonation will cause the formation of toxic residual oxidants. In coldwater RAS, ozonation requirements to achieve complete disinfection of the process flow have been determined (Summerfelt et al. 2009).

Ozonation improves microscreen filter performance and minimizes the accumulation of dissolved matter affecting the water colour (Summerfelt et al. 2009).

However, excessive ozonation may severely impact farmed fish by causing adverse effects including histopathologic tissue damage (Richardson et al. 1983; Reiser et al. 2010) and alterations in feeding behaviour (Reiser et al. 2010) as well as oxidative stress (Ritola et al. 2000, 2002; Livingstone 2003). Additionally, ozonation by-products may be harmful. Bromate is one of these and is potentially toxic. Tango and Gagnon (2003) showed that ozonated marine RAS have concentrations of bromate that are likely to impair fish health. Chronic, sublethal ozone-produced oxidants (OPO) toxicity was investigated in juvenile turbot by Reiser et al. (2011), while rainbow trout health and welfare were assessed in ozonated and non-ozonated RAS by Good et al. (2011). Raising rainbow trout to market size in ozonated RAS improved fish performance without significantly impacting their health and welfare while high OPO doses affect welfare of juvenile turbot.

3.3.5 Denitrification

In most recirculating aquaculture systems, nitrate, the end product of nitrification, tends to accumulate. Such accumulation is commonly controlled by dilution (introducing new water in the system). The control of nitrate by dilution may be a limiting factor to a RAS operation due to environmental regulations, poor availability of new water, the cost of treating the incoming and effluent water streams or the costs associated with chilling or heating the new water.

Biological nitrate removal in RAS can be achieved by facultative anaerobic bacteria using a dissimilatory pathway to convert nitrate to nitrogen gas in the presence of carbon and nitrate as electron donors (van Rijn et al. 2006). Denitrification reactors are thus biological reactors which are typically operated in anaerobic conditions and generally dosed with some type of carbon source such as ethanol, methanol, glucose, molasses, etc. Denitrification technology has been under development since the 1990s (van Rijn and Riviera 1990), but its popularity among the recirculating aquaculture industry has only increased over the past years, offering innovative denitrification reactor solutions.

One of the most notable applications of denitrification systems in aquaculture is the 'zero exchange' RAS (Yogev et al. 2016), which employ anaerobic digestion of biosolids produced in the system to produce volatile fatty acids (VFA) which are then used by denitrifiers as a carbon source. Klas at al. (2006) developed a 'single-sludge' denitrification system, where production of VFA from biosolids and denitrification occur in a single, mixed reactor. Suhr et al. (2014) developed the single-sludge concept further, adapting it for end-of-pipe treatment of fish farming effluents and adding an extra step which separates VFA production from the denitrification reactor in a hydrolysis tank. These works have provided valuable information on the possibilities of using aquacultural biosolids instead of expensive inorganic carbon sources for denitrification. Furthermore, Christianson et al. (2015) studied the effectiveness of autotrophic, sulphur-based denitrification reactors as an alternative to conventional heterotrophic denitrification reactors. Autotrophic reactors produce less biomass (solids) and can be supplied with sulphur particles, which are cheaper than conventional inorganic carbon sources.

VFAs are also the precursor component in the production of biopolymers such as Polyhydroalkanoates (PHAs), used to produce biodegradable plastics (Pittmann and Steinmetz 2013). This could hold potential for fish farms employing anaerobic activated sludge processes to be part of the 'biorefinery' concept applied to wastewater treatment plants.

3.3.6 Microbial Control

Microbial communities are important constituents of the aquatic ecosystem. In aquaculture production systems, they play significant roles in nutrient recycling, degradation of organic matter and treatment and control of disease (Zeng et al. 2017). Developing efficient, productive, biologically secure and disease-free RAS requires a thorough understanding of all life support processes from physical and chemical (gas transfer, thermal treatment, ozonation, UV irradiation, pH and salinity adjustments) to biological processes (nitrification, denitrification and aerobic heterotrophic activity). While physical and chemical processes can be controlled, biological filtration systems rely on the interaction of microbial communities with each other and their environment as a consequence of nutrient input (fish waste output) and, as such, are not as easily controlled (Schreier et al. 2010). Recent studies using molecular tools have not only allowed for evaluating microbial diversity in RAS but have also provided some insight into their activities that should lead to a better understanding of microbial community interactions. These approaches are certain to provide novel RAS process arrangements as well as insight into new processes and tools to enhance and monitor these systems (Schreier et al. 2010). Current understanding of RAS biofilter microbial diversity in both freshwater and marine systems is based on studies using 16S rRNA and functional gene-specific probes or 16S rRNA gene libraries rather than culture-based techniques (Table 3.1).

Insights into the temporal and spatial dynamics of microbiota in RAS are also still limited (Schreier et al. 2010), and potential solutions to maintain or restore beneficial microbial communities in RAS are lacking (Rurangwa and Verdegem 2015). Besides a microbial community that purifies the water, microbiota in RAS can also harbour pathogens or produce off-flavour-causing compounds (Guttman and van Rijn 2008). Given the difficulty to treat disease during operation without negatively affecting beneficial microbiota, microbial management in RAS is rather a necessity from the start-up through the whole production process. Microorganisms are introduced into RAS through different pathways: make-up water, air, animal vectors, feed, fish stocking, dirty equipment and via staff or visitors (Sharrer et al. 2005; Blancheton et al. 2013). Specific microbes can also, on the other hand, be applied intentionally to steer microbial colonization to improve system performance or animal health (Rurangwa and Verdegem 2015).

Table 3.1 Primary activities associated with RAS biofiltration units and participating microorganisms. (From Schreier et al. 2010)

Process	Reaction	Microorganism	
		Freshwater	Marine
Nitrification			
Ammonium oxidation	$NH_4^+ + 1.5O_2 \rightarrow NO_2^- + 2H^+ + H_2O$	*Nitrosomonas oligotropha*	*Nitrosomonas sp.*
			Nitrosomonas cryotolerans
			Nitrosomonas europaea
			Nitrosomonas cinnybus/ nitrosa
			Nitrosococcus mobilis
Nitrite oxidation	$NO_2^- + H_2O \rightarrow NO_3^- + 2H^+ + 2e^-$	*Nitrospira spp.*	
		Nitrospira marina[a]	*Nitrospira marina*[a]
		Nitrospira moscoviensis[a]	*Nitrospira moscoviensis*[a]
Denitrification			
Autotrophic	$S_2^- + 1.6NO_3^- + 1.6H^+ \rightarrow$		*Thiomicrosporia denitrificans*
(sulfide-dependent)	$SO_4^{2-} + 0.8N_2(g) + 0.8H_2O$		*Thiothrix disciformis*[a]
			Rhodobacter litoralis[a]
			Hydrogenophaga sp.
Heterotrophic	$5CH_3COO^- + 8NO_3^- + 3H^+ \rightarrow$		*Pseudomonas fluorescens*
	$10HCO_3^- + 4N_2(g) + 4H_2O$	*Pseudomonas sp.*	*Pseudomonas stutzeri*
		Comamonas sp.	*Pseudomonas sp.*
			Paracoccus denitrificans
Dissimilatory nitrate	$NO_3^- + 2H^+ + 4H_2 \rightarrow NH_4^+ + 3H_2O$		Various *Proteobacteria* and *Firmicutes*
Reduction to ammonia (DNRA)			
Anaerobic ammonium	$NH_4^+ + NO_2^- \rightarrow N_2(g) + 2H_2O$		*Planctomycetes spp.*
oxidation (Anammox)			*Brocadia sp.*[a]
Sulfate reduction	$SO_4^{2-} + CH_3COO^- + 3H^+ \rightarrow$		*Desulfovibrio sp.*,
	$HS^- + 2HCO_3^- + 3H^+$		*Dethiosulfovibrio sp.*,
			Fusibacter sp., *Bacteroides* sp.
Sulfide oxidation	$HS^- + 2O_2 \rightarrow SO_4^{2-} + H^+$		*Thiomicrospira sp.*

(continued)

Table 3.1 (continued)

Process	Reaction	Microorganism	
		Freshwater	Marine
Methanogenesis	$4H_2 + H^+ + HCO_3^- \rightarrow CH_4(g) + 3H_2O$		Methanogenic Archaea [Mirzoyan and Gross, unpublished]

[a]Microorganisms identified solely on the basis of partial 16S rRNA gene or functional gene sequences

One of the approaches for inhibiting pathogen colonization is the use of probiotic bacteria that may compete for nutrients, produce growth inhibitors, or, quench cell-to-cell communication (quorum sensing) that allows for settling within biofilms (Defoirdt et al. 2007, 2008; Kesarcodi-Watson et al. 2008). Probiotic bacteria include *Bacillus*, *Pseudomonas* (Kesarcodi-Watson et al. 2008) and *Roseobacter* spp. (Bruhn et al. 2005), and bacteria related to these have also been identified in RAS biofilters (Schreier et al. 2010) (Table 3.1). To obtain the information needed to manage microbial stability in RAS, Rojas-Tirado et al. (2017) have identified the factors affecting changes in the bacterial dynamics in terms of their abundance and activity. Their studies show that bacterial activity was not a straightforward predictable parameter in the water phase as nitrate-N levels in identical RAS showed unexpected sudden changes/fluctuations within one of the systems. Suspended particles in RAS provide surface area that can be colonized by bacteria. More particles accumulate as the intensity of recirculation increases, thus potentially increasing the bacterial carrying capacity of the systems. Pedersen et al. (2017) explored the relationship between total particle surface area (TSA) and bacterial activity in freshwater RAS. They indicated a strong, positive, linear correlation between TSA and bacterial activity in all systems with low to moderate recirculation intensity. However, the relationship apparently ceased to exist in the systems with the highest recirculation intensity. This is likely due to the accumulation of dissolved nutrients sustaining free-living bacterial populations, and/or accumulation of suspended colloids and fine particles less than 5 μm in diameter, which were not characterized in their study but may provide significant surface area.

In RAS, various chemical compounds (mainly nitrates and organic carbon) accumulate in the rearing water. These chemical substrata regulate the ecophysiology of the bacterial communities on the biofilter and have an impact on its nitrification efficiency and reliability. Michaud et al. (2014) investigated the shift of the bacterial community structure and major taxa relative abundance in two different biological filters and concluded that the dynamics and flexibility of the bacterial community to adapt to influent water changes seemed to be linked with the biofilter performance. One of the key aspects for improving the reliability and sustainability of RAS is the appropriate management of the biofilter bacterial populations, which is directly linked to the C (carbon) availability (Avnimelech 1999). It should be noted that RAS have properties that may actually contribute to microbial stabilization, including long water retention time and a large surface area of biofilters for bacterial growth, which could potentially limit the chances of proliferation of opportunistic microbes in the rearing water (Attramadal et al. 2012a).

Attramadal et al. (2012a) compared the development of the microbial community in a RAS with moderate ozonation (to 350 mV) to that of a conventional flow-through system (FTS) for the same group of Atlantic cod, *Gadus morhua*. They found less variability in bacterial composition between replicate fish tanks of the RAS than between tanks of the FTS. The RAS had a more even microbial community structure with higher species diversity and periodically a lower fraction of opportunists. The fish in RAS performed better than their control in the FTS, despite being exposed to an apparent inferior physico-chemical water quality. While researching the effects of moderate ozonation or high-intensity UV irradiation on the microbial environment in RAS for marine fish larvae, Attramadal et al. (2012b) emphasized that a RAS for such larvae should probably not include strong disinfection because it leads to a reduction in bacterial numbers, which is likely to result in a destabilization of the microbial community. Furthermore, their results support the hypothesis of RAS as a microbial control strategy during the first feeding of fish larvae.

RAS and microbial maturation as tools for K-selection of microbial communities was the subject of the study by Attramadal et al. (2014) in which they hypothesized that fish larvae that are reared in water dominated by K-strategists (mature microbial communities) will perform better, because they are less likely to encounter opportunistic (R-selected) microbes and develop detrimental host–microbe interactions. The results of their experiment showed a high potential for increasing fish survival by using K-selection of bacteria, which is a cheap and easy method that can be used in all kinds of new or existing aquaculture systems. Small changes in the management (organic load and maturation of water) of water treatment give significantly different microbiota in fish tanks (Attramadal et al. 2016). On the other hand, humic substances (HS) are natural organic compounds, comprising a wide array of pigmented polymers of high organic weight. They are end products in the degradation of complex organic compounds and, when abundant, produce a typical brown to dark-brownish colour of the soil and water (Stevenson 1994). In a zero-discharge aquaculture system, HS-like substances were detected in the culture water as well as in the fish blood (Yamin et al. 2017a). A protective effect of HS was reported in fish exposed to toxic metal (Peuranen et al. 1994; Hammock et al. 2003) and toxic ammonia and nitrite concentrations (Meinelt et al. 2010). Furthermore, evidence was provided for their fungistatic effect against the fish pathogen, *Saprolengia parasitica* (Meinelt et al. 2007). In common carp (*Cyprinus carpio*) exposed to (a) humic-rich water and sludge from a recirculating system, (b) a synthetic humic acid and (c) a Leonardite-derived humic-rich extract, infection rates were reduced to 14.9%, 17.0% and 18.8%, respectively, as compared to a 46.8% infection rate in the control treatment (Yamin et al. 2017b). Likewise, the exposure of guppy fish (*Poecilia reticulata*), infected with the monogenea *Gyrodactylus turnbulli* and *Dactylogyrus* sp. to humic-rich culture water and feed, reduced both the infection prevalence (% of infected fish) and the infection intensity (parasites per fish) of the two parasites (Yamin et al. 2017c).

It is believed that the fundamental research in the area of microbial ecology of the nitrification/denitrification reactor systems in RAS may provide innovations which may alter and/or improve the reactor performance in RAS drastically. Up until now, the microbial community in reactors is still difficult to control (Leonard et al. 2000, 2002; Michaud et al. 2006, 2009; Schreier et al. 2010; Rojas-Tirado et al. 2017) and many of the inefficiencies of the system originate from this (Martins et al. 2010b).

3.3.7 Energy Efficiency

Economic viability of fish production in a recirculating aquaculture system depends, in part, on minimizing the energy requirements of operating such facilities. RAS require a higher technical infrastructure than open systems, thus energy costs in RAS have already been rated as major constraints which may prevent this technology from widespread application (Singh and Marsh 1996). Of all the costs associated with the electricity use in RAS, ventilation and water cooling are generally the most important. In indoor RAS, building ventilation is important to control humidity and carbon dioxide levels. Poor humidity control may result in a rapid deterioration of building structures, while atmospheric carbon dioxide accumulation will affect carbon dioxide stripping processes operating in the RAS and cause dizziness in workers. In order to keep an acceptable atmosphere inside the facilities, ventilation or air conditioning plants are widely in operation (Gehlert et al. 2018). These ventilation systems may be fitted with measures to reduce energy use. Furthermore, in order to develop an environmentally sustainable RAS, energy may be assumed as a key driving parameter, and in particular, energy can be considered an important indicator. Energetic performance analysis of the RAS has been performed by Kucuk et al. (2010) to contribute to the energy management in the RAS. In order to improve the energetic performance of the RAS, they recommended that operating conditions of the components, particularly, the pumps should be optimized and improved based on the fish production capacity of the system.

To increase the efficiency, RAS managers need guidelines and tools to optimize production. Energy audits can provide real data that can be used for decision-making. Badiola et al. (2014) investigated the total energy consumption (kWh) of a RAS cod system continuously for 14 months and identified the heat pump as a top energy consumer of rearing fish requiring high water thermal treatment. Gehlert et al. (2018) concluded that ventilation units offered a significant potential for energy savings in the RAS. Most of the time, when climate parameters in the facility stay within a desired range, air flow rates can be kept at low levels for saving energy. Additionally, energy saving measures in the RAS may include: software with energy performance data, alternative energy sources to heat the water and the use of frequency converters (Badiola et al. 2014).

3.4 Animal Welfare Issues

3.4.1 Introduction

During the last decade, fish welfare has attracted a lot of attention, and this has led to the aquaculture industry incorporating a number of husbandry practices and technologies specifically developed to improve this aspect. The neocortex, which in humans is an important part of the neural mechanism that generates the subjective experience of suffering, is lacking in fish and non-mammalian animals, and it has been argued that its absence in fish indicates that fish cannot suffer. A strong alternative view, however, is that complex animals with sophisticated behaviours, such as fish, probably have the capacity for suffering, though this may be different in degree and kind from the human experience of this state (Huntingford et al. 2006).

The UK government's Farm and Animal Welfare Committee (FAWC) has based their guidelines on the 'Five Freedoms' framework, which defines ideal states rather than specific levels of acceptable welfare (FAWC 2014). Freedom from hunger and thirst, discomfort, pain, injury, disease, fear and distress, as well as the freedom to express normal behaviour, provides us with a defined framework with which to assess welfare issues. Physical health is the most universally accepted measure of welfare and is undoubtedly a necessary requirement for good welfare. In a competitive, expanding and emerging industry, aquaculturists who incorporate welfare considerations into their daily husbandry practices can gain a competitive advantage and added price premium (Olesen et al. 2010) through improved consumer perception and confidence in their products. Grimsrud et al. (2013) provided evidence that there is a high willingness to pay, among all Norwegian households, to improve the welfare of farmed Atlantic salmon through increased resistance to diseases and salmon lice, which may imply less use of medicines and chemicals in the production process.

In intensive RAS, animal welfare is tightly connected to the performance of the systems. Over the past few years, animal welfare in the RAS has been mostly studied from the perspective of water quality and fish crowding effects on growth performance, stress bioindicators or the development of health disorders. The main goal of animal welfare research in the RAS has been to build and operate systems that maximize productivity and minimize stress and mortalities. Topics of interest have been stocking density limits (Calabrese et al. 2017), concentration limits of nitrogenous compounds in the rearing water (Davidson et al. 2014), concentration limits for dissolved carbon dioxide (Good et al. 2018), the effects of ozonation (Good et al. 2011; Reiser et al. 2011) and to a lesser extent, the accumulation of recalcitrant compounds in the RAS (van Rijn and Nussinovitch 1997) with limited water exchanges and noise (Martins et al. 2012; Davidson et al. 2017).

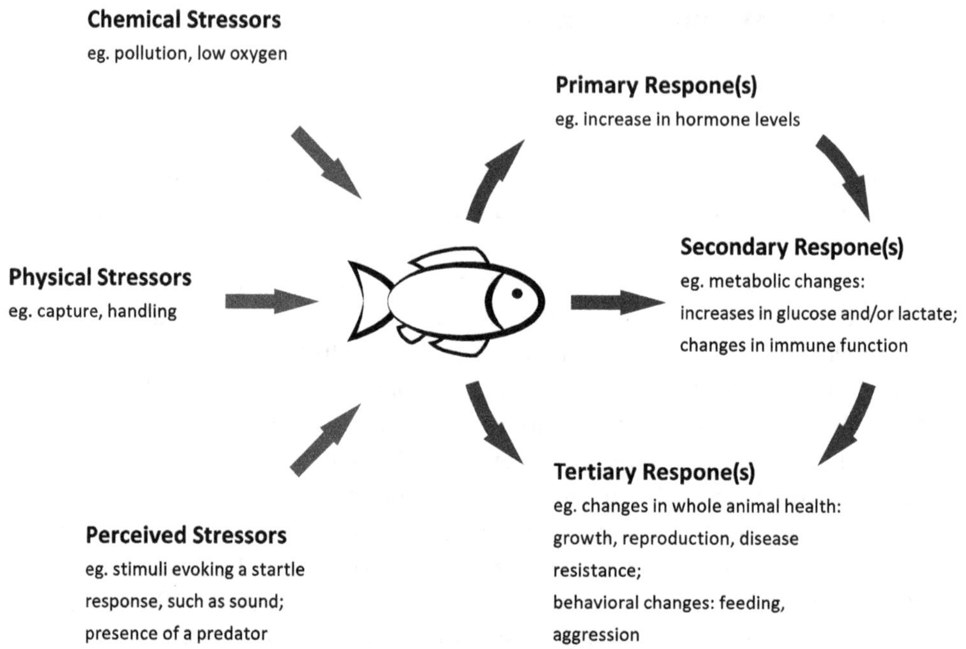

Fig. 3.6 Physical, chemical and other perceived stressors can affect fish and cause primary, secondary and/or whole-body responses. (After Barton 2002)

3.4.2 Stress

The stress response in fish is an adaptive function in the face of a perceived threat to homeostasis and stress physiology does not necessarily equate to suffering and diminished welfare (Ashley 2007) (Fig. 3.6). Stress responses serve a very important function to preserve the individual. Welfare measures in aquaculture are, therefore, largely associated with the tertiary effects of stress response that are generally indicative of prolonged, repeated or unavoidable stress (Conte 2004).

Stocking density is a pivotal factor affecting fish welfare in the aquaculture industry, especially RAS where high densities in confined environments are aimed at high productivity. Although rarely defined, stocking density is the term normally used to refer to the weight of fish per unit volume or per unit volume in unit time of water flow through the holding environment (Ellis et al. 2001). The concept of minimum space for a fish is more complex than for terrestrial species as fish utilize a three-dimensional medium (Conte 2004).

Beyond providing for the physiological needs, the FAWC (2014) recommends that fish 'need sufficient space to show most normal behaviour with minimal pain, stress and fear'. Stocking density is, therefore, an area that illustrates both the significance of species differences and the existence of a complex web of interacting factors that affect fish welfare. Calabrese et al. (2017) have researched stocking density limits for post-smolt Atlantic salmon (*Salmosalar* L.) with emphasis on production performance and welfare wherein fin damage and cataracts were observed in stocking densities of 100 kg m^{-3} and above. However, the effect of stocking density on measures of welfare varies between species. For instance, sea bass (*Dicentrarchus labrax*) showed higher stress levels at high densities, as indicated by cortisol, innate immune response and expression of stress-related genes (Vazzana et al. 2002; Gornati et al. 2004). High stocking densities in juvenile gilthead sea bream (*S. aurata*) also produce a chronic stress situation, reflected by high cortisol levels, immunosuppression and altered metabolism (Montero et al. 1999). In contrast, Arctic charr (*Salvelinus alpinus*) feed and grow well when stocked at high densities while showing a depressed food intake and growth rates at low densities (Jorgensen et al. 1993).

Diet may also play an important role in stress sensitivity. African catfish (*Clarias gariepinus*) receiving a diet with a high supplementation of ascorbic acid (vitamin C) during early development showed a lower stress sensitivity (Merchie et al. 1997). On the other hand, common carp (*Cyprinus carpio*), fed large doses of vitamin C, showed a more pronounced cortisol (a steroid hormone released with stress) increase in response to stress when compared to fish fed recommended levels of the vitamin (Dabrowska et al. 1991). Tort et al. (2004) have shown that a modified diet providing a supplementary dosage of vitamins and trace minerals to assist the immune system may help to co-reduce some of the effects of winter disease syndrome. Other common aquaculture diseases regarding animal welfare and stress are reviewed in Ashley (2007).

3.4.3 Accumulation of Substances in the Process Water

Intensive and 'zero-discharge' RAS offer significant environmental advantages. However, culturing of fish in continuously recycled water raises the question of whether substances released by the fish into the water may accumulate, resulting in decreased growth rates and impaired welfare. The existence of growth retardation in Nile tilapia (*Oreochromis niloticus*) by comparing the growth, feeding behaviour and stress response of fish cultured in the RAS with different levels of substances accumulated (TAN, NO_2-N and NO_3-N, orthophosphate-P) was investigated by Martins et al. (2010a). Results showed that the large individuals had a trend towards growth retardation in the highest accumulation RAS while small individuals, on the contrary, seem to grow better in such systems based on high levels of blood glucose as a stress indicator. A similar study done by the same author on carp embryos and larvae (Martins et al. 2011) found results that suggest that the concentration of substances (orthophosphate-P, nitrate, arsenic and copper) were likely to affect the development. Despite these findings, the authors claim that overall, the percentage of mortalities and deformities recorded in the study were relatively low compared to other studies. In both studies, the authors used systems with very limited water exchange rates with the aid of denitrification reactors (30 litres of new water per Kg of feed per day). Similarly the accumulation of hormones in coldwater salmonid RAS has been studied by Good et al. 2014, 2017). Their research in 2014 found neither a relationship between water exchange rate and hormone accumulation (except for testosterone) nor a link between hormone accumulation and precocious maturation in Atlantic salmon, but further study was suggested. Their study in 2017 focused on the use of ozonation for the reduction of hormones in the same RAS, with inconclusive results regarding the steroid hormone accumulation, but with a positive reduction of oestradiol by ozone.

On the other hand, the accumulation of humic substances in 'zero exchange' RAS has shown to have a protective effect against bacterial infections (Yamin et al. 2017a) and ectoparasites (Yamin et al. 2017b). Humic acids have also been shown to reduce ammonia and nitrite toxicity (Meinelt et al. 2010). This has implications for RAS operated with ozonation, as ozone may improve water quality while sacrificing the apparently beneficial effects of humic substances.

3.4.4 Health and Behaviour

The fundamental characteristics of good welfare are good health and absence of disease and, with respect to aquaculture, good productivity (Turnbull and Kadri 2007; Volpato et al. 2007). While the physical health of an animal is fundamental for good welfare (Ashley 2007; Duncan 2005), the fact that an animal is healthy does not necessarily mean that its welfare status is adequate. Thus, welfare is a broader and more overarching concept than the concept of health. Physiological and behavioural measures are intrinsically linked and are dependent on one another for a correct interpretation with regard to welfare (Dawkins 1998).

The behaviour of animals and in our case fish, represents a reaction to the environment as fish perceive it and behaviour is therefore a key element of fish welfare. Changes in foraging behaviour, gill ventilation activity, aggression, individual and group swimming behaviour, stereotypic and abnormal behaviours have been linked with acute and chronic stressors in aquaculture and can therefore be regarded as likely indicators of poor welfare (Martins et al. 2011). Behavioural welfare indicators have the advantage of being fast and easy to observe and therefore are good candidates for use 'on-farm'. Examples of behaviour that are commonly used as an indicators of welfare are changes in food-anticipatory behaviour, feed intake, swimming activity and ventilation rates (Huntingford et al. 2006). However, Barreto and Volpato (2004) caution the use of ventilation frequency as an indicator of stress in fish because although ventilation frequency is a very sensitivity response to disturbance, it is of limited use because it does not reflect the severity of the stimulus.

3.4.5 *Noise*

Farmed fish are cultured over long periods of time in the same tanks of the same colour(s) and the same shape and exposed to the same, potentially harmful, background noises (Martins et al. 2012). Intensive aquaculture systems and particularly recirculating systems utilize equipment such as aerators, air and water pumps, blowers and filtration systems that inadvertently increase noise levels in fish culture tanks. Sound levels and frequencies measured within intensive aquaculture systems are within the range of fish hearing, but species-specific effects of aquaculture production noise are not well defined (Davidson et al. 2009).

Bart et al. (2001) found that mean broadband sound pressure levels (SPL) differed across various intensive aquaculture systems. In his study, a sound level of 135 dB re 1μPa was measured in an earthen pond near an operating aerator, whereas large fiberglass tanks (14 m diameter) within a recirculating system had the highest SPLs of 153 dB re 1μPa.

Field and laboratory studies have shown that fish behaviour and physiology can be negatively impacted by intense sound. Terhune et al. (1990) have observed decreased growth and smoltification rates of Atlantic salmon, *Salmo salar*, in fiberglass tanks that had underwater sound levels 2–10 dB re 1μPa higher at 100–500 Hz than in concrete tanks. Therefore, chronic exposure to aquaculture production noise could cause increased stress, reduced growth rates and feed conversion efficiency and decreased survival. However, Davidson et al. (2009) found that after 5 months of noise exposure, no significant differences were identified between treatments for mean weight, length, specific growth rates, condition factor, feed conversion or survival of rainbow trout, *Oncorhynchus mykiss*. Similar findings are described by Wvysocki et al. (2007). However, these findings should not be generalized across all cultured fish species, because many species, including catfish and cyprinids, have much greater hearing sensitivity than rainbow trout and could be affected differently by noise. For instance, Papoutsoglou et al. (2008) provided initial evidence that music transmission under specific rearing conditions could have enhancing effects on *S. aurata* growth performance, at least at specific fish sizes. Moreover, the observed music effects on several aspects of fish physiology (e.g. digestive enzymes, fatty acid composition and brain neurotransmitters) imply that certain music could possibly provide even further enhancement in growth, quality, welfare and production.

3.5 Scalability Challenges in RAS

RAS are capital-intensive operations, requiring high funding expenditure on equipment, infrastructure, influent and effluent treatment systems, engineering, construction and management. Once the RAS farm is built, working capital is also needed until harvests and successful sales are achieved. Operational expenditures are also substantial and are mostly comprised of fixed costs such as rent, interest on loans, depreciation and variable costs such as fish feed, seed (fingerlings or eggs), labour, electricity, technical oxygen, pH buffers, electricity, sales/marketing, maintenance costs, etc.

When comparing productivity and economics, RAS will invariably compete with other forms of fish production and even other sources of protein production for human consumption. This competition is likely to exert a downwards pressure on the sale price of fish, which in turn must be high enough to make a RAS business profitable. As in other forms of aquaculture production, reaching higher economies of scale is generally a way to reduce the cost of production and thus obtain access to markets. Some examples of reduction in costs of production that can be achieved with larger facilities are:

1. Reduced transportation costs on bulk orders of feed, chemicals, oxygen.
2. Discounts on the purchase of larger quantities of equipment.
3. Access to industrial electricity rates.
4. Automation of farm processes such as process monitoring and control, feeding, harvesting, slaughter and processing.
5. Maximization of the use of labour: The same manpower was needed to take care of 10 tons of fish as was needed to take care of 100 tons of fish or more.

Following the increase of economies of scale in the net pen aquaculture sector, larger RAS are being developed on scales not considered a decade ago. The last decade has seen the construction of facilities with production capacities of thousands of tons per year, and this sheer size increase of RAS facilities is bringing new technical challenges, which are explored in the next section.

3.5.1 Hydrodynamics and Water Transport

Proper control of the hydrodynamic conditions in fish tanks is essential to ensure uniform water quality and adequate solids transport towards the tank outlets (Masaló 2008; Oca and Masalo 2012). Tanks which are not capable of flushing metabolites quickly enough will have less carrying capacity. Ensuring proper hydrodynamic performance in fish tanks is an important aquacultural engineering research topic which has helped the industry design and operate tanks of different shapes and sizes. However, increasing tank sizes used in commercial RAS are posing new engineering challenges to designers and operators. Recent investigations are underway to optimize the hydrodynamic characteristics in large octagonal tanks used for salmon smolt production (Gorle et al. 2018), by studying the effect of fish biomass, geometry and inlet and outlet structures in large tanks used in Norwegian smolt facilities. Similarly, Summerfelt et al. (2016) found a trend towards a decreasing feed loading rate per unit flow in modern tanks compared to tanks built more than a decade ago in Norwegian smolt facilities. A lowered feed burden effectively results in improved tank water quality as the recirculating water is treated at a faster rate, preventing the accumulation of metabolites and the depletion of oxygen in the tank even further compared to older tanks which operated at higher feed burdens. Future work will likely provide more information on the hydrodynamics of tanks with more than 1000 m^3 in volume. Other examples of enormous tanks which are currently being used are the tanks used in the RAS 2020 systems (Kruger, Denmark) or the Niri concept (Niri, Norway). Adoption of these new concepts using larger tanks will play a vital role in their profitability, as long as proper hydrodynamic conditions are achieved.

3.5.2 Stock Loss Risk

In RAS, the intensive rearing conditions can lead to sudden and catastrophic loss of fish if the system fails. Sources of system failure may include mechanical failure of pumping systems and RAS equipment, power outages, loss of oxygenation/aeration systems, hydrogen sulphide build-up and release, operational accidents and more. These risks and solutions to them need to be identified and incorporated into operational procedures.

The increasing size of RAS operations may also mean an increased risk of financial loss if catastrophic loss of fish occurs. On the other hand, risk-mitigation measures and system redundancy may also increase the cost of a RAS project and thus, designers and engineers must strike a balance between these elements.

Aside from industry and media reports, little academic research has been done on the risk of commercial RAS ventures. Badiola et al. (2012) surveyed RAS farms and analysed the main technical issues, finding that poor system design, water quality problems and mechanical problems were the main risk elements affecting the viability of RAS.

3.5.3 Economics

Debate on the economic viability of RAS focuses mostly on the high capital start-up costs of recirculating aquaculture farms and the long lead time before fish are ready to be marketed, as well as the perception that RAS farms have high operating costs. De Ionno et al. (2007) studied the commercial performance of RAS farms, concluding that economic viability increases with the scale of the operation. According to this study, farms smaller than 100 tons per year of production capacity are only marginally profitable in the Australian context where the study took place. Timmons and Ebeling (2010) also provide a case for achieving large economies of scale (in the order of magnitude of thousands of tons of production per year) which allow for the production cost reductions through vertical integration projects such as the inclusion of processing facilities, hatcheries or feed mills. Liu et al. (2016) studied the economic performance of a theoretical RAS farm with a capacity of 3300 tons per year, compared to a traditional net pen farm of the same capacity. At this scale the RAS operation reaches similar production costs compared to the net pen farm, but the higher capital investment doubles the payback period in comparison, even when the fish from the RAS farm are sold at a premium price. In the future, costly and strict licensing requiring good environmental performance may increase the viability of RAS as competitive option for the production of Atlantic salmon.

3.5.4 Fish Handling

On land-based farms, fish handling is often required for various reasons: to separate fish into weight classes, to reduce stocking densities, to transport fish across growing departments (i.e. from a nursery to an on-growing department) or to harvest fish when they are market ready. According to Lekang (2013), fish are handled most effectively with active methods such as fish pumps and also with passive methods such as the use of visual or chemical signals that allow the fish to move themselves from one place on the farm to the next.

Summerfeltet al. (2009) studied several means to crowd and harvest salmonids from large circular tanks using Cornell-type dual drains. Strategies included crowding fish with purse seines, clam-shell bar crowders and herding fish between tanks taking advantage to their innate avoidance response to carbon dioxide. Harvesting techniques included extracting the fish through the sidewall discharge port of a Cornell-type dual-drain tank or using an airlift to lift the crowded fish to a dewatering box. AquaMAOF (Israel) employs swimways and tanks sharing a common wall to passively transfer the fish through the farm, with harvesting taking place using a pescalator (Archimedes screw pump) at the end of a swimway. The RAS2020 concept from Kruger (Denmark) uses bar graders/crowders permanently installed in a donut-shaped or circular raceway tank to move and crowd the fish without the need for fish pumps.

Despite continuing developments on this topic, the increasing size of RAS farms will keep challenging designers and operators on how to handle fish safely, economically and without stress. The expanding range of designs, species under production and operation intensity of RAS farms may result in various and novel fish transport and harvest technologies.

3.6 RAS and Aquaponics

Aquaponic systems are a branch of recirculating aquaculture technology in which plant crops are included to either diversify the production of a business, to provide extra water filtration capacity, or a combination of the two.

As a branch of RAS, aquaponic systems are bound to the same physical, chemical and biological phenomena that occur in RAS. Therefore, the same fundamentals of water ecology, fluid mechanics, gas transfer, water depuration etc. apply in more or less equal terms to aquaponics with the exception of water quality control, as plants and fish may have specific and different requirements.

The fundamental economic realities of RAS and aquaponics are also related. Both technologies, are capital intensive and highly technical and are affected by economies of scale, appropriate design of the components, reliance on market conditions and the expertise of operators.

3.6.1 Welfare

In aquaponic systems, the uptake of nutrients should be maximized for the healthy production of plant biomass but without neglecting the best welfare conditions for the fish in terms of water quality (Yildiz et al. 2017). Measures to reduce the risks of the introduction or spread of diseases or infection and to increase biosecurity in aquaponics are also important. The possible impacts of allelochemicals, i.e. chemicals released by the plants, should be also taken into account. Moreover, the effect of diet digestibility, faeces particle size and settling ratio on water quality should be carefully considered. There is still a lack of knowledge regarding the relationship between the appropriate levels of minerals needed by plants, and fish metabolism, health and welfare (Yildiz et al. 2017) which requires further research.

3.6.2 Microbial Diversity and Control

As mentioned earlier in the chapter, aquaponics combines a recirculating aquaculture system with a hydroponic unit. One of its most important features is the reliance on bacteria and their metabolic products. Also, Sect. 3.2.6.discussed the importance of microbial communities and its control in RAS. Bacteria serve as the bridge that connects fish excrements, which are high in ammonium concentration, to plant fertilizer, which should be a combination of low ammonium and high nitrate (Somerville et al. 2014). As aquaponic systems can have different subunits, i.e. fish tanks, biofilter, drum filter, settler tanks and hydroponic units, each having different possible designs and different optimal conditions, the microbial communities in these components may differ considerably. This provides an interesting topic of research with the ultimate goal of improving system management processes. Schmautz et al. (2017) attempted to characterize the microbial community in different areas of aquaponic systems. They concluded that fish faeces contained a separate community dominated by bacteria of the genus *Cetobacterium*, whereas the samples from plant roots, biofilter and periphyton were more similar to each other, with more diverse bacterial communities. The biofilter samples contained large numbers of *Nitrospira* (3.9% of total community) that were found only in low numbers in the periphyton or the plant roots. On the other hand, only small percentages of *Nitrosomonadales* (0.64%) and *Nitrobacter* (0.11%) were found in the same samples. This second group of organisms are commonly tested for their presence in aquaponics systems as they are mainly held responsible for nitrification (Rurangwa and Verdegem 2015; Zou et al. 2016); *Nitrospira* has only recently been described as a total nitrifier (Daims et al. 2015), being able to directly convert ammonium to nitrate in the system. The dominance of *Nitrospira* is thus a novelty in such systems and might be correlated with a difference in the basic setup (Graber et al. 2014).

Schmautz et al. (2017) also emphasized that the increased presence of *Nitrospira* does not necessarily correlate to larger activity of these organisms in the system, as its metabolic activities were not measured. In addition, many species of bacteria and coliforms are inherently present in aquaponic recirculating biofilters carrying out transformations of organic matter and fish waste. This implies the presence of many microorganisms that can be pathogens for plants and fish, as well as for people. For this purpose, some microorganisms have been considered safety indicators for products and water quality in the system (Fox et al. 2012). Some of these safety indicators are *Escherichia coli* and *Salmonella* spp. Much needed research has thus recently been carried out in order to ascertain microbial safety of aquaponic products (Fox et al. 2012; Sirsat and Neal 2013). One future direction for the analysis of microbial activity in aquaponics has been identified by Munguia-Fragozo et al. (2015), who reviewed the *Omic* technologies for microbial community analysis. They concluded that metagenomics and metatranscriptomics analysis will be crucial in future studies of microbial diversity in aquaponic biosystems.

- From a period of technological consolidation to a new era of industrial implementation, RAS technology has considerably developed over the past two decades. The last few years have seen an increase in the number and scale of recirculating aquaculture farms. With the increase in acceptance of the technology, improvements over traditional engineering approaches, innovations and new technical challenges keep emerging.
- Aquaponics combines a recirculating aquaculture system with a hydroponic unit. RAS are complex aquatic production systems that involve a range of physical, chemical and biological interactions.
- Dissolved oxygen (DO) is generally the most important water quality parameter in intensive aquatic systems. However, addition of sufficient oxygen to the rearing water can be achieved relatively simply and thus, the control of other water parameters become more challenging.
- High dissolved carbon dioxide (CO_2) concentrations have a negative effect in fish growth. The removal of CO_2 from water to concentrations below 15 mg/L is challenging due to its high solubility and the limited efficiency of degassing equipment.
- Ammonia has been traditionally treated in recirculation systems with nitrifying biofilters. Some emerging technologies are being explored as alternatives to ammonia removal.

- Biosolids in RAS originate from fish feed, faeces and biofilms and are one of the most critical and difficult water quality parameters to control. A multi-step treatment system where solids of different sizes and removed via different mechanisms, is the most common approach.
- Ozone, as a powerful oxidizer, can be used in RAS to eliminate microorganisms, nitrite and humic substances. Ozonation improves microscreen filter performance and minimizes the accumulation of dissolved matter affecting the water colour.
- Denitrification reactors are biological reactors which are typically operated under anaerobic conditions and generally dosed with some type of carbon source such as ethanol, methanol, glucose and molasses. One of the most notable applications of denitrification systems in aquaculture is the 'zero exchange' RAS.
- In aquaculture production systems microbial communities play significant roles in nutrient recycling, degradation of organic matter and treatment and control of disease. The role of water disinfection in RAS is being challenged by the idea of using microbially mature water to control opportunistic pathogens.
- In intensive RAS, animal welfare is tightly connected to the performance of the systems. The main goal of animal welfare research in RAS has been to build and operate systems that maximize productivity and minimize stress and mortalities.

References

Altinok I, Grizzle JM (2004) Excretion of ammonia and urea by phylogenetically diverse fish species in low salinities. Aquaculture 238:499–507. https://doi.org/10.1016/j.aquaculture.2004.06.020

Ashley PJ (2007) Fish welfare: current issues in aquaculture. Appl Anim Behav Sci 104 (2–4):199–235. https://doi.org/10.1016/j.applanim.2006.09.001

Ashley KI, Mavinic DS, Hall KJ (2008) Oxygenation performance of a laboratory-scale Speece Cone hypolimnetic aerator: preliminary assessment. Can J Civ Eng 35:663–675. https://doi.org/10.1139/L08-011

Attramadal K, Salvesen I, Xue R, Øie G, Størseth TR, Vadstein O, Olsen Y (2012a) Recirculation as a possible microbial control strategy in the production of marine larvae. Aquac Eng 46:27–39. https://doi.org/10.1016/j.aquaeng.2011.10.003

Attramadal K, Øie G, Størseth TR, Morten OA, Vadstein O, Olsen Y (2012b) The effects of moderate ozonation or high intensity UV-irradiation on the microbial environment in RAS for marine larvae. Aquaculture 330–333:121–129. https://doi.org/10.1016/j.aquaculture.2011.11.042

Attramadal KJK, Truong TMH, Bakke I, Skjermo J, Olsen Y, Vadstein O (2014) RAS and microbial maturation as tools for K-selection of microbial communities improve survival in cod larvae. Aquaculture 432:483–490. https://doi.org/10.1016/j.aquaculture.2014.05.052

Attramadal K, Minniti G, Øie G, Kjørsvik E, Østensen M-A, Bakke I, Vadstein O (2016) Microbial maturation of intake water at different carrying capacities affects microbial control in rearing tanks for marine fish larvae. Aquaculture 457:68–72. https://doi.org/10.1016/j.aquaculture.2016.02.015

Avnimelech Y (1999) Carbon/nitrogen ratio as a control element in aquaculture systems. Aquaculture 176(3–4):227–235. https://doi.org/10.1016/S0044-8486(99)00085-X

Badiola M, Mendiola D, Bostock J (2012) Recirculating Aquaculture Systems (RAS) analysis: main issues on management and future challenges. Aquac Eng 51:26–35

Badiola M, Basurko OC, Gabiña G, Mendiola D (2014) Energy audits in Recirculating Aquaculture Systems (RAS): as a way forward to guarantee sustainability. AE2014 conference paper. https://doi.org/10.13140/2.1.1218.5604

Barreto RE, Volpato GL (2004) Caution for using ventilatory frequency as an indicator of stress in fish. Behav Process 66:43–51. https://doi.org/10.1016/j.beproc.2004.01.001

Bart AN, Clark J, Young J, Zohar Y (2001) Underwater ambient noise measurements in aquaculture systems: a survey. Aquac Eng 25:99–110. https://doi.org/10.1016/S0144-8609(01)00074-7

Barton B (2002) Stress in fishes: a diversity of responses with particular reference to changes in circulating corticosteroids. Integr Comp Biol 42:517–525

Becke C, Steinhagen D, Schumann M, Brinker A (2016) Physiological consequences for rainbow trout (*Oncorhynchus mykiss*) of short-term exposure to increased suspended solid load. Aquac Eng. https://doi.org/10.1016/j.aquaeng.2016.11.001

Blancheton JP, Attramadal KJK, Michaud L, Roque d'Orbcastel E, Vadstein O (2013) Insight into bacterial population in aquaculture systems and its implication. Aquac Eng 53:30–39. https://doi.org/10.1016/j.aquaeng.2012.11.009

Bock E, Sundermeyer-Klinger H, Stackerbrandt E (1983) New facultative lithoatotrophic nitrinte-oxidizing bacteria. Arch Microbiol 136:281–284. https://doi.org/10.1007/BF00425217

Bovendeur J, Eding EH, Henken AM (1987) Design and performance of a water recirculation system for high-density culture of the African catfish, *Clarias-gariepinus* (Burchell 1822). Aquaculture 63:329–353. https://doi.org/10.1016/0044-8486(87)90083-4

Browdy CL, Ray AJ, Leffler JW, Avnimelech Y (2012) Biofloc-based aquaculture systems. In: Tidwell JH (ed) Aquaculture production systems. Wiley, Oxford

Bruhn JB, Nielsen KF, Hjelm M, Hansen M, Bresciani J, Schulz S, Gram L (2005) Ecology, inhibitory activity, and morphogenesis of a marine antagonistic bacterium belonging to the *Roseobacter clade*. Appl Environ Microbiol 71:7263–7270. https://doi.org/10.1128/AEM.71.11.7263-7270.2005

Calabrese S, Nilsen TO, Kolarevic J, Ebbesson LOE, Pedrosa C, Fivelstad S, Hosfeld C, Stefansson SO, Terjesen BF, Takle H, Martins CIM, Sveier H, Mathisen F, Imsland AK, Handeland SO (2017) Stocking density limits for post-smolt Atlantic salmon (*Salmo salar* L.) with emphasis on production performance and welfare. Aquaculture 468(1):363–370. https://doi.org/10.1016/j.aquaculture.2016.10.041

Camargo JA, Alonso A (2006) Ecological and toxicological effects of inorganic nitrogen pollution in aquatic ecosystems: a global assessment. Environ Int 32:831–849. https://doi.org/10.1016/j.envint.2006.05.002

Chen S, Coffin DE, Malone RF (1994) Suspended solids control in recirculating aquaculture systems. In: Timmons MB, Losordo TM (eds) Aquaculture water reuse systems: engineering design and management. Elsevier, Amsterdam, pp 61–100

Christianson L, Lepine C, Tsukuda S, Saito K, Summerfelt S (2015) Nitrate removal effectiveness of fluidized sulfur-based autotrophic denitrification biofilters for recirculating aquaculture systems. Aquac Eng:10–18. https://doi.org/10.1016/j.aquaeng.2015.07.002

Colt J (1991) Aquacultural production systems. J Anim Sci 69:4183–4192. https://doi.org/10.2527/1991.69104183x

Colt J (2006) Water quality requirements for reuse systems. Aquac Eng 34(3):143–156. https://doi.org/10.1016/j.aquaeng.2005.08.011

Colt J, Bouck G (1984) Design of packed columns for degassing. Aquac Eng 3:251–273. https://doi.org/10.1016/0144-8609(84)90007-4

Colt JE, Watten BJ (1988) Applications of pure oxygen in fish culture. Aquac Eng 7:397–441. https://doi.org/10.1016/0144-8609(88)90003-9

Conte FS (2004) Stress and the welfare of cultured fish. Appl Anim Behav Sci 86:205–223. https://doi.org/10.1016/j.applanim.2004.02.003

Couturier M, Trofimencoff T, Buil JU, Conroy J (2009) Solids removal at a recirculating salmon-smolt farm. Aquac Eng 41:71–77. https://doi.org/10.1016/j.aquaeng.2009.05.001

Cripps SJ, Bergheim A (2000) Solids management and removal for intensive land-based aquaculture production systems. Aquac Eng 22:33–56. https://doi.org/10.1016/S0144-8609(00)00031-5

Crossley PL (2004) Sub-irrigation in wetland agriculture. Agric Hum Values 21(2/3):191–205. https://doi.org/10.1023/B:AHUM.0000029395.84972.5e. Accessed 18 Dec 2017

Dabrowska H, Dabrowski K, Meyerburgdorff K, Hanke W, Gunther KD (1991) The effect of large doses of Vitamin-C and magnesium on stress responses in Common Carp, *Cyprinus-Carpio*. Comp Biochem Physiol A 99:681–685. https://doi.org/10.1016/0300-9629(91)90150-B

Daims H, Lebedeva EV, Pjevac P, Han P, Herbold C, Albertsen M, Jehmlich N, Palatinszky M, Vierheilig J, Bulaev A, Kirkegaard RH, von Bergen M, Rattei T, Bendinger B, Nielsen PH, Wagner M (2015) Complete nitrification by *Nitrospira* bacteria. Nature 528:504–509. https://doi.org/10.1038/nature16461

Davidson J, Summerfelt S (2004) Solids flushing, mixing, and water velocity profiles within large (10 and 150 m3) circular "Cornell-type" dual-drain tanks. Aquac Eng 32:245–271. https://doi.org/10.1016/j.aquaeng.2004.03.009

Davidson J, Bebak J, Mazik P (2009) The effects of aquaculture production noise on the growth, condition factor, feed conversion, and survival of rainbow trout, *Oncorhynchus mykiss*. Aquaculture 288(3–4):337–343. https://doi.org/10.1016/j.aquaculture.2008.11.037

Davidson J, Good C, Welsh C, Summerfelt S (2014) Comparing the effects of high vs. low nitrate on the health, performance, and welfare of juvenile rainbow trout *Oncorhynchus mykiss* within water recirculating aquaculture systems. Aquac Eng 59:30–40. https://doi.org/10.1016/j.aquaeng.2014.01.003

Davidson J, Good C, Williams C, Summerfelt S (2017) Evaluating the chronic effects of nitrate on the health and performance of post-smolt Atlantic salmon *Salmo salar* in freshwater recirculation aquaculture systems. Aquac Eng 79:1–8

Dawkins MS (1998) Evolution and animal welfare. Q Rev Biol 73:305–328. https://doi.org/10.1086/420307

De Carvalho RAPLF, Lemos DEL, Tacon AGJ (2013) Performance of single-drain and dual-drain tanks in terms of water velocity profile and solids flushing for in vivo digestibility studies in juvenile shrimp. Aquac Eng 57:9–17. https://doi.org/10.1016/j.aquaeng.2013.05.004

De Ionno PN, Wines GL, Jones PL, Collins RO (2007) A bioeconomic evaluation of a commercial scale recirculating finfish growout system—An Australian perspective. Aquaculture 259(1):315–327. https://doi.org/10.1016/j.aquaculture.2006.05.047

Defoirdt T, Boon N, Sorgeloos P, Verstraete W, Bossier P (2007) Alternatives to antibiotics to control bacterial infections: luminescent vibriosis in aquaculture as an example. Trends Biotechnol 25:472–479. https://doi.org/10.1016/j.tibtech.2007.08.001

Defoirdt T, Boon N, Sorgeloos P, Verstraete W, Bossier P (2008) Quorum sensing and quorum quenching in *Vibrio harveyi*: lessons learned from in vivo work. ISME J 2:19–26. https://doi.org/10.1038/ismej.2007.92

Díaz V, Ibáñez R, Gómez P, Urtiaga AM, Ortiz I (2012) Kinetics of nitrogen compounds in a commercial marine Recirculating Aquaculture System. Aquac Eng 50:20–27. https://doi.org/10.1016/j.aquaeng.2012.03.004

Dolan E, Murphy N, O'Hehir M (2013) Factors influencing optimal micro-screen drum filter selection for recirculating aquaculture systems. Aquac Eng 56:42–50. https://doi.org/10.1016/j.aquaeng.2013.04.005

Drennan DG, Hosler KC, Francis M, Weaver D, Aneshansley E, Beckman G, Johnson CH, Cristina CM (2006) Standardized evaluation and rating of biofilters. II. Manufacturer's and user's perspective. Aquac Eng 34:403–416. https://doi.org/10.1016/j.aquaeng.2005.07.001

Duncan IJ (2005) Science-based assessment of animal welfare: farm animals. Rev Sci Tech 24:483–492

Ebeling JM, Timmons MB (2012) Recirculating aquaculture systems. In: Tidwell JH (ed) Aquaculture production systems. Wiley, Oxford

Ebeling JM, Timmons MB, Bisogni JJ (2006) Engineering analysis of the stoichiometry of photoautotrophic, autotrophic, and heterotrophic removal of ammonia-nitrogen in aquaculture systems. Aquaculture 257:346–358. https://doi.org/10.1016/j.aquaculture.2006.03.019

EIFAC/ICES (1980) World conference on flow-through and recirculation systems, Stavanger, Norway 1980 and the 1981 World Aquaculture conference, Venice, Italy

Ellis T, Scott AP, Bromage N, North B, Porter M (2001) What is stocking density? Trout News 32:35–37

Emparanza EJM (2009) Problems affecting nitrification in commercial RAS with fixed-bed biofilters for salmonids in Chile. Aquac Eng 41:91–96. https://doi.org/10.1016/j.aquaeng.2009.06.010

Eshchar M, Mozes N, Fediuk M (2003) Carbon dioxide removal rate by aeration devices in marine fish tanks. Isr J Aquac-Bamidgeh 55:79–85

FAO (2001) Integrated agriculture-aquaculture: a primer 407. ISBN 9251045992

FAO (2004) Aquaculture management and conservation service (FIMA) 2004–2018. Cultured aquatic species information programme. *Carassius carassius*. Cultured aquatic species information programme. Weimin M. Accessed 5 Jan 2018

Fast AW, Tan EC, Stevens DF, Olson JC, Qin J, Barclay DK (1999) Paddlewheel aerator oxygen transfer efficiencies at three salinities. Aquac Eng 19:99–103. https://doi.org/10.1016/S0144-8609(98)00044-2

FAWC (2014) Farm animal welfare committee. Welfare of farmed fish. Available on www.defra.gov.uk/fawc. Accessed 18 Jan 2018

Fernandes P, Pedersen L-F, Pedersen PB (2015) Microscreen effects on water quality in replicated recirculating aquaculture systems. Aquac Eng 65:17–26. https://doi.org/10.1016/j.aquaeng.2014.10.007

Fox BK, Tamaru CS, Hollyer J et al (2012) A preliminary study of microbial water quality related to food safety in recirculating aquaponic fish and vegetable production systems. College of Tropical Agriculture and Human Resources, University of Hawaii at Manoa. Food Safety and Technology

Gehlert G, Griese M, Schlachter M, Schulz C (2018) Analysis and optimisation of dynamic facility ventilation in recirculating aquaculture systems. Aquac Eng 80:1–10. https://doi.org/10.1016/j.aquaeng.2017.11.003

Gendel Y, Lahav O (2013) A novel approach for ammonia removal from fresh-water recirculated aquaculture systems, comprising ion exchange and electrochemical regeneration. Aquac Eng 52:27–38. https://doi.org/10.1016/j.aquaeng.2012.07.005

Gomez RC (2011) Integrated fish farming systems. Available at: http://www.fao.org/fileadmin/templates/FCIT/Meetings/World_Water_Day_2011/5-integrated_aquaculture.pdf. Accessed 12 Feb 2018.

Gonçalves AA, Gagnon GA (2011) Ozone application in recirculating aquaculture system: an overview. Ozone Sci Eng 33:345–367. https://doi.org/10.1080/01919512.2011.604595

Good C, Davidson J, Welsh C, Snekvik K, Summerfelt S (2011) The effects of ozonation on performance, health and welfare of rainbow trout *Oncorhynchus mykiss* in low-exchange water recirculation aquaculture systems. Aquac Eng 44:97–102. https://doi.org/10.1016/j.aquaeng.2011.04.003

Good C, Davidson J, Earley RL, Lee E, Summerfelt S (2014) The impact of water exchange rate and treatment processes on water-borne hormones in recirculation aquaculture systems containing sexually maturing Atlantic salmon *Salmo salar*. J Aquac Res Development 5:260. https://doi.org/10.4172/2155-9546.1000260

Good C, Davidson J, Earley RL, Styga J, Summerfelt S (2017) The effects of ozonation on select waterborne steroid hormones in reicirculation aquaculture systems containing sexually mature Atlantic salmon *Salmo salar*. Aquac Eng 79:9–16. https://doi.org/10.1016/j.aquaeng.2017.08.004

Good C, Davidson J, Terjesen BF, Talke H, Kolarevic J, Baeverfjord G, Summerfelt S (2018) The effects of long-term 20mg/l carbon dioxide exposure on the health and performance of Atlantic salmon *Salmo salar* post-smolts in water recirculation aquaculture systems. Aquac Eng 81:1–9. https://doi.org/10.1016/j.aquaeng.2018.01.003

Gorle JMR, Terjesen BF, Mota VC, Summerfelt S (2018) Water velocity in commercial RAS culture tanks for Atlantic salmon smolt production. Aquac Eng 81:89–100

Gornati R, Papis E, Rimoldi S, Terova G, Saroglia M, Bernardini G (2004) Rearing density influences the expression of stress-related genes in sea bass (*Dicentrarchus labrax* L.). Gene 341:111–118. https://doi.org/10.1016/j.gene.2004.06.020

Graber A, Antenen N, Junge R (2014) The multifunctional aquaponic system at ZHAW used as research and training lab. In: Maček Jerala M, Maček MA (eds) Conference VIVUS: transmission of innovations, knowledge and practical experience into everyday practice, Strahinj, Slovenia

Grady CPL, Lim HC (1980) Biological wastewater treatment: theory and applications. Marcel Dekker, New York

Grimsrud KM, Nielsen HM, Olesen I (2013) Households' willingness-to-pay for improved fish welfare in breeding programs for farmed Atlantic salmon. Aquaculture 372–375:19–27. https://doi.org/10.1016/j.aquaculture.2012.10.009

Gutierrez-Wing MT, Malone RF (2006) Biological filters in aquaculture: trends and research directions for freshwater and marine applications. Aquac Eng 34:163–171. https://doi.org/10.1016/j.aquaeng.2005.08.003

Guttman L, Rijn J (2008) Identification of conditions underlying production of geosmin and 2-methylisoborneol in a recirculating system. Aquaculture 279:85–91. https://doi.org/10.1016/j.aquaculture.2008.03.047

Hammock D, Huang CC, Mort G, Swinehart JH (2003) The effect of humic acid on the uptake of mercury(II), cadmium(II), and zinc(II) by Chinook salmon (*Oncorhynchus tshawytscha*) eggs. Arch Environ Contam Toxicol 44:83–88. https://doi.org/10.1007/s00244-002-1261-9

Hulata G, Simon Y (2011) An overview on desert aquaculture in Israel. In: Crespi V, Lovatelli, A (eds) Aquaculture in desert and arid lands: development constraints and opportunities. FAO technical workshop. Hermosillo, Mexico. FAO fisheries and aquaculture proceedings no. 20. Rome, FAO. 6–9 July 2010, pp 85–112

Huntingford FA, Adams C, Braithwaite VA, Kadri S, Pottinger TG, Sandøe P, Turnbull JF (2006) Current issues in fish welfare. J Fish Biol 68:332–372. https://doi.org/10.1111/j.0022-1112.2006.001046.x

Jorgensen EH, Christiansen JS, Jobling M (1993) Effects of stocking density on food intake, growth performance and oxygen consumption in Arctic charr (*Salvelinus alpinus*). Aquaculture 110:191–204. https://doi.org/10.1016/0044-8486(93)90272-Z

Kesarcodi-Watson A, Kaspar H, Lategan MJ, Gibson L (2008) Probiotics in aquaculture: the need, principles and mechanisms of action and screening processes. Aquaculture 274:1–14. https://doi.org/10.1016/j.aquaculture.2007.11.019

Klas S, Mozes N, Lahav O (2006) Development of a single-sludge denitrification method for nitrate removal from RAS effluents: lab-scale results vs. model prediction. Aquaculture 259:342–353. https://doi.org/10.1016/j.aquaculture.2006.05.049

Krom MD, van Rijn J (1989) Water quality processes in fish culture systems: processes, problems, and possible solutions. In: de Pauw N, Jaspers E, Ackerfors H, Wilkins N (eds) Aquaculture-a biotechnology in progress, vol 2. European aquaculture society, Bredene, pp 1091–1111

Kucuk H, Midilli A, Özdemir A, Çakmak E, Dincer I (2010) Exergetic performance analysis of a recirculating aquaculture system. Energy Convers Manag 51(5):1033–1043. https://doi.org/10.1016/j.enconman.2009.12.007

Lahav O, Ben Asher R, Gendel Y (2015) Potential applications of indirect electrochemical ammonia oxidation within the operation of freshwater and saline-water recirculating aquaculture systems. Aquac Eng 65:55–64. https://doi.org/10.1016/j.aquaeng.2014.10.009

Lee J (2014) Separation of fine organic particles by a low-pressure hydrocyclone (LPH). Aquac Eng 63:32–38. https://doi.org/10.1016/j.aquaeng.2014.07.002

Lekang OI (2013) Aquaculture engineering, 2nd edn. Wiley, Ames

Leonard N, Blancheton JP, Guiraud JP (2000) Populations of heterotrophic bacteria in an experimental recirculating aquaculture system. Aquac Eng 22:109–120. https://doi.org/10.1016/S0144-8609(00)00035-2

Leonard N, Guiraud JP, Gasset E, Cailleres JP, Blancheton JP (2002) Bacteria and nutrients – nitrogen and carbon – in a recirculating system for sea bass production. Aquac Eng 26:111–127. https://doi.org/10.1016/S0144-8609(02)00008-0

Lewis WM, Wehr LW (1976) A fish-rearing system incorporating cages, water circulation, and sewage removal. Prog Fish Cult 38(2):78–81. https://doi.org/10.1577/1548-8659(1976)38[78:AFSICW]2.0.CO;2

Lewis WM, Yopp JH, Schramm HL Jr, Brandenburg AM (1978) Use of hydroponics to maintain quality of recirculated water in a fish culture system. Trans Am Fish Soc 107(1):92–99. https://doi.org/10.1577/1548-8659(1978)107<92:UOHTMQ>2.0.CO;2

Li X, Przybyla C, Triplet S, Liu Y, Blancheton JP (2015) Long-term effects of moderate elevation of oxidation-reduction potential on European seabass (*Dicentrarchus labrax*) in recirculating aquaculture systems. Aquac Eng 64:15–19. https://doi.org/10.1016/j.aquaeng.2014.11.006

Liu Y, Rosten T, Henriksen K, Hognes E, Summerfelt S, Vinci B (2016) Comparative economic performance and carbon footprint of two farming models for producing Atlantic salmon (*Salmo salar*): land-based closed containment system in freshwater and open net pen in seawater. Aquac Eng 71:1–12

Livingstone DR (2003) Oxidative stress in aquatic organisms in relation to pollution and aquaculture. Rev Med Vet-Toulouse 154:427–430

Loyless JC, Malone RE (1998) Evaluation of air-lift pump capabilities for water delivery, aeration, and degasification for application to recirculating aquaculture systems. Aquac Eng 18 (2):117–133. https://doi.org/10.1016/S0144-8609(98)00025-9

Malone RF, Pfeiffer TJ (2006) Rating fixed film nitrifying biofilters used in recirculating aquaculture systems. Aquac Eng 34:389–402. https://doi.org/10.1016/j.aquaeng.2005.08.007

Manan H, Moh JHZ, Kasan NA et al (2017) Identification of biofloc microscopic composition as the natural bioremediation in zero water exchange of pacific white shrimp, Penaeus vannamei, cultured in close hatchery system. Appl Water Sci 7:2437–2446. https://doi.org/10.1007/s13201-016-0421-4

Mancinelli RL (1996) The nature of nitrogen: an overview. Life Support Biosph Sci 3(1–2):17–24

Martins CIM, Eding EH, Schneider O, Rasmussen R, Olesen B, Plesner L, Verreth JAJ (2005) Recirculation aquaculture systems in Europe. CONSENSUS. Oostende, Belgium, Consensus working group. Eur Aquacult Soc 31

Martins CIM, Ochola D, Ende SSW, Eding EH, Verreth JAJ (2010a) Is growth retardation present in Nile tilapia *Oreochromis niloticus* cultured in low water exchange recirculating aquaculture systems. Aquaculture 298:43–50. https://doi.org/10.1016/j.aquaculture.2009.09.030

Martins CIM, Eding EH, Verdegem MCJ, Heinsbroek LTN, Schneider O, Blancheton JP, Roque d'Orbcastel E, Verreth JAJ (2010b) New developments in recirculating aquaculture systems in Europe: a perspective on environmental sustainability. Aquac Eng 43(3):83–93. https://doi.org/10.1016/j.aquaeng.2010.09.002

Martins CIM, Eding EH, Verreth JAJ (2011) Stressing fish in recirculating aquaculture systems (RAS): does stress induced in one group of fish affect the feeding motivation of other fish sharing the same RAS? Aquac Res 42:1378–1384. https://doi.org/10.1111/j.1365-2109.2010.02728.x

Martins CIM, Galhardo L, Noble C et al (2012) Behavioural indicators of welfare in farmed fish. Fish Physiol Biochem 38:17. https://doi.org/10.1007/s10695-011-9518-8

Masalo I (2008) Hydrodynamic characterization of aquaculture tanks and design criteria for improving self-cleaning properties. Ph.D thesis. Universitat Politechnica de Catalunya BARCELONATECH, Spain

Meinelt T, Paul A, Phan TM, Zwirnmann E, Krüger A, Wienke A, Steinberg CEW (2007) Reduction in vegetative growth of the water mold *Saprolegnia parasitica* (Coker) by humic substances of different qualities. Aquat Toxicol 83:93–103. https://doi.org/10.1016/j.aquatox.2007.03.013

Meinelt T, Kroupova H, Stüber A, Rennert B, Wienke A, Steinberg CEW (2010) Can dissolved aquatic humic substances reduce the toxicity of ammonia and nitrite in recirculating aquaculture systems? Aquaculture 306:378–383. https://doi.org/10.1016/j.aquaculture.2010.06.007

Merchie G, Lavens P, Verreth J, Ollevier F, Nelis H, DeLeenheer A, Storch V, Sorgeloos P (1997) The effect of supplemental ascorbic acid in enriched live food for *Clarias gariepinus* larvae at start feeding. Aquaculture 151:245–258. https://doi.org/10.1016/S0044-8486(96)01505-0

Michaud L, Blancheton JP, Bruni V, Piedrahita R (2006) Effect of particulate organic carbon on heterotrophic bacterial populations and nitrification efficiency in biological filters. Aquac Eng 34:224–233. https://doi.org/10.1016/j.aquaeng.2005.07.005

Michaud L, Lo Giudice A, Troussellier M, Smedile F, Bruni V, Blancheton JP (2009) Phylogenetic characterization of the heterotrophic bacterial communities inhabiting a marine recirculating aquaculture system. J Appl Microbiol 107:1935–1946. https://doi.org/10.1111/j.1365-2672.2009.04378.x

Michaud L, Lo Giudice A, Interdonato F, Triplet S, Ying L, Blancheton JP (2014) C/N ratio-induced structural shift of bacterial communities inside lab-scale aquaculture biofilters. Aquac Eng 58:77–87. https://doi.org/10.1016/j.aquaeng.2013.11.002

Montero D, Izquierdo MS, Tort L, Robaina L, Vergara JM (1999) High stocking density produces crowding stress altering some physiological and biochemical parameters in gilthead seabream, *Sparus aurata*, juveniles. Fish Physiol Biochem 20:53–60. https://doi.org/10.1023/A:1007719928905

Moran D (2010) Carbon dioxide degassing in fresh and saline water. I: degassing performance of a cascade column. Aquac Eng 43:29–36. https://doi.org/10.1016/j.aquaeng.2010.05.001

Munguia-Fragozo P, Alatorre-Jacome O, Rico-Garcia E et al (2015) Perspective for aquaponic systems: "Omic" technologies for microbial community analysis. Biomed Res Int 2015, Article ID 480386, 10 pages. https://doi.org/10.1155/2015/480386

Murray RG, Watson SW (1965) Structure of *Nitrocystis* oceanus and comparison with *Nitrosomonas* and *Nitrobacter*. J Bacteriol 89(6):1594–1609

Murray F, Bostock J, Fletcher M (2014) Review of RAS technologies and their commercial application. Final report. Available at http://www.hie.co.uk

Naegel LCA (1977) Combined production of fish and plants in recirculating water. Aquaculture 10 (1):17–24. https://doi.org/10.1016/0044-8486(77)90029-1

Noga E (2010) Fish disease: diagnosis and treatment, 2nd edn. Wiley, Ames

Oca J, Masalo I (2012) Flow pattern in aquaculture circular tanks: influence of flow rate, water depth, and water inlet & outlet features. Aquac Eng 52:65–72

Olesen I, Alfnes F, Bensze Røra M, Kolstad K (2010) Eliciting consumers' willingness to pay for organic and welfare-labelled salmon in a non-hypothetical choice experiment. Livest Sci 127 (2–3):218–226. https://doi.org/10.1016/j.livsci.2009.10.001

Palm HW, Knaus U, Appelbaum S, Goddek S, Strauch SM, Vermeulen T, Haïssam JM, Kotzen B (2018) Towards commercial aquaponics: a review of systems, designs, scales and nomenclature. Aquac Int 26:813. https://doi.org/10.1007/s10499-018-0249-z

Papoutsoglou SE, Karakatsouli N, Batzina A, Papoutsoglou ES, Tsopelakos A (2008) Effect of Mozart's music stimulus on gilthead seabream (*Sparus aurata* L.) physiology under different light intensity in a recirculating water system. J Fish Biol 73:980–1004. https://doi.org/10.1111/j.1095-8649.2008.02001.x

Park J, Kim PK, Lim T, Daniels HV (2013) Ozonation in seawater recirculating systems for black seabream *Acanthopagrus schlegelii* (Bleeker): effects on solids, bacteria, water clarity, and color. Aquac Eng 55:1–8. https://doi.org/10.1016/j.aquaeng.2013.01.002

Park J, Kim PK, Park S, Daniels HV (2015) Effects of two different ozone doses on total residual oxidants, nitrogen compounds and nitrification rate in seawater recirculating systems for black seabream *Acanthopagrus schlegelii* (Bleeker). Aquac Eng 67:1–7. https://doi.org/10.1016/j.aquaeng.2015.05.003

Pedersen PB, von Ahnen M, Fernandes P, Naas C, Pedersen L-F, Dalsgaard J (2017) Particle surface area and bacterial activity in recirculating aquaculture systems. Aquac Eng 78A:18–23. https://doi.org/10.1016/j.aquaeng.2017.04.005

Peuranen S, Vuorinen PJ, Vuorinen M, Hollender A (1994) The effects of iron, humic acids and low pH on the gills and physiology of brown trout (*Salmo trutta*). Ann Zool Fenn 31:389–396

Pittmann T, Steinmetz H (2013) Development of a process to produce bioplastic at municipal wastewater treatment plants. Conference paper

Prosser JI (1989) Autotrophic nitrification in bacteria. Adv Microb Physiol 30:125–181

Randall DJ, Tsui TKN (2002) Ammonia toxicity in fish. Marine Poll Bull 45:17–23

Reiser S, Schroeder JP, Wuertz S, Kloas W, Hanel R (2010) Histological and physiological alterations in juvenile turbot (*Psetta maxima*, L.) exposed to sublethal concentrations of ozone-produced oxidants in ozonated seawater. Aquaculture 307:157–164. https://doi.org/10.1016/j.aquaculture.2010.07.007

Reiser S, Wuertz S, Schroeder JP, Kloas W, Hanel R (2011) Risks of seawater ozonation in recirculation aquaculture – effects of oxidative stress on animal welfare of juvenile turbot (*Psetta maxima*, L.). Aquat Toxicol 105:508–517. https://doi.org/10.1016/j.aquatox.2011.08.004

Richardson LB, Burton DT, Block RM, Stavola AM (1983) Lethal and sublethal exposure and recovery effects of ozone-produced oxidants on adult white perch (*Morone americana* Gmelin). Water Res 17:205–213. https://doi.org/10.1016/0043-1354(83)90101-X

Ritola O, Lyytikainen T, Pylkko P, Molsa H, Lindstrom-Seppa P (2000) Glutathione-dependant defence system and monooxygenase enzyme activities in Arctic charr *Salvelinus alpinus* (L.) exposed to ozone. Aquaculture 185:219–233. https://doi.org/10.1016/S0044-8486(99)00355-5

Ritola O, Livingstone DR, Peters LD, Lindstrom-Seppa P (2002) Antioxidant processes are affected in juvenile rainbow trout (*Oncorhynchus mykiss*) exposed to ozone and oxygen-supersaturated water. Aquaculture 210:1–9. https://doi.org/10.1016/S0044-8486(01)00823-7

Rojas-Tirado P, Bovbjerg Pedersen P, Pedersen L-F (2017) Bacterial activity dynamics in the water phase during start-up of recirculating aquaculture systems. Aquac Eng 78A:24–31. https://doi.org/10.1016/j.aquaeng.2016.09.004

Rosenthal H (1980) Recirculation systems in western Europe. World symposium on aquaculture in heated effluents and recirculation system, Stavanger, Institut für Kuesten- und Binnefischerei, Bundesforschungsanstalt Hamburg, BRD

Rosenthal H, Black EA (1993) Recirculation systems in aquaculture. In: Wang J-K (ed) Techniques for modem aquaculture. ASAE, St. Joseph, pp 284–294

Rurangwa E, Verdegem MCJ (2015) Microorganisms in recirculating aquaculture systems and their management. Rev Aquac 7:117–130. https://doi.org/10.1111/raq.12057

Rusten B, Eikebrokk B, Ulgenes Y, Lygren E (2006) Design and operations of the Kaldnes moving bed biofilm reactors. Aquac Eng 34:322–331. https://doi.org/10.1016/j.aquaeng.2005.04.002

Schmautz Z, Graber A, Jaenicke S, Goesmann A, Junge R, Smits THM (2017) Microbial diversity in different compartments of an aquaponics system. Arch Microbiol 199:613. https://doi.org/10.1007/s00203-016-1334-1

Schreier HJ, Mirzoyan N, Saito K (2010) Microbial diversity of biological filters in recirculating aquaculture systems. Curr Opin Biotechnol 21:318–325. https://doi.org/10.1016/j.copbio.2010.03.011

Schroeder JP, Croot PL, Von Dewitz B, Waller U, Hanel R (2011) Potential and limitations of ozone for the removal of ammonia, nitrite, and yellow substances in marine recirculating aquaculture systems. Aquac Eng 45:35–41. https://doi.org/10.1016/j.aquaeng.2011.06.001

Sharrer MJ, Summerfelt ST, Bullock GL, Gleason LE, Taeuber J (2005) Inactivation of bacteria using ultraviolet irradiation in a recirculating salmonid culture system. Aquac Eng 33:135–149. https://doi.org/10.1016/j.aquaeng.2004.12.001

Singh S, Marsh LS (1996) Modelling thermal environment of a recirculating aquaculture facility. Aquaculture 139:11–18. https://doi.org/10.1016/0044-8486(95)01164-1

Sirsat S, Neal J (2013) Microbial profile of soil-free versus insoil grown lettuce and intervention methodologies to combat pathogen surrogates and spoilage microorganisms on lettuce. Foods 2 (4):488–498. https://doi.org/10.3390/foods2040488

Somerville C, Cohen M, Pantanella E, Stankus A, Lovatelli A (2014) Small-scale aquaponic food production: integrated fish and plant farming. In: FAO fisheries and aquaculture technical paper food and agriculture organization of the United Nations, Rome, Italy, p 262

Stevenson FJ (1994) Humus Chemistry: genesis, composition, reactions, 2nd edn. Wiley, New York, p 496. https://doi.org/10.1021/ed072pA93.6

Suhr KI, Pedersen LF, Nielsen JL (2014) End-of-pipe single-sludge denitrification in pilot-scale recirculating aquaculture systems. Aquac Eng 62:28–35. https://doi.org/10.1016/j.aquaeng.2014.06.002

Summerfelt ST (2003) Ozonation and UV irradiation – An introduction and examples of current applications. Aquac Eng 28:21–36. https://doi.org/10.1016/S0144-8609(02)00069-9

Summerfelt ST (2006) Design and management of conventional fluidized-sand biofilters. Aquac Eng 34:275–302. https://doi.org/10.1016/j.aquaeng.2005.08.010

Summerfelt ST, Hochheimer JN (1997) Review of Ozone processes and applications as an oxidizing agent in Aquaculture. Progress Fish-Culturist 59:94–105. https://doi.org/10.1577/1548-8640(1997)059<0094:ROOPAA>2.3.CO;2

Summerfelt S, Davidson JW, Waldrop TB, Tsukuda S, Bebak-Williams J (2004) A partial-reuse system for coldwater aquaculture. Aquac Eng 157–181. https://doi.org/10.1016/j.cell.2013.02.048

Summerfelt ST, Sharrer MJ, Tsukuda SM, Gearheart M (2009) Process requirements for achieving full-flow disinfection of recirculating water using ozonation and UV irradiation. Aquac Eng 40:17–27. https://doi.org/10.1016/j.aquaeng.2008.10.002

Summerfelt ST, Zulke A, Kolarevic J, Megard Reiten BK, Selset R, Guiterrez X, Terjesen BF (2015) The effects of alkalinity on ammonia removal, carbon dioxide stripping, and pH in semi-commercial scale water recirculating aquaculture systems operated with moving bed bioreactors. Aquac Eng 65:46–54. https://doi.org/10.1016/j.aquaeng.2014.11.002

Summerfelt ST, Mathisen F, Buran Holan A, Terjesen BF (2016) Survey of large circular and octagonal tanks operated at Norwegian commercial smolt and post-smolt sites. Aquac Eng 74:105–110. https://doi.org/10.1016/j.aquaeng.2016.07.004

Tal Y, Watts JEM, Schreier HJ (2006) Anaerobic ammonium-oxidizing (Anammox) bacteria and associated activity in fixed-film biofilters of a marine recirculating aquaculture system. Appl Environ Microbiol 72:2896–2904. https://doi.org/10.1128/AEM.72.4.2896-2904.2006

Tal Y, Schreier HJ, Sowers KR, Stubblefield JD, Place AR, Zohar Y (2009) Environmentally sustainable land-based marine aquaculture. Aquaculture 286:28–35. https://doi.org/10.1016/j.aquaculture.2008.08.043

Tango MS, Gagnon GA (2003) Impact of ozonation on water quality in marine recirculation systems. Aquac Eng 29:125–137. https://doi.org/10.1016/S0144-8609(03)00061-X

Terhune JM, Friars GW, Bailey JK, O'Flynn FM (1990) Noise levels may influence Atlantic salmon smolting rates in tanks. J Fish Biol 37:185–197. https://doi.org/10.1111/j.1095-8649.1990.tb05939.x

Timmons MB, Ebeling JM (2010) Recirculating aquaculture, 2nd edn. Cayuga Aqua Ventures, Ithaca

Timmons MB, Losordo TM (1994) Aquaculture water reuse systems – engineering design and management. Elsevier, Amsterdam

Tomita-Yokotani K, Anilir S, Katayama N, Hashimoto H, Yamashita M (2009) Space agriculture for habitation on mars and sustainable civilization on earth. Recent Adv Space Technol:68–69. https://doi.org/10.1109/RAST.2009.5158276

Tort L, Rotllant J, Liarte C, Acerete L, Hernandez A, Ceulemans S, Coutteau P, Padros F (2004) Effects of temperature decrease on feeding rates, immune indicators and histopathological changes of gilthead sea bream S*parus aurata* fed with an experimental diet. Aquaculture 229:55–65. https://doi.org/10.1016/S0044-8486(03)00403-4

Turnbull JF, Kadri S (2007) Safeguarding the many guises of farmed fish welfare. Dis Aquat Org 75:173–182. https://doi.org/10.3354/dao075173

Vadstein O, Mo TA, Bergh Ø (2004) Microbial interactions, prophylaxis and diseases. In: Moksness E, Kjørsvik E, Olsen Y (eds) Culture of cold-water marine fishes. Blackwell Publishing, Bath, pp 28–72. https://doi.org/10.1016/j.aquaeng.2005.04.004

van Rijn J (1996) The potential for integrated biological treatment systems in recirculating fish culture—a review. Aquaculture 139(3–4):181–201. https://doi.org/10.1016/0044-8486(95)01151-X

van Rijn J (2013) Waste treatment in recirculating aquaculture systems. Aquac Eng 53:49–56. https://doi.org/10.1016/j.aquaeng.2012.11.010

van Rijn J, Nussinovitch A (1997) An empirical model for predicting degradation of organic solids in fish culture systems based on short-term observations. Aquaculture 154(2):173–179. https://doi.org/10.1016/S0044-8486(97)00048-3

van Rijn J, Rivera G (1990) Aerobic and anaerobic biofiltration in an aquaculture unit: nitrite accumulation as a result of nitrification and denitrification. Aquac Eng 9:1–18

van Rijn J, Tal Y, Schreier HJ (2006) Denitrification in recirculating systems: theory and applications. Aquac Eng 34(3):364–376

Vazzana M, Cammarata M, Cooper EL, Parrinello N (2002) Confinement stress in sea bass (*Dicentrarchus labrax*) depresses peritoneal leukocyte cytotoxicity. Aquaculture 210:231–243. https://doi.org/10.1016/S0044-8486(01)00818-3

Veerapen JP, Lowry BJ, Couturier MF (2005) Design methodology for the swirl separator. Aquac Eng 33:21–45. https://doi.org/10.1016/j.aquaeng.2004.11.001

Verreth JAJ, Eding EH (1993) European farming industry of African catfish (*Clarias gariepinus*) facts and figures. J World Aquacult Soc 24:6–13

Volpato GL, Goncalves-de-Freitas E, Fernandes-de-Castilho M (2007) Insights into the concept of fish welfare. Dis Aquat Org 75:165–171. https://doi.org/10.3354/dao075165

Wagner EJ, Bosakowski T, Miller SA (1995) Evaluation of the absorption efficiency of the Low Head Oxygenation System. Aquac Eng 14:49–57. https://doi.org/10.1016/0144-8609(94)P4426-C

Wold P-A, Holan AB, Øie G, Attramadal K, Bakke I, Vadstein O, Leiknes TO (2014) Effects of membrane filtration on bacterial number and microbial diversity in marine recirculating aquaculture system (RAS) for Atlantic cod (*Gadus morhua* L.) production. Aquaculture 422–423:69–77. https://doi.org/10.1016/j.aquaculture.2013.11.019

Wong KB, Piedrahita RH (2000) Settling velocity characterization of aquacultural solids. Aquac Eng 21:233–246. https://doi.org/10.1016/S0144-8609(99)00033-3

Wysocki LE, Davidson JW III, Smith ME, Frankel AS, Ellison TE, Mazik PM, Popper AN, Bebak J (2007) The effects of aquaculture production noise on hearing, growth, and disease resistance of rainbow trout, *Oncorhynchus mykiss*. Aquaculture 272:687–697. https://doi.org/10.1016/j.aquaculture.2007.07.225

Yamin G, Borisover M, Cohen E, van Rijn J (2017a) Accumulation of humic-like and proteinaceous dissolved organic matter in zero-discharge aquaculture systems as revealed by fluorescence EEM spectroscopy. Water Res 108:412–421. https://doi.org/10.1016/j.watres.2016.11.028

Yamin G, Falk R, Avtalion RR, Shoshana N, Ofek T, Smirnov R, Rubenstein G, van Rijn J (2017b) The protective effect of humic-rich substances on atypical *Aeromonas salmonicida subsp. salmonicida* infection in common carp (*Cyprinus carpio* L.). J Fish Dis 40(12):1783–1790. https://doi.org/10.1111/jfd.12645

Yamin G, Zilberg D, Levy G, van Rijn J (2017c) The protective effect of humic-rich substances from monogenean parasites infecting the guppy (*Poecilia reticulata*). Aquaculture 479:487–489. https://doi.org/10.1016/j.aquaculture.2017.06.022

Yildiz HY, Robaina L, Pirhonen J, Mente E, Domínguez D, Parisi G (2017) Fish welfare in aquaponic systems: its relation to water quality with an emphasis on feed and faeces—a review. Water 9(1):13. https://doi.org/10.3390/w9010013

Yogev U, Sowers KR, Mozes N, Gross A (2016) Nitrogen and carbon balance in a novel near-zero exchange saline recirculating aquaculture system. Aquaculture 467:118–126

Zeng Q, Tian X, Wang L (2017) Genetic adaptation of microbial populations present in high-intensity catfish production systems with therapeutic oxytetracycline treatment. Sci Rep 7:17491. https://doi.org/10.1038/s41598-017-17640-3

Zhu S, Chen S (2002) The impact of temperature on nitrification rate in fixed film biofilters. Aquac Eng 26(4):221–237. https://doi.org/10.1016/S0144-8609(02)00022-5

Zou Y, Hu Z, Zhang J, Xie H, Guimbaud C, Fang Y (2016) Effects of pH on nitrogen transformations in media-based aquaponics. Bioresour Technol 210:81–87. https://doi.org/10.1016/j.biortech.2015.12.079

4

Hydroponic Types and Disinfection Methods

Carmelo Maucieri, Carlo Nicoletto, Erik van Os, Dieter Anseeuw, Robin Van Havermaet, and Ranka Junge

Abstract Hydroponics is a method to grow crops without soil, and as such, these systems are added to aquaculture components to create aquaponics systems. Thus, together with the recirculating aquaculture system (RAS), hydroponic production forms a key part of the aqua-agricultural system of aquaponics. Many different existing hydroponic technologies can be applied when designing aquaponics systems. This depends on the environmental and financial circumstances, the type of crop that is cultivated and the available space. This chapter provides an overview of different hydroponic types, including substrates, nutrients and nutrient solutions, and disinfection methods of the recirculating nutrient solutions.

Keywords Hydroponics · Soilless culture · Nutrients · Grow media · Aeroponics · Aquaponics · Nutrient solution

---

C. Maucieri (✉) · C. Nicoletto
Department of Agriculture, Food, Natural Resources, Animals and Environment, University of Padova, Campus of Agripolis, Legnaro, Italy
e-mail: carmelo.maucieri@unipd.it; carlo.nicoletto@unipd.it

E. v. Os
Wageningen University and Research, Business Unit Greenhouse Horticulture, Wageningen, The Netherlands
e-mail: erik.vanos@wur.nl

D. Anseeuw
Inagro, Roeselare, Belgium
e-mail: info@inagro.be

R. V. Havermaet
Provinciaal Proefcentrum voor de Groenteteelt Oost-Vlaanderen, Kruishoutem, Belgium
e-mail: robin@pcgroenteteelt.be

R. Junge
Institute of Natural Resource Sciences Grüental, Zurich University of Applied Sciences, Wädenswil, Switzerland
e-mail: ranka.junge@zhaw.ch

4.1 Introduction

In horticultural crop production, the definition soilless cultivation encompasses all the systems that provide plant production in soilless conditions in which the supply of water and of minerals is carried out in nutrient solutions with or without a growing medium (e.g. stone wool, peat, perlite, pumice, coconut fibre, etc.). Soilless culture systems, commonly known as hydroponic systems, can further be divided into open systems, where the surplus nutrient solution is not recycled, and closed systems, where the excess flow of nutrients from the roots is collected and recycled back into the system (Fig. 4.1).

Soilless culture systems have evolved as one possible solution to avoid soil-borne diseases that have always been a problem in the greenhouse cultivation industry.

Nowadays, soilless growing systems are common in horticultural practice in most European countries, although not in every country does this occur on a large scale. The advantages of soilless systems compared to soil grown crops are:

- Pathogen-free start with the use of substrates other than soil and/or easier control of soil-borne pathogens.
- Growth and yield are independent of the soil type/quality of the cultivated area.
- Better control of growth through a targeted supply of nutrient solution.
- The potential for reusing the nutrient solution allowing for maximizing resources.
- Increased quality of produce gained by the better control of other environmental parameters (temperature, relative humidity) and pests.

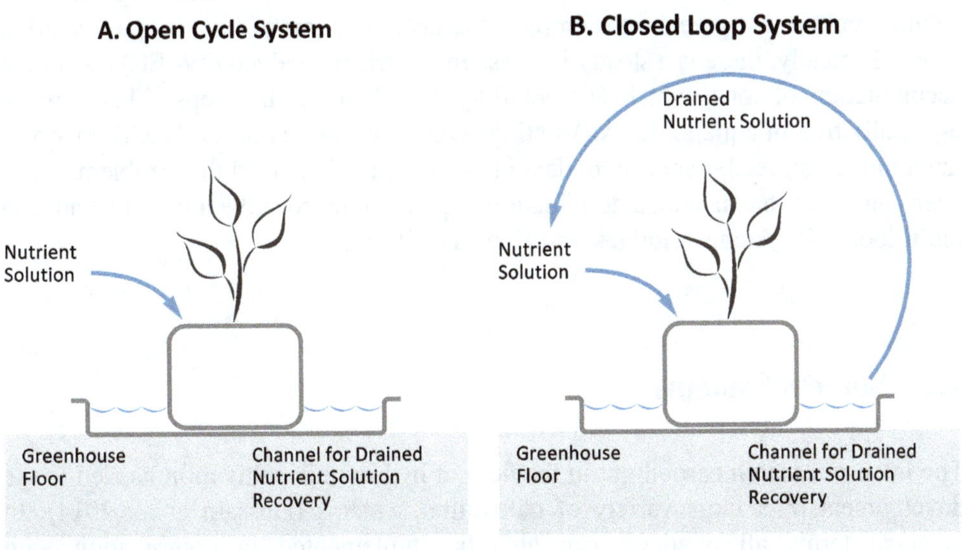

Fig. 4.1 Scheme of open cycle (**a**) and closed loop systems (**b**)

In most cases, open loop or run-to-waste systems rather than closed loop or recirculation systems are adopted, although in more and more European countries the latter are mandatory. In these open systems, the spent and/or superfluous nutrient solution is deposited into the ground and surface water bodies, or it is used in open field cultivation. However, regarding economics and environmental concerns, soilless systems should be as closed as possible, i.e. where recirculation of the nutrient solution occurs, where the substrate is reused and where more sustainable materials are used.

The advantages of closed systems are:

- A reduction in the amount of waste material.
- Less pollution of ground and surface water.
- A more efficient use of water and fertilizers.
- Increased production because of better management options.
- Lower costs because of the savings in materials and higher production.

There are also a number of disadvantages such as:

- The required high water quality.
- High investments.
- The risk of rapid dispersal of soil-borne pathogens by the recirculating nutrient solution.
- Accumulation of potential phytotoxic metabolites and organic substances in the recirculating nutrient solution.

In commercial systems, the problems of pathogen dispersal are tackled by disinfecting the water through physical, chemical and/or biological filtration techniques. However, one of the main factors that hinder the use of recirculating nutrient solution culture for greenhouse crops is the accumulation of salts in the irrigation water. Typically, there is a steady increase in electrical conductivity (EC) due to the accumulation of ions, which are not fully absorbed by the crops. This may be especially true in aquaponic (AP) settings where sodium chloride (NaCl), incorporated in the fish feed, may accumulate in the system. To amend this problem, it has been suggested that an added desalination step could improve the nutrient balance in multi-loop AP systems (Goddek and Keesman 2018).

4.2 Soilless Systems

The intense research carried out in the field of hydroponic cultivation has led to the development of a large variety of cultivation systems (Hussain et al. 2014). In practical terms all of these can also be implemented in combination with

Table 4.1 Classification of hydroponic systems according to different aspects

Characteristic	Categories	Examples
Soilless system	No substrate	NFT (nutrient film technique)
		Aeroponics
		DFT (deep flow technique)
	With substrate	Organic substrates (peat, coconut fibre, bark, wood fibre, etc.)
		Inorganic substrates (stone wool, pumice, sand, perlite, vermiculite, expanded clay)
		Synthetic substrates (polyurethane, polystyrene)
Open/closed systems	Open or run-to-waste systems	The plants are continuously fed with "fresh" solution without recovering the solution drained from the cultivation modules (Fig. 4.1a)
	Closed or recirculation systems	The drained nutrient solution is recycled and topped up with lacking nutrients to the right EC level (Fig. 4.1b)
Water supply	Continuous	NFT (nutrient film technique) DFT (deep flow technique)
	Periodical	Drip irrigation, ebb and flow, aeroponics

aquaculture; however, for this purpose, some are more suitable than others (Maucieri et al. 2018). The great variety of systems that may be used necessitates a categorization of the different soilless systems (Table 4.1).

4.2.1 Solid Substrate Systems

At the start of soilless cultivation in the 1970s, many substrates were tested (Wallach 2008; Blok et al. 2008; Verwer 1978). Many failed for reasons such as being too wet, too dry, not sustainable, too expensive and releasing of toxic substances. Several solid substrates survived: stone wool, perlite, coir (coconut fibre), peat, polyurethane foam and bark. Solid substrate systems can be divided as follows:

Fibrous Substrates These may be organic (e.g. peat, straw and coconut fibre) or inorganic (e.g. stone wool). They are characterized by the presence of fibres of different sizes, which give the substrate a high water-retention capacity (60–80%) and a modest air capacity (free porosity) (Wallach 2008). A high percentage of the retained water is easily available for the plant, which is directly reflected in the minimum volume of substrate per plant required to guarantee a sufficient water supply. In these substrates there are no obvious water and salinity gradients along the profile, and, consequently, the roots tend to grow faster, evenly and abundantly, using the entire available volume.

Granular Substrates They are generally inorganic (e.g. sand, pumice, perlite, expanded clay) and are characterized by different particle sizes and thus textures; they have high porosity and are free draining. Water-holding capacity is rather poor (10–40%), and much of the water retained is not easily available to the plant (Maher et al. 2008). Therefore, the required volume of substrate per plant is higher compared to the fibrous ones. In granular substrates, a marked gradient of moisture is observed along the profile and this causes the roots to develop mainly on the bottom of containers. Smaller particle sizes, increase in the capacity for water retention, moisture homogeneity and greater EC and a lower volume of the substrate are required for the plant.

Substrates are usually enveloped in plastic coverings (so-called grow bags or slabs) or inserted in other types of containers of various sizes and of synthetic materials.

Before planting the substrate should be saturated in order to:

- Provide adequate water and nutrients supply in the entire substrate slab.
- Achieve uniform EC and pH levels.
- Expel the presence of air and make a homogeneous wetting of the material.

It is equally important for a substrate dry phase after planting to stimulate the plants to evolve homogeneous substrate exploration by roots to obtain an abundant and well-distributed root system at the various levels and to expose the roots to air. Using a substrate for the second time by rewetting may be a problem because saturation is not possible due to drain holes in the plastic envelope. In an organic substrate (such as coir), adopting short and frequent irrigation turns, it is possible to recover the water-retention capability to use it for the second time, more easily than inert substrates (stone wool, perlite) (Perelli et al. 2009).

4.2.2 Substrates for Medium-Based Systems

A substrate is necessary for the anchorage of the roots, a support for the plant and also as a water-nutritional mechanism due to its microporosity and cation exchange capacity.

Plants grown in soilless systems are characterized by an unbalanced shoot/root ratio, demands for water, air and nutrients that are much greater than in open field conditions. In the latter case, the growth rates are slower, and the quantities of substrate are theoretically unlimited. To satisfy these requirements, it is necessary to resort to substrates which, alone or in mixture, ensure optimal and stable chemical–physical and nutritional conditions. An array of materials with different characteristics and costs can be used as substrates as illustrated in Fig. 4.2. However, as yet, there is no one substrate that can be used universally in all cultivation situations.

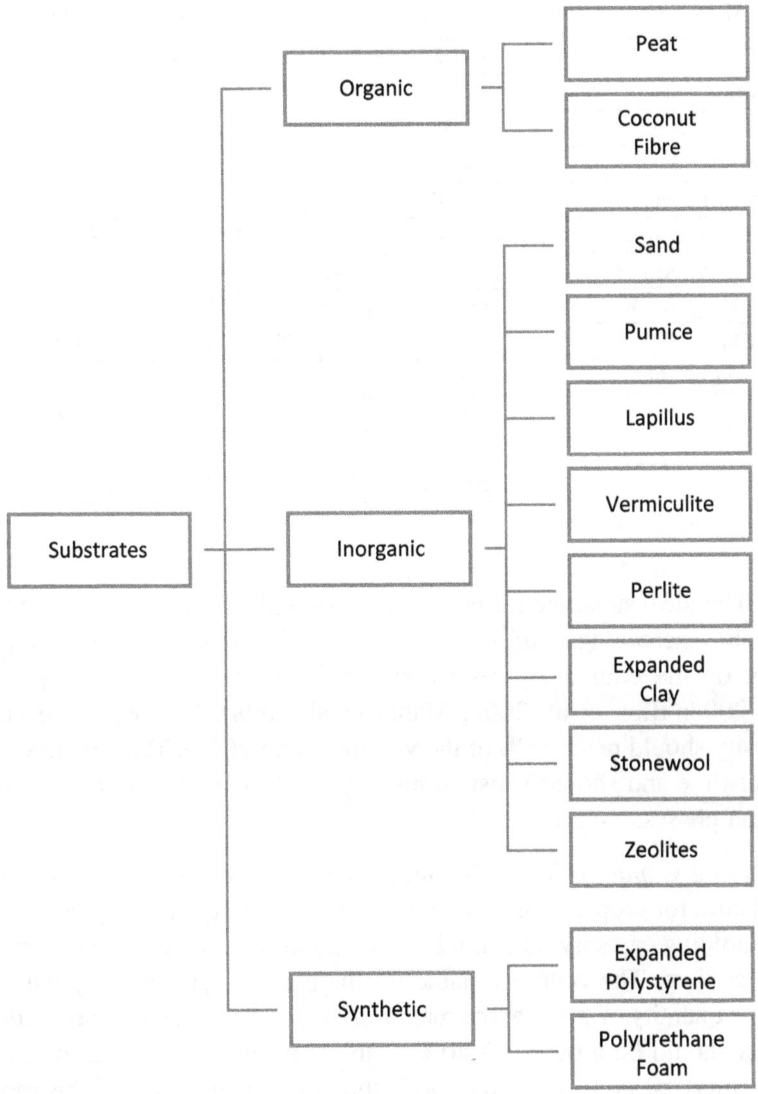

Fig. 4.2 Materials usable as substrates in soilless systems

4.2.3 Characterization of Substrates

Bulk Density (BD) BD is expressed by the dry weight of the substrate per unit of volume. It enables the anchorage of the roots and offers plant support. The optimum BD for crops in a container varies between 150 and 500 kg m^{-3} (Wallach 2008). Some substrates, because of their low BD and their looseness, as is the case of perlite (ca. 100 kg m^{-3}), polystyrene in granules (ca. 35 kg m^{-3}) and non-compressed sphagnum peat (ca. 60 kg m^{-3}), are not suitable for use alone, especially with plants that grow vertically.

Table 4.2 Main chemical–physical characteristics of peats and coconut fibre. (dm = dry matter)

	Raised bogs		Fen bogs	
Characteristics	Blond	Brown	Black	Coconut fibre (coir)
Organic matter (% dm)	94–99	94–99	55–75	94–98
Ash (% dm)	1–6	1–6	23–30	3–6
Total porosity (% vol)	84–97	88–93	55–83	94–96
Water-retention capacity (% vol)	52–82	74–88	65–75	80–85
Free porosity (% vol)	15–42	6–14	6–8	10–12
Bulk density (kg m^{-3})	60–120	140–200	320–400	65–110
CEC (meq%)	100–150	120–170	80–150	60–130
Total nitrogen (% dm)	0.5–2.5	0.5–2.5	1.5–3.5	0.5–0.6
C/N	30–80	20–75	10–35	70–80
Calcium (% dm)	<0.4	<0.4	>2	–
pH (H_2O)	3.0–4.0	3.0–5.0	5.5–7.3	5.0–6.8

Source: Enzo et al. (2001)

Porosity The ideal substrate for potted crops should have a porosity of at least 75% with variable percentages of macropores (15–35%) and micropores (40–60%) depending on the cultivated species and the environmental and crop conditions (Wallach 2008; Blok et al. 2008; Maher et al. 2008). In small-sized containers, total porosity should reach 85% of the volume (Bunt 2012). The structure should be stable over time and should resist compaction and the reduction of volume during dehydration phases.

Water-Holding Capacity Water-holding capacity ensures adequate levels of substrate moisture for crops, without having to resort to frequent irrigations. However, the water-holding capacity must not be too high in order to avoid root asphyxia and too much cooling. The water available for the plant is calculated by the difference between the quantity of water at the retention capacity and that retained at the wilting point. This should be around 30–40% of the apparent volume (Kipp et al. 2001). Finally, it must be considered that with the constant increase of the root system biomass during growth, the free porosity in the substrate is gradually reduced and the hydrological characteristics of the substrate are modified.

Cation Exchange Capacity (CEC) CEC is a measure of how many cations can be retained on substrate particle surfaces. In general, organic materials have a higher CEC and a higher buffer capability than mineral ones (Wallach 2008; Blok et al. 2008) (Table 4.2).

pH A suitable pH is required to suit the needs of the cultivated species. Substrates with a low pH are more suitable for crops in containers, as they are more easily modified towards the desired levels by adding calcium carbonate and also because they meet the needs of a wider number of species. Moreover, during cultivation the pH value tends to rise due to irrigation with water rich in carbonates. The pH may also vary in relation to the type of fertilizer used. It is more difficult to correct an alkaline substrate. This can however be achieved by adding sulphur or physiologically acid fertilizers (ammonium sulphate, potassium sulphate) or constitutionally acid fertilizers (mineral phosphate).

Electrical Conductivity (EC) Substrates should have a known nutrient content and low EC values, (see also Table 4.4). It is often preferable to use a chemically inert substrate and to add the nutrients in relation to the specific crop requirements. Particular attention has to be paid to the EC levels. High EC levels indicate the presence of ions (e.g. Na^+) that, although not being important as nutrients, can play a decisive role in the suitability of the substrate.

Health and Safety Health in the systems and safety for operatives are provided by the absence of pathogens (nematodes, fungi, insects), potentially phytotoxic substances (pesticides) and weed seeds. Some industrially produced materials (expanded clay, perlite, stone wool, vermiculite and polystyrene) guarantee high levels of sterility due to the high temperatures applied during their processing.

Sustainability Another important characteristic of a substrate is its sustainability profile. Many commonly used substrates face ecological challenges relating to their provenance, production process and/or subsequent processing and end-of-life footprint. In this regard, substrates originating from materials with a low ecological footprint (modified in an environmentally friendly way and ultimately biodegradable) are an extra characteristic to consider. Reusability of the substrate can also be an important aspect of the sustainability of a substrate.

Cost Last but not least, the substrate must be inexpensive or at least cost-effective, readily available and standardized from the chemical–physical point of view.

4.2.4 Type of Substrates

The choice of substrates ranges from products of organic or mineral origin which are present in nature and which are subjected to special processing (e.g. peat, perlite, vermiculite), to those of organic origin derived from human activities (e.g. waste or by-products of agricultural, industrial and urban activities) and of industrial origin obtained by synthesis processes (e.g. polystyrene).

4.2.4.1 Organic Materials

This category includes natural organic substrates, including residues, waste and by-products of organic nature derived from agricultural (manure, straw, etc.) or, for example, industrial, by-products of the wood industry, etc. or from urban settlements, e.g. sewage sludge, etc. These materials can be subjected to additional processing, such as extraction and maturation.

All the materials that can be used in hydroponics can also be used in AP. However, as the bacterial load in an AP solution may be higher than in conventional hydroponic solutions, it can therefore be expected that organic substrates may be prone to an increased decomposition rate, causing substrate compaction and root aeration problems. Therefore, organic materials can be considered for crops with a shorter growth cycle, whilst mineral substrates may be preferred for crops with a long growth cycle.

Peat

Peat, used alone or with other substrates, is currently the most important material of organic origin for substrate preparation. The term peat refers to a product derived from residues of bryophytes (*Sphagnum*), Cyperaceae (*Trichophorum*, *Eriophorum*, *Carex*) and others (*Calluna*, *Phragmites*, etc.) transformed in anaerobic conditions.

Raised bogs are formed in cold and very rainy environments. Rainwater, without salts, is retained on the surface by mosses and vegetable residues, creating a saturated environment. In raised bogs we can distinguish a deeper, much decomposed layer of dark colour (*brown peat*) and a slightly decomposed, shallower layer of a light colour (*blond peat*). Both of the peats are characterized by good structural stability, very low availability of nutrients and acidic pH whilst they mainly differ in their structure (Table 4.2).

Brown peats, with very small pores, have a higher water-retention capacity and less free porosity for air and have higher CEC and buffer capability. Physical characteristics vary in relation to the particle size that allows water absorption from 4 up to 15 times its own weight. Raised bogs usually satisfy the requirements needed for a good substrate. Moreover, they have constant and homogenous properties, and so they can be industrially exploited. However, the use of these peats requires pH corrections with, e.g. calcium carbonate ($CaCO_3$). Generally, for a sphagnum peat with a pH 3–4, 2 kg m^{-3} of $CaCO_3$ should be added to increase the pH for one unit. Attention must be paid to avoid the complete drying of the substrate. It should also be taken into account that peat is subjected to microbiological decomposition processes which, over time, may increase water-retention capacity and reduce the free porosity.

Fen bogs are mainly present in temperate areas (e.g. Italy and western France), where *Cyperaceae*, *Carex* and *Phragmites* are dominant. These peats are formed in the presence of stagnant water. The oxygen, salts and calcium content in the water allow for a faster decomposition and humification, compared to that which occurs in the raised bogs. This results in a very dark, brown to black peat with a higher nutrient content, in particular nitrogen and calcium, a higher pH, higher bulk density and much lower free porosity (Table 4.2). They are rather fragile in the dry state, and have a remarkable plasticity in the humid state, which confers high susceptibility to compression and deformation. The carbon/nitrogen (C/N) ratio is generally between 15 and 48 (Kuhry and Vitt 1996; Abad et al. 2002). Because of its properties, black peat is of low value and is not suitable as a substrate, but can be mixed with other materials.

It should be noted that in some countries there is a drive to reduce peat use and extraction to reduce environmental effects and various peat substitutes have been identified with varied success.

Coconut Fibre

Coconut fibre (coir) is obtained from removing the fibrous husks of coconuts and is a by-product of the copra (coconut oil production) and fibre extraction industry, and is composed almost exclusively of lignin. Before use, it is composted for 2–3 years, and then it is dehydrated and compressed. Prior to its use, it must be rehydrated by adding up to 2–4 times of its compressed volume with water. Coconut fibre possesses chemical–physical characteristics that are similar to blond peat (Table 4.2), but with the advantages of having a higher pH. It also has a lower environmental impact than peat (excessive exploitation of peat bogs) and stone wool where there are problems with disposal. This is one of the reasons why it is increasingly preferred in soilless systems (Olle et al. 2012; Fornes et al. 2003).

Wood-Based Substrates

Organic substrates which are derived from wood or its by-products, such as bark, wood chips or saw dust, are also used in global commercial plant production (Maher et al. 2008). Substrates based on these materials generally possess good air content and high saturated hydraulic conductivities. The disadvantages can include low water-retention capacities, insufficient aeration caused by microbial activity, inappropriate particle-size distribution, nutrient immobilization or negative effects due to salt and toxic compound accumulations (Dorais et al. 2006).

4.2.4.2 Inorganic Materials

This category includes natural materials (e.g. sand, pumice) and mineral products derived from industrial processes (e.g. vermiculite, perlite) (Table 4.3).

Sand

Sands are natural inorganic material with particles between 0.05 and 2.0 mm diameter, originating from the weathering of different minerals. The chemical composition of sands may vary according to origin, but in general, it is constituted by 98.0–99.5% silica (SiO_2) (Perelli et al. 2009). pH is mainly related to the carbonate content. Sands with lower calcium carbonate content and pH 6.4–7.0 are better suited as substrate material because they do not influence the solubility of phosphorus and some microelements (e.g. iron, manganese). Like all mineral-origin substrates, sands have a low CEC and low buffering capability (Table 4.3). Fine sands (0.05–0.5 mm) are the most suitable for use in hydroponic systems in mixtures 10–30% by volume with organic materials. Coarse sands (>0.5 mm) can be used in order to increase the drainage capacity of the substrate.

Pumice

Pumice comprises aluminium silicate of volcanic origin, being very light and porous, and may contain small amounts of sodium and potassium and traces of calcium, magnesium and iron depending on the place of origin. It is able to retain calcium, magnesium, potassium and phosphorus from the nutrient solutions and to gradually release these to the plant. It usually has a neutral pH, but some materials may have

Table 4.3 Main chemical–physical characteristics of inorganic substrates used in soilless systems

Substrates	Bulk density (kg m^{-3})	Total porosity (%vol)	Free porosity (%vol)	Water-retention capacity (%vol)	CEC (meq %)	EC (mS cm^{-1})	pH
Sand	1400–1600	40–50	1–20	20–40	20–25	0.10	6.4–7.9
Pumice	450–670	55–80	30–50	24–32	–	0.08–0.12	6.7–9.3
Volcanic tuffs	570–630	80–90	75–85	2–5	3–5	–	7.0–8.0
Vermiculite	80–120	70–80	25–50	30–55	80–150	0.05	6.0–7.2
Perlite	90–130	50–75	30–60	15–35	1.5–3.5	0.02–0.04	6.5–7.5
Expanded clay	300–700	40–50	30–40	5–10	3–12	0.02	4.5–9.0
Stone wool	85–90	95–97	10–15	75–80	–	0.01	7.0–7.5
Expanded Polystyrene	6–25	55	52	3	–	0.01	6.1

Source: Enzo et al. (2001)

excessively high pH, good free porosity but low water-retention capacity (Table 4.3). The structure however tends to deteriorate fairly quickly, due to the easy breaking up of the particles. Pumice, added to peat, increases the drainage and aeration of the substrate. For horticulture use, pumice particles from 2 to 10 mm in diameter are preferred (Kipp et al. 2001).

Volcanic Tuffs

Tuffs derive from volcanic eruptions, with particles ranging between 2 and 10 mm diameter. They may have a bulk density ranging between 850 and 1100 kg m^{-3} and a water-retention capacity between 15% and 25% by volume (Kipp et al. 2001).

Vermiculite

Vermiculite comprises hydrous phyllosilicates of magnesium, aluminium and iron, which in the natural state have a thin lamellar structure that retains tiny drops of water. Exfoliated vermiculite is commonly used in the horticultural industry and is characterized by a high buffer capability and CEC values similar to those of the best peats (Table 4.3), but, compared to these, it has a higher nutrient availability (5–8% potassium and 9–12% magnesium) (Perelli et al. 2009). NH_4^+ is especially strongly retained by vermiculite; the activity of the nitrifying bacteria, however, allows the recovery of part of the fixed nitrogen. Similarly, vermiculite binds over 75% of phosphate in an irreversible form, whereas it has low absorbent capacity for Cl^-, NO_3^- and SO_4^-. These characteristics should be carefully assessed when vermiculite is used as a substrate. The vermiculite structure is not very stable because of a low compression resistance and tends to deteriorate over time, reducing water drainage. It can be used alone; however, it is preferable to mix it with perlite or peat.

Perlite

Perlite comprises aluminium silicate of volcanic origin containing 75% SiO_2 and 13% Al_2O_3. The raw material is crushed, sieved, compressed and heated to 700–1000 °C. At these temperatures, the little water contained in the raw material turns into vapour by expanding the particles into small whitish-grey aggregates which, unlike vermiculite, have a closed cell structure. It is very light and possesses high free porosity even after the soaking. It contains no nutrients, has negligible CEC and is virtually neutral (Table 4.3) (Verdonk et al. 1983). pH, however, can vary easily, because the buffer capacity is insignificant. pH ought to be controlled via the quality of the irrigation water and should not fall below 5.0 in order to avoid the phytotoxic effects of the aluminium. The closed cell structure allows water to be held only on the surface and in the spaces between the agglomerations, so the water-retention capacity is variable in relation to the dimensions of the agglomerations. It is marketed in different sizes, but the most suitable for horticulture are 2–5 mm diameter. It can be used as a substrate in rooting beds, because it ensures good aeration. In mixtures with organic materials, it enhances the softness, permeability and aeration of the substrate. Perlite can be reused for several years as long as it is sterilized between uses.

Expanded Clay

Expanded clay is obtained by treating clay powder at about 700 °C. Stable aggregates are formed, and, depending on the used clay material, they have variable values with regard to CEC, pH and bulk density (Table 4.3). Expanded clay can be used in mixtures with organic materials in the amounts of about 10–35% by volume, to which it provides more aeration and drainage (Lamanna et al. 1990). Expanded clays with pH values above 7.0 are not suitable for use in soilless systems.

Stone Wool

Stone wool is the most used substrate in soilless cultivation. It originates from the fusion of aluminium, calcium and magnesium silicates and carbon coke at 1500–2000 °C. The liquefied mixture is extruded in 0.05 mm diameter strands and, after compression and addition of special resins, the material assumes a very light fibrous structure with a high porosity (Table 4.3).

Stone wool is chemically inert and, when added to a substrate, it improves its aeration and drainage and also offers an excellent anchorage for plant roots. It is used alone, as a sowing substrate and for soilless cultivation. The slabs used for the cultivation can be employed for several production cycles depending on quality, as long as the structure is able to guarantee enough porosity and oxygen availability for root systems. Usually, after several crop cycles, the greater part of substrate porosity is filled with old, dead roots, and this is due to the compaction of the substrate over time. The result is a then a reduced depth of substrate where irrigation strategies may need adaptation.

Zeolites

Zeolites comprise hydrated aluminium silicates characterized by the capacity to absorb gaseous elements; they are high in macro- and microelements, they have high absorbent power and they have high internal surface (structures with 0.5 mm pores). This substrate is of great interest as it absorbs and slowly releases K^+ and NH_4^+ ions, whilst it is not able to absorb Cl^- and Na^+, which are hazardous to plants. Zeolites are marketed in formulations which differ in the N and P content and which can be used in seed sowing, for the rooting of cuttings or during the cultivation phase (Pickering et al. 2002).

4.2.4.3 Synthetic materials

Synthetic materials include both low-density plastic materials and ion-exchange synthetic resins. These materials, called "expanded", because they are obtained by a process of dilation at high temperatures, are not yet widely used, but they possess physical properties suitable to balance the characteristics of other substrates.

Expanded Polystyrene

Expanded polystyrene is produced in granules of 4–10 mm diameter with a closed cell structure. It does not decompose, is very light and has a very high porosity but with an extremely low water-retention capacity (Table 4.3). It has no CEC and virtually zero buffer capability, so it is added to the substrate (e.g. peat) exclusively to improve its porosity and drainage. The preferred particle size is 4–5 mm (Bunt 2012).

Polyurethane Foam

Polyurethane foam is a low-density material (12–18 kg m^{-3}) with a porous structure that allows absorption of water equal to 70% of its volume. It is chemically inert, has an almost neutral pH (6.5–7.0), does not contain useful nutrients available to plants and does not undergo decomposition (Kipp et al. 2001). In the market it is possible to find it in the form of granules, rooting cubes or blocks. Like a stone wool, it can also be used for soilless cultivation.

4.2.5 Preparation of Mixed Cultivation Substrates

Mixed substrates can be useful to reduce overall substrate costs and/or to improve some characteristics of the original materials. For example, peat, vermiculite and coir can be added to increase water-retention capacity; perlite, polystyrene, coarse sand and expanded clay to increase free porosity and drainage; blonde peat to raise the acidity; higher quantities of organic material or suitable amounts of clay soil to increase CEC and buffer capability; and low decomposable substrates for increased durability and stability. The characteristics of the mixtures rarely represent the average of the components because with the mixing the structures are modified between the individual particles and consequently the relationship of physical and chemical characteristics. In general, mixtures with a low nutrient content are preferable, in order to be able to better manage cultivation. The right relationship among the different constituents of a mixture also varies with the environmental conditions in which it operates. At high temperatures it is rational to use components that possess a higher water-retention capacity and do not allow fast evaporation (e.g. peat) and, at the same time, are resilient to decomposition. In contrast, in humid environments, with low solar radiation, the components characterized by high porosity are preferred to ensure good drainage. In this case, it will be necessary to add coarse substrates such as sand, pumice, expanded clay and expanded polystyrene (Bunt 2012).

4.3 Types of Hydroponic Systems According to Water/Nutrient Distribution

4.3.1 Deep Flow Technique (DFT)

Deep flow technique (DFT), also known as deep water technique, is the cultivation of plants on floating or hanging support (rafts, panels, boards) in containers filled with 10–20 cm nutrient solution (Van Os et al. 2008) (Fig. 4.3). In AP this can be up to 30 cm. There are different forms of application that can be distinguished mainly by the depth and volume of the solution, and the methods of recirculation and oxygenation.

One of the simplest systems comprises 20–30 cm deep tanks, which can be constructed of different materials and waterproofed with polyethylene films. The tanks are equipped with floating rafts (several types are available from suppliers) that serve to support the plants above the water whilst the plants' roots penetrate the water. The system is particularly interesting as it minimizes costs and management. For example, there is a limited need for the automation of the control and correction

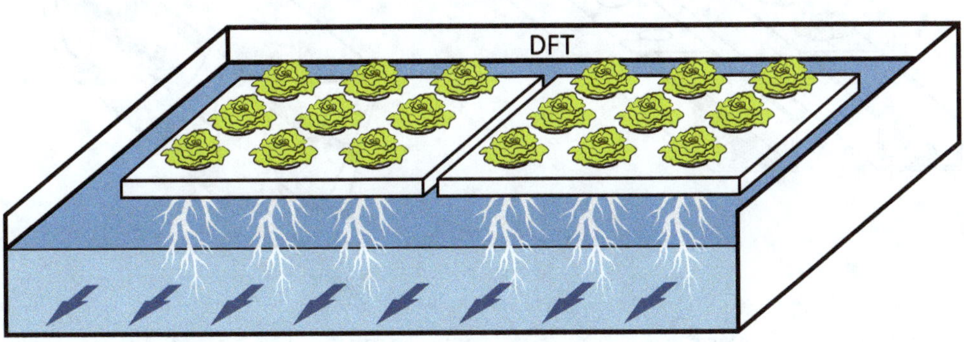

Fig. 4.3 Illustration of a DFT system with floating panels

of the nutrient solution, particularly in short duration crops such as lettuce, where the relatively high volume of solution facilitates the replenishment of the nutrient solution only at the end of each cycle, and only the oxygen content needs to be monitored periodically. Oxygen levels should be above 4–5 mg L^{-1}; otherwise, nutrient deficiencies may appear due to root systems uptake low performance. Circulation of the solution will normally add oxygen, or Venturi systems can be added which dramatically increase air into the system. This is especially important when water temperatures are greater than 23 °C, as such high temperatures may stimulate lettuce bolting.

4.3.2 Nutrient Film Technique (NFT)

The NFT technique is used ubiquitously and can be considered the classic hydroponic cultivation system, where a nutrient solution flows along and circulates in troughs with a 1–2 cm layer of water (Cooper 1979; Jensen and Collins 1985; Van Os et al. 2008) (Fig. 4.4). The recirculation of the nutrient solution and the absence of substrate represent one of the main advantages of the NFT system. An additional advantage is its great potential for automation to save on labour costs (planting & harvesting) and the opportunity to manage the optimal plant density during crop cycle. On the other hand, the lack of substrate and low water levels makes the NFT vulnerable to the failure of pumps, due to e.g. clogging or a failure in the power supply. Temperature fluctuations in the nutrient solution can cause plant stress followed by diseases.

Fig. 4.4 Illustration of NFT system (left) and a multilayer NFT trough, developed and marketed by New Growing Systems (NGS), Spain (right)

The development of the root system, part of which remains suspended in air above the nutrient flow and which is exposed to an early ageing and loss of functionality, represents a major constraint as it prevents the production of long-cycle crops (over 4–5 months). Because of its high susceptibility to temperature variations, this system is not suitable for cultivation environments characterized by high levels of irradiation and temperature (e.g. southern areas of the Mediterranean basin). However, in response to these challenges, a multilayer NFT trough has been designed which allows for longer production cycles without clogging problems (NGS). It is made of a series of interconnected layers placed in a cascade, so that even in strong rooting plant species, such as tomatoes, the nutrient solution will still find its way to the roots by by-passing the root-clogged layer via a lower positioned layer.

4.3.3 Aeroponic Systems

The aeroponic technique is mainly aimed at smaller horticultural species, and has not yet been widely used due to the high investment and management costs. Plants are supported by plastic panels or by polystyrene, arranged horizontally or on inclined tops of growing boxes. These panels are supported by a structure made with inert materials (plastic, steel coated with plastic film, polystyrene boards), in order to form closed boxes where the suspended root system can develop (Fig. 4.5).

The nutrient solution is directly sprayed on the roots, which are suspended in the box in air, with static sprinklers (sprayers), inserted on pipes housed inside the box

Fig. 4.5 Illustration of the aeroponics technique

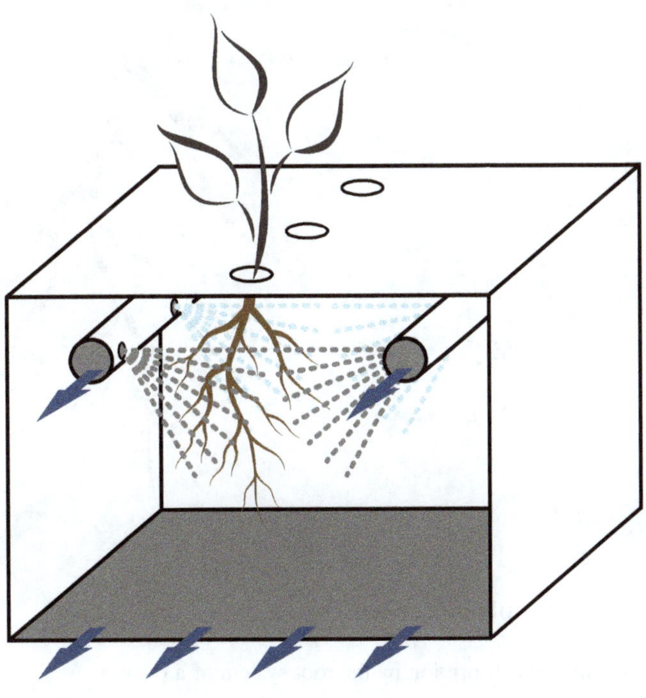

module. The spray duration is from 30 to 60 s, whilst the frequency varies depending on the cultivation period, the growth stage of the plants, the species and the time of day. Some systems use vibrating plates to create micro droplets of water which form a steam which condenses on the roots. The leachate is collected on the bottom of the box modules and conveyed to the storage tank, for reuse.

4.4 Plant Physiology

4.4.1 Mechanisms of Absorption

Amongst the main mechanisms involved in plant nutrition, the most important is the absorption which, for the majority of the nutrients, takes place in ionic form following the hydrolysis of salts dissolved in the nutrient solution.

Active roots are the main organ of the plant involved in nutrient absorption. Anions and cations are absorbed from the nutrient solution, and, once inside the plant, they cause the protons (H^+) or hydroxyls (OH^-) to exit which maintains the balance between the electric charges (Haynes 1990). This process, whilst maintaining the ionic equilibrium, can cause changes in the pH of the solution in relation to the quantity and quality of the nutrients absorbed (Fig. 4.6).

The practical implications of this process for the horticulturist are two-fold: to provide adequate buffer capability to the nutrient solution (adding bicarbonates if needed) and to induce slight pH changes with the choice of fertilizer. The effect of fertilizers on the pH relates to the different chemical forms of the used compounds.

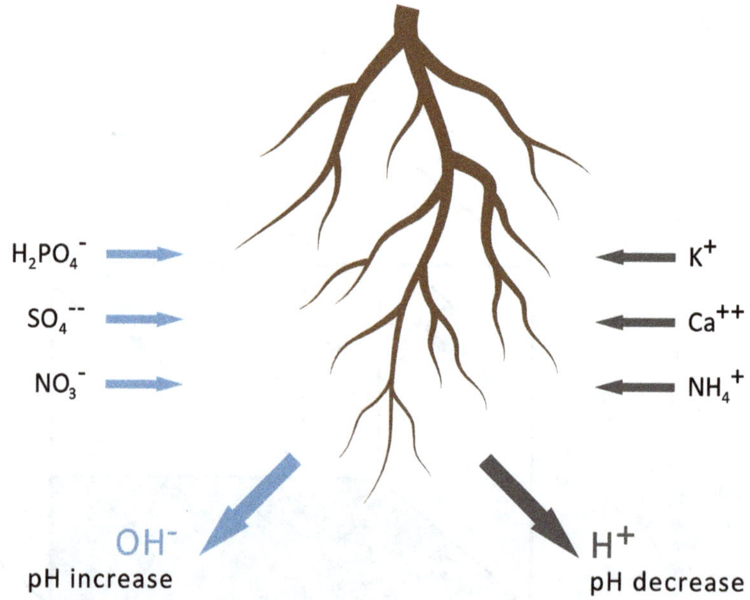

Fig. 4.6 Ion absorption by the root system of a plant

In the case of N, for example, the most commonly used form is nitric nitrogen (NO_3^-), but when the pH should be lowered, nitrogen can be supplied as ammonium nitrogen (NH_4^+). This form, when absorbed, induces the release of H^+ and consequently an acidification of the medium.

Climatic conditions, especially air and substrate temperature and relative humidity, exert a major influence on the absorption of nutrients (Pregitzer and King 2005; Masclaux-Daubresse et al. 2010; Marschner 2012; Cortella et al. 2014). In general, the best growth occurs where there are few differences between substrate and air temperature. However, persistently high temperature levels in the root system have a negative effect. Sub-optimal temperatures reduce the absorption of N (Dong et al. 2001). Whilst NH_4^+ is effectively used at optimum temperatures, at low temperatures, the bacterial oxidation is reduced, causing accumulation within the plant that can produce symptoms of toxicity and damage to the root system and the aerial biomass. Low temperatures at the root level also inhibit the assimilation of K and P, as well as the P translocation. Although the available information regarding the effect of low temperatures on the absorption of micronutrients is less clear, it appears that Mn, Zn, Cu, and Mo uptake are most affected (Tindall et al. 1990; Fageria et al. 2002).

4.4.2 *Essential Nutrients, Their Role and Possible Antagonisms*

The appropriate management of plant nutrition must be based on basic aspects that are influenced by uptake and use of macro, and micro-nutrients (Sonneveld and Voogt 2009). Macro-nutrients are needed in relatively large amounts, whilst micro-nutrients or trace elements are needed in small amounts. Furthermore, nutrient availability to the plant in the case of the soilless systems presents more or less consistent phenomena of synergy and antagonism (Fig. 4.7).

Nitrogen (N) Nitrogen is absorbed by plants to produce amino acids, proteins, enzymes and chlorophyll. The most used nitrogen forms for plant fertilization are nitrate and ammonium. Nitrates are quickly absorbed by the roots, are highly movable inside the plants and can be stored without toxic effects. Ammonium can be absorbed by plants only in low quantities and cannot be stored at high quantities because it exerts toxic effects. Quantities higher than 10 mg L^{-1} inhibit plant calcium and copper uptake, increase the shoot growth compared to root growth and result in a strong green colour of the leaves. Further excesses in ammonia concentration result in phytotoxic effects such as chlorosis along the leaves' margins. Excess in nitrogen supply causes high vegetative growth, increase of crop cycle length, strong green leaf colour, low fruit set, high content of water in the tissues, low tissue lignification and high tissue nitrate accumulation. Commonly, nitrogen deficiency is characterized by a pale green colour of the older leaves (chlorosis), reduced growth and senescence advance.

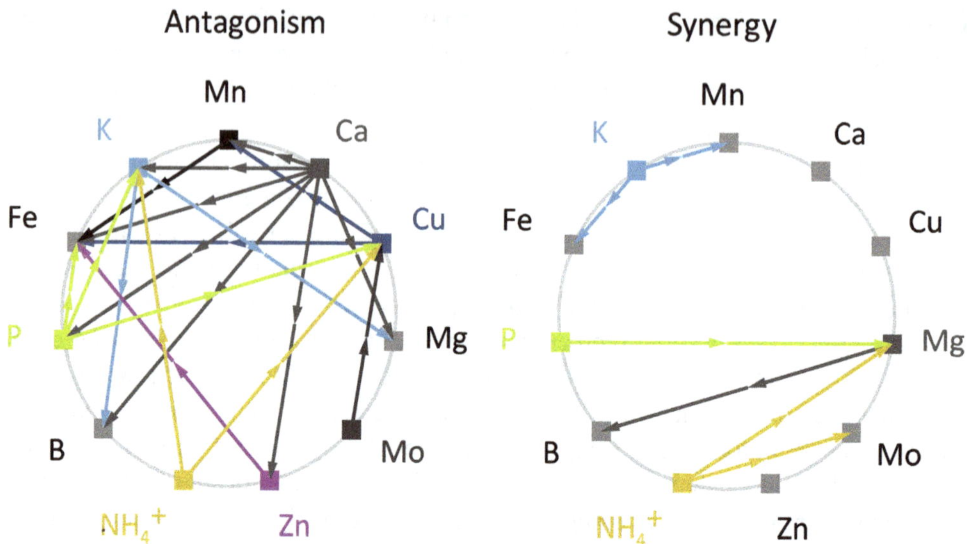

Fig. 4.7 Nutrients synergies and antagonisms amongst ions. Connected ions present synergistic or antagonistic relationship according to the direction of the arrow

Potassium (K) Potassium is fundamental for cell division and extension, protein synthesis, enzyme activation and photosynthesis and also acts as a transporter of other elements and carbohydrates through the cell membrane. It has an important role in keeping the osmotic potential of the cell in equilibrium and regulating the stomatal opening. The first signs of deficiency are manifested in the form of yellowish spots that very quickly necrotize on the margins of the older leaves. Potassium deficient plants are more susceptible to sudden temperature drops, water stress and fungal attacks (Wang et al. 2013).

Phosphorus (P) Phosphorus stimulates roots development, the rapid growth of buds and flower quantity. P is absorbed very easily and can be accumulated without damage to the plant. Its fundamental role is linked to the formation of high-energy compounds (ATP) necessary for plant metabolism. The average quantities requested by plants are rather modest (10–15% of the needs of N and K) (Le Bot et al. 1998). However, unlike what occurs in soil, P is easily leachable in soilless crops. The absorption of P appears to be reduced by low substrate temperatures (< 13 °C) or at increasing pH values (> 6.5) which can lead to deficiency symptoms (Vance et al. 2003). Under these conditions a temperature increase and/or pH reduction is more effective than additional amendments of phosphorus fertilizers. P excess can reduce or block the absorption of some other nutrients (e.g. K, Cu, Fe) (Fig. 4.7). Phosphorus deficiency manifests in a green-violet colour of the older leaves, which may follow chlorosis and necrosis in addition to the stunted growth of the vegetative apex. However, these symptoms are non-specific and make P deficiencies difficult to be identified (Uchida 2000).

Calcium (Ca) Calcium is involved in cell wall formation, membrane permeability, cell division and extension. Good availability gives the plant greater resistance to fungal attacks and bacterial infections (Liu et al. 2014). The absorption is very closely linked to the water flow between roots and aerial parts. Its movement occurs through the xylem and is therefore particularly influenced by low temperatures at the root level, by reduced water supply (drought or salinity of the solution) or by excessive relative humidity of the air. As Ca is not mobile within the plant, deficiencies start from the most recently formed parts (Adams 1991; Adams and Ho 1992; Ho et al. 1993). The main symptoms are plant growth being stunted, deformation of the margins of the younger leaves, light green or sometimes chlorotic colouring of new tissues and a stunted root system without fine roots. The deficiencies are displayed in different ways, e.g. apical rot in tomato and/or marginal browning of leaves in lettuce.

Magnesium (Mg) Magnesium is involved in the constitution of chlorophyll molecules. It is immobilized at pH values below 5.5 and enters into competition with the absorption of K and Ca (Fig. 4.7). Symptoms of deficiency are yellowing between leaf veins and internal chlorosis of the basal leaves. As Mg can be easily mobilized, magnesium-deficient plants will first break down chlorophyll in the older leaves and transport the Mg to younger leaves. Therefore, the first sign of magnesium deficiency is the interveinal chlorosis in older leaves, contrary to iron deficiency where interveinal chlorosis first appears in the youngest leaves (Sonneveld and Voogt 2009).

Sulphur (S) Sulphur is required by the plant in quantities comparable to those of phosphorus, and in order to optimize its absorption, it must be present in a 1:10 ratio with nitrogen (McCutchan et al. 2003). It is absorbed as sulphate. The deficiencies are not easily detected, as the symptoms can be confused with those of nitrogen deficiency, except that the deficiency of nitrogen begins to manifest itself from the older leaves, whilst that of sulphur from the youngest ones (Schnug and Haneklaus 2005). S nutrition has a significant role in ameliorating the damages in photosynthetic apparatus caused by Fe-deficiency (Muneer et al. 2014).

Iron (Fe) Iron is one of the most important micro-nutrients because it is key in many biological processes such as photosynthesis (Briat et al. 2015; Heuvelink and Kierkels 2016). To improve its absorption, the nutrient solution pH should be around 5.5–6.0, and the Mn content should not be allowed to become too high because the two elements subsequently enter into competition (Fig. 4.7). The optimal ratio of Fe–Mn is around 2:1 for most crops (Sonneveld and Voogt 2009). At low temperatures, the assimilation efficiency is reduced. The deficiency symptoms are characterized by interveinal chlorosis from the young leaves towards the older basal ones, and by reduced root system growth. Symptoms of deficiency are not always due to the low presence of Fe in the nutrient solution, but often they are due to the Fe unavailability for the plant. The use of chelating agents guarantees constant availability of Fe for the plant.

Chlorine (Cl) Chlorine has been recently considered a micro-nutrient, even if its content in plants (0.2–2.0% dw) is quite high. It is easily absorbed by the plant and is very mobile within it. It is involved in the photosynthetic process and the regulation of the stomata opening. Deficiencies, which are rather infrequent, occur with typical symptoms of leaves drying out, especially at the margins. Much more widespread is the damage due to an excess of Cl that leads to conspicuous plant shrinkage which is relative to the different sensitivities of different species. To avoid crop damage, it is always advisable to check the Cl content in the water used to prepare nutrient solutions and choose suitable fertilizers (e.g. K_2SO_4 rather than KCl).

Sodium (Na) Sodium, if in excess, is harmful to plants, as it is toxic and interferes with the absorption of other ions. The antagonism with K (Fig. 4.7), for example, is not always harmful because in some species (e.g. tomatoes), it improves the fruit taste, whereas in others (e.g. beans), it can reduce plant growth. Similar to Cl, it is important to know the concentration in the water used to prepare the nutrient solution (Sonneveld and Voogt 2009).

Manganese (Mn) Manganese forms part of many coenzymes and is involved in the extension of root cells and their resistance to pathogens. Its availability is controlled by the pH of the nutrient solution and by competition with other nutrients (Fig. 4.7). Symptoms of deficiency are similar to those of the Fe except for the appearance of slightly sunken areas in the interveinal areas (Uchida 2000). Corrections can be made by adding $MnSO_4$ or by lowering the pH of the nutrient solution.

Boron (B) Boron is essential for fruit setting and seed development. The absorption methods are similar to those already described for Ca with which it can compete. The pH of the nutrient solution must be below 6.0 and the optimal level seems to be between 4.5 and 5.5. Symptoms of deficiency can be detected in the new structures that appear dark green, the young leaves greatly increase their thickness and have a leathery consistency. Subsequently they can appear chlorotic and then necrotic, with rusty colouring.

Zinc (Zn) Zinc plays an important role in certain enzymatic reactions. Its absorption is strongly influenced by the pH and the P supply of the nutrient solution. pH values between 5.5 and 6.5 promote the absorption of Zn. Low temperature and high P levels reduce the amount of zinc absorbed by the plant. Zinc deficiencies occur rarely, and are represented by chlorotic spots in the interveinal areas of the leaves, very short internodes, leaf epinasty and poor growth (Gibson 2007).

Copper (Cu) Copper is involved in respiratory and photosynthetic processes. Its absorption is reduced at pH values higher than 6.5, whilst pH values lower than 5.5 may result in toxic effects (Rooney et al. 2006). High levels of ammonium and phosphorus interact with Cu reducing the availability of the latter. The excessive presence of Cu interferes with the absorption of Fe, Mn and Mo. The deficiencies are manifested by interveinal chlorosis which leads to the collapse of the leaf tissues that look like desiccated (Gibson 2007).

Molybdenum (Mo) Molybdenum is essential in protein synthesis and in nitrogen metabolism. Contrary to other micro-nutrients, it is better available at neutral pH values. Symptoms of deficiency start with chlorosis and necrosis along the main rib of old leaves, whilst the young leaves appear deformed (Gibson 2007).

4.4.3 Nutrient Management in Relation to the Requirements of Plants

Since the development of soilless horticulture systems in the 1970s (Verwer 1978; Cooper 1979), different nutrient solutions have been developed and adjusted according to the growers' preferences (Table 4.4; De Kreij et al. 1999). All mixes follow the principles of excess availability of all elements to prevent deficiencies and balance between (bivalent) cations to avoid competition between cations in plant nutrient uptake (Hoagland and Arnon 1950; Steiner 1961; Steiner 1984; Sonneveld and Voogt 2009). Commonly, the EC is allowed to rise in the root zone to a limited degree. In tomatoes, for example, the nutrient solution typically has an EC of ca. 3 dS m^{-1}, whilst in the root zone in the stone wool slabs, the EC may rise to 4–5 dS m^{-1}. However, in northern European countries, for the first irrigation of new stone wool slabs at the beginning of the production cycle, the nutrient solution may have an EC as high as 5 dS m^{-1}, saturating the stone wool substrate with ions up to an EC of 10 dS m^{-1}, which will subsequently be flushed after 2 weeks. To provide sufficient flushing of the root zone, in a typical drip-irrigation stone wool slab system, about 20–50% of the dosed water is collected as drainage water. The drainage water is then recycled, filtered, mixed with fresh water and topped up with nutrients for use in the next cycle (Van Os 1994).

In tomato production, increasing the EC can be applied to enhance lycopene synthesis (promoting the bright red coloration of the fruits), total soluble solids (TSS) and fructose and glucose content (Fanasca et al. 2006; Wu and Kubota 2008). Furthermore, tomato plants have higher absorption rates for N, P, Ca and Mg and low absorption of K during the early (vegetative) stages. Once the plants start developing fruits, leaf production is slowed down leading to a reduction in N and Ca requirements, whilst K requirement increases (e.g. Zekki et al. 1996; Silber, Bar-Tal 2008). In lettuce, on the other hand, an increased EC may promote tip-burn disease during hot growing conditions. Huett (1994) showed a significant decrease in the number of leaves with tip-burn disease per plant when the EC was dropped from 3.6 to 0.4 dS m^{-1}, as well as when the nutrient formulation K/Ca was reduced from 3.5:1 to 1.25:1. In AP the management of nutrients is more difficult than in hydroponics since they mainly depend on fish stock density, feed type and feeding rates.

4.4.4 Nutrient Solution Properties

Phosphorus is an element which occurs in forms that are strongly dependent on environmental pH. In the root zone, this element can be found as PO_4^{3-}, HPO_4^{2-} and $H_2PO_4^-$ ions, where the last two ions are the main forms of P taken up by the plants. Thus, when the pH is slightly acidic (pH 5–6), the largest amount of P is presented in a nutrient solution (De Rijck and Schrevens 1997).

Potassium, calcium and magnesium are available to plants in a wide range of pH. However, the presence of other ions may interfere in their plant availability due to the formation of compounds with different grades of solubility. At a pH above 8.3, Ca^{2+} and Mg^{2+} ions easily precipitate as carbonates by reacting with CO_3^{2-}. Also sulphate forms relatively strong complexes with Ca^{2+} and Mg^{2+} (De Rijck and Schrevens 1998). As pH increases from 2 to 9, the amount of SO_4^{2-} forming soluble complexes with Mg^{2+} as $MgSO_4$ and with K^+ as KSO_4^- increases (De Rijck and Schrevens 1999). In general, nutrient availability for plant uptake at pH above 7 may be restricted due to a precipitation of Boron, Fe^{2+}, Mn^{2+}, PO_4^{3-}, Ca^{2+} and Mg^{2+} due to insoluble and unavailable salts. The most appropriate pH values of the nutrient solution for the development of crops lie between 5.5 and 6.5 (Sonneveld and Voogt 2009).

4.4.5 Water Quality and Nutrients

The quality of the supplied water is extremely important in hydroponic and AP systems. For long-term recirculation, the chemical composition should be well known and monitored frequently to avoid an imbalance in nutrient supply but also to avoid the accumulation of certain elements leading to toxicity. De Kreij et al. (1999) made an overview of the chemical demands on water quality for hydroponic systems.

Before starting, an analysis of the water supply has to be made on the macro- and microelements. Based on the analysis, a scheme for the nutrient solution can be made. For example, if rainwater is used, special attention has to be made for Zn when collection takes place via untreated gutters. In tap water, problems may appear with Na, Ca, Mg, SO_4 and HCO_3. Furthermore, surface and bore hole water may be used which may also contain amounts of Na, Cl, K, Ca, Mg, SO_4 and Fe but also microelements as Mn, Zn, B and Cu. It should be noted that all valves and pipes should be made of synthetic materials such as PVC and PE, and not containing Ni or Cu parts.

It often happens that water supplies contain a certain amount of Ca and Mg; therefore, the contents have to be subtracted from the amount in the nutrient solution to avoid accumulation of these ions. HCO_3 has to be compensated preferably by nitric acid, about 0.5 mmol L^{-1} that can be maintained as a pH buffer in the nutrient solution. Phosphoric and sulphuric acid can also be possibly used to compensate pH, but both will rapidly give a surplus of $H_2PO_4^-$ or SO_4^{2-} in the nutrient solution. In AP systems nitric acid (HNO_3) and potassium hydroxide (KOH) can be also used to regulate pH and at same time supply macronutrients in the system (Nozzi et al. 2018).

4.4.5.1 Water Quality Management

For the formulation of nutrient solutions, simple fertilizers (granular, powder or liquid) and substances (e.g. acid compounds) that affect the pH are preferably used. The integration of the nutrient elements into the solution takes into account the optimum values of the quantities of each element. This has to be made in relation to the requirements of the species and its cultivars considering phenological phases and substrate. The calculation of nutrient supplements must be carried out considering the conditions of the water used, according to a *strict set of priorities*. On the priority scale, magnesium and sulphates are positioned at the bottom, at the same level, because they have less nutritional importance and the plants do not present damage even if their presence is abundant in the nutrient solution. This characteristic has an advantageous practical feedback as it allows an exploitation of the two elements in order to balance the nutritional composition with respect to other macronutrients whose deficiency or excess may be negative for production. As an example, we can consider a nutrient solution where an integration of only potassium or only nitrate is required. The salts to be used, in this case, are respectively potassium sulphate or magnesium nitrate. In fact, if the most common potassium nitrate or calcium nitrate were used, the levels of nitrate, in the first case, and calcium, in the second case, would automatically increase. Moreover, when the analysis of the water used shows an imbalance between cations and anions, and in order to be able to calculate a nutrient solution with the EC in equilibrium, the correction of the water values is carried out reducing the levels of magnesium and/or sulphates.

The following points provide guidelines for the formulation of nutrient solutions:

1. Definition of the species and cultivar requirements. Consideration of the cultivation environment and the characteristics of water need to be taken into account. In order to satisfy the needs of the plants in warm periods and with intense radiation, the solution must possess a lower EC and K content, which contrasts to a higher quantity of Ca. Instead, when the temperature and brightness reach sub-optimal levels, it is advisable to raise the values of the EC and K by reducing those of the Ca. It is important to note regarding cultivars that there are substantial variations, especially for the values of the NO_3^-, due to the different vegetative vigorousness of the cultivars. For tomatoes, in fact, 15 mmol L^{-1} of NO_3^- is used on average (Table 4.4), and in the case of cultivars characterized by low vegetative vigour and in certain phenological phases (e.g. fruit setting of the fourth trusses), up to 20 mM L^{-1} of NO_3^- is adopted. In case that some elements such as Na are

Table 4.4 Nutrient solutions in hydroponic cultivation of lettuce (DFT) tomato, pepper and cucumber (stone wool slabs with drip irrigation) in the Netherlands (De Kreij et al. 1999)

	pH	EC	NH_4	K	Ca	Mg	NO_3	SO_4	P	Fe	Mn	Zn	B	Cu	Mo
		$dS\ m^{-1}$	$mmol\ L^{-1}$	$mmol\ L^{-1}$	$mmol\ L^{-1}$	$mmol\ L^{-1}$	$mmol\ L^{-1}$	$mmol\ L^{-1}$	$mmol\ L^{-1}$	$mmol\ L^{-1}$	$mmol\ L^{-1}$	$mmol\ L^{-1}$	$mmol\ L^{-1}$	$mmol\ L^{-1}$	$mmol\ L^{-1}$
Lettuce (Wageningen UR)	5.9	1.7	1.0	4.4	4.5	1.8	10.6	1.5	1.5	28.1	1.5	6.4	47.0	1.0	0.7
Lettuce	5.8	1.2	0.7	4.8	2.3	0.8	8.9	0.8	1.0	35.1	4.9	3.0	18.4	0.5	0.5
Lettuce	5.8	1.2		3.0	2.5	1.0	7.5	1.0	0.5	50.0	3.7	0.6	4.8	0.5	0.01
Tomato generative	5.5	2.6–3.0	1.2	13.0	4.2	1.9	15.4	4.7	1.5	15.0	10.0	5.0	30.0	0.8	0.5
Tomato vegetative	5.5	2.6	1.2	8.3	5.7	2.7	15.4	4.7	1.5	15.0	10.0	5.0	30.0	0.8	0.5
Cucumber	5.5	3.2	1.2	10.4	6.7	2.0	23.3	1.5–2.0	1.5–2.0	15.0	10.0	5.0	25.0	0.8	0.5
Pepper	5.6	2.5–3.0	1.2	5–7	4–5	2.0	17.0	1.8–2.0	1.5–2.5	25.0	10.0	7.0	30.0	1.0	0.5
Plant propagation	5.5	2.3	1.2	6.8	4.5	3.0	16.8	2.5	1.3	25.0	10.0	5.0	35.0	1.0	0.5

Adopted and modified from Vermeulen (2016, personal communication)

present in the water, in order to reduce its effect, which is particularly negative for some crops, it will be necessary to increase the amount of NO_3^- and Ca and possibly decrease the K, keeping the EC at the same level.

2. Nutrient requirement calculations should be obtained by subtracting the values of the chemical elements of the water from the chemical elements defined above. For example, the established need for Mg of peppers (*Capsicum* sp.) is 1.5 mM L^{-1}, having the water at 0.5 mM L^{-1}, and 1.0 mM L^{-1} of Mg should be added to the water (1.5 requirement – 0.5 water supply = 1.0).
3. Choice and calculation of fertilizers and acids to be used. For example, having to provide Mg, as in the example of point 2 above, $MgSO_4$ or $Mg(NO_3)_2$ can be used. A decision will be made taking into account the collateral contribution of sulphate or nitrate as well.

4.4.6 Comparison Between Hydroponic and Aquaponic Production

During their life cycle, plants need several essential macro- and microelements for regular development (boron, calcium, carbon, chlorine, copper, hydrogen, iron, magnesium, manganese, molybdenum, nitrogen, oxygen, phosphorous, potassium, sulphur, zinc), usually absorbed from the nutrient solution (Bittsanszky et al. 2016). The nutrient concentration and ratio amongst them are the most important variables capable to influence plant uptake. In AP systems fish metabolic wastes contain nutrients for the plants, but it must be taken into account, especially at commercial scales, that the nutrient concentrations supplied by the fish in AP systems are significantly lower and unbalanced for most nutrients compared to hydroponic systems (Nicoletto et al. 2018). Usually, in AP, with appropriate fish stocking rates, the levels of nitrate are sufficient for good plant growth, whereas the levels of K and P are generally insufficient for maximum plant growth. Furthermore, calcium and iron could also be limited. This can reduce the crop yield and quality and so nutrient integration should be carried out to support an efficient nutrient reuse. Microbial communities play a crucial role in the nutrient dynamic of AP systems (Schmautz et al. 2017), converting ammonium to nitrate, but also contributing to the processing of particulate matter and dissolved waste in the system (Bittsanszky et al. 2016). Plant uptake of N and P represents only a fraction of the amount removed from the water (Trang and Brix 2014), indicating that microbial processes in the root zone of the plants, and in the substrate (if present) and throughout the whole system, play a major role.

The composition of fish feeds depends on the type of fish and this influences nutrient release from fish's metabolic output. Typically, fish feed contains an energy source (carbohydrates and/or lipids), essential amino acids, vitamins, as well as other organic molecules that are necessary for normal metabolism but some that the fish's cells cannot synthesize. Furthermore, it must be taken into account that a plant's nutritional requirements vary with species (Nozzi et al. 2018), variety, life cycle stage, day length and weather conditions and that recently (Parent et al. 2013; Baxter 2015), the Liebig's law (plant growth is controlled by the scarcest resource) has been superseded by complex algorithms that consider the interactions between the individual nutrients. Both these aspects do not allow a simple evaluation of the effects of changes in nutrient concentrations in hydroponic or AP systems.

The question thus arises whether it is necessary and effective to add nutrients to AP systems. As reported by Bittsanszky et al. (2016), AP systems can only be operated efficiently and thus successfully, if special care is taken through the continuous monitoring of the chemical composition of the recirculating water for adequate concentrations and ratios of nutrients and of the potentially toxic component, ammonium. The necessity to add nutrients depends on plant species and growth stage. Frequently, although fish density is optimal for nitrogen supply, the addition of P and K with mineral fertilizers, at least, should be carried out (Nicoletto et al. 2018). In contrast to, for example, lettuce, tomatoes which need to bear fruit, mature and ripen, need supplemental nutrients. In order to calculate these needs, a software can be used, such as *HydroBuddy* which is a free software (Fernandez 2016) that is used to calculate the amount of required mineral nutrient supplements.

4.5 Disinfection of the Recirculating Nutrient Solution

To minimize the risk of spreading soil-borne pathogens, disinfection of the circulating nutrient solution is required (Postma et al. 2008). Heat treatment (Runia et al. 1988) was the first method used. Van Os (2009) made an overview for the most important methods and a summary is given below. Recirculating of the nutrient solution opens possibilities to save on water and fertilizers (Van Os 1999). The big disadvantage of the recirculation of the nutrient solution is the increasing risk of spreading root-borne pathogens all over the production system. To minimize such risks, the solution should be treated before reuse. The use of pesticides for such a treatment is limited as effective pesticides are not available for all such pathogens, and if available, resistance may appear, and environmental legislation restricts discharge of water with pesticides (and nutrients) into the environment (European Parliament and European Council 2000). In addition, in AP systems, the use of pesticides exerts negative effects on fish health and cannot be carried out, even if hydroponic and AP parts of the system are in different rooms, because spraying of chemicals may enter the nutrient solution via condensation water or via direct spraying on the substrate slabs. In view of this, a biological control approach can be adopted to manage pest diseases, and this can be accessed via the EU Aquaponics Hub Fact Sheet (EU Aquaponics Hub). At the same time, similar problems can be observed for fish treatment using veterinary drugs that are not compatible with the plant's cycle.

4.5.1 *Description of Disinfection Methods*

Disinfection of the circulating nutrient solution should take place continuously. All drain returned (10–12 h during daytime) has to be treated within 24 h. For a greenhouse of 1000 m^2 in a substrate cultivation (stone wool, coir, perlite), a disinfection capability of about 1–3 m^3 per day is needed to disinfect an estimated needed surplus of 30% of the water supplied with drip irrigation to tomato plants during a 24-h period in summer conditions. Because of the variable return rate of drain water, a sufficiently large catchment tank for drain water is needed in which the water is stored before it is pumped to the disinfection unit. After disinfection another tank is required to store the clean water before adjusting EC and pH and blending with new water to supply to the plants. Both tanks have an average size of 5 m^3 per 1000 m^2. In a nutrient film system (NFT), about 10 m^3 per day should be disinfected daily. It is generally considered that such a capacity is uneconomical to disinfect (Ruijs 1994). DFT requires similar treatment. This is the main reason why NFT and DFT production units do not normally disinfect the nutrient solution. Disinfection is carried out either by non-chemical or chemical methods as follows:

4.5.1.1 Non-chemical Methods

In general these methods do not alter the chemical composition of the solution, and there is no build-up of residuals:

1. *Heat treatment.* Heating the drain water to temperatures high enough to eradicate bacteria and pathogens is the most reliable method for disinfection. Each type of organism has its own lethal temperature. Non-spore-forming bacteria have lethal temperatures between 40 and 60 °C, fungi between 40 and 85 °C, nematodes between 45 and 55 °C and viruses between 80 and 95°C (Runia et al. 1988) at an exposure time of 10 s. Generally, the temperature set point of 95 °C is high enough to kill most of the organisms that are likely to cause diseases with a minimum time of 10 s. Whilst this may seem very energy intensive, it should be noted that the energy is recovered and reused with heat exchangers. Availability of a cheap energy source is of greater importance for practical application.
2. *UV radiation.* UV radiation is electromagnetic radiation with a wavelength between 200 and 400 nm. Wavelengths between 200 and 280 nm (UV-C), with an optimum at 254 nm, has a strong killing effect on micro-organisms, because it minimizes the multiplication of DNA chains. Different levels of radiation are needed for different organisms so as to achieve the same level of efficacy. Runia (1995) recommends a dose which varies from 100 mJ cm^{-2} for eliminating bacteria and fungi to 250 mJ cm^{-2} for eliminating viruses. These relatively high doses are needed to compensate for variations in water turbidity and variations in penetration of the energy into the solution due to low turbulence around the UV lamp or variations in output from the UV lamp. Zoschke et al. (2014) reviewed that UV irradiation at 185 and 254 nm offers water organic

contaminant control and disinfection. Moreover, Moriarty et al. (2018) reported that UV radiation efficiently inactivated coliforms in AP systems.

3. *Filtration.* Filtration can be used to remove any undissolved material out of the nutrient solution. Various types of filters are available relative to the range of particle sizes. Rapid sand filters are often used to remove large particles from the drain water before adding, measuring and control of EC, pH and application of new fertilizers. After passing the fertilizer unit, often a fine synthetic filter (50–80 um) is built in the water flow to remove undissolved fertilizer salts or precipitates to avoid clogging of the irrigation drippers. These synthetic filters are also used as a pretreatment for disinfection methods with heat treatment, ozone treatment or UV radiation. With reductions in filtration pore size, the flow is inhibited, so that removal of very small particles requires a combination of adequate filters and high pressure followed by frequent cleaning of the filter(s). Removal of pathogens requires relatively small pore sizes (<10 μm; so-called micro-, ultra- or nanofiltration).

4.5.1.2 Chemical Methods

1. *Ozone (O_3).* Ozone is produced from dry air and electricity using an ozone-generator (converting $3O_2 \rightarrow 2O_3$). The ozone-enriched air is injected into the water that is being sanitized and stored for a period of 1 h. Runia (1995) concluded that an ozone supply of 10 g per hour per m^3 drain water with an exposure time of 1 h is sufficient to eliminate all pathogens, including viruses. The reduction of microbial populations in vegetable production in soilless systems managed with ozone has also been observed by Nicoletto et al. (2017). Human exposure to the ozone that vents from the system or the storage tanks should be avoided since even a short exposure time of a concentration of 0.1 mg L^{-1} of ozone may cause irritation of mucous membranes. A drawback of the use of ozone is that it reacts with iron chelate, as UV does. Consequently, higher dosages of iron are required and measures need to be taken to deal with iron deposits in the system. Recent research (Van Os 2017) with contemporary ozone installations looks promising, where complete elimination of pathogens and breakdown of remaining pesticides is achieved, with no safety problems.
2. *Hydrogen peroxide (H_2O_2).* Hydrogen peroxide is a strong, unstable oxidizing agent that reacts to form H_2O and an $O^{\cdot}$- radical. Commercially so-called activators are added to the solution to stabilize the original solution and to increase efficacy. Activators are mostly formic acid or acetic acid, which decrease pH in the nutrient solution. Different dosages are recommended (Runia 1995) against *Pythium* spp. (0.005%), other fungi (0.01%), such as *Fusarium*, and viruses (0.05%). The 0.05% concentration is also harmful to plant roots. Hydrogen peroxide is especially helpful for cleaning the watering system, whilst the use for disinfection has been taken over by other methods. The method is considered inexpensive, but not efficient.

3. *Sodium hypochlorite (NaOCl).* Sodium hypochlorite is a compound having different commercial names (e.g. household bleach) with different concentrations but with the same chemical structure (NaOCl). It is widely used for water treatment, especially in swimming pools. The product is relatively inexpensive. When added to water, sodium hypochlorite decomposes to HOCl and $NaOH^-$ and depending on the pH to OCl^-; the latter decomposes to Cl^- and $O^\cdot$ for strong oxidation. It reacts directly with any organic substance, and if there is enough hypochlorite, it also reacts with pathogens. Le Quillec et al. (2003) showed that the tenability of hypochlorite depends on the climatic conditions and the related decomposing reactions. High temperatures and contact with air cause rapid decomposition, at which $NaClO_3$ is formed with phytotoxic properties. Runia (1995) showed that hypochlorite is not effective for eliminating viruses. Chlorination with a concentration of 1–5 mg Cl L^{-1} and an exposure time of 2 h achieved a reduction of 90–99.9% of *Fusarium oxysporum*, but some spores survived at all concentrations. Safety measures have to be taken for safe storage and handling. Hypochlorite might work against a number of pathogens, but not all, but at the same time, Na^+ and Cl^- concentration is increased in a closed growing system which will also lead to levels which decrease productivity of the crop and at which time the nutrient solution has to be leached. Despite the above-mentioned drawbacks, the product is used and recommended by commercial operatives as a cheap and useful method.

4.5.2 Chemical Versus Non-chemical Methods

Growers prefer disinfection methods with excellent performance in combination with low costs. A good performance can be described by eliminating pathogens with a reduction of 99.9% (or a log 3 reduction) combined with a clear, understandable and controllable process. Low costs are preferably combined with low investments, low maintenance costs and no need for the grower to perform as a laboratory specialist. Heat treatment, UV radiation and ozone treatment show a good performance. However, investments in ozone treatment are very high, resulting in high annual costs. Heat treatment and UV radiation also have high annual costs, but investments are lower, whilst the eliminating process is easy to control. The latter two methods are most popular among growers, especially at nurseries larger than 1 or 2 ha. Slow sand filtration is less perfect in performance but has considerably lower annual costs. This method could be recommended for producers smaller than 1 ha and for growers with lower investment capital, as sand filters can be constructed by the grower themselves. Sodium hypochlorite and hydrogen peroxide are also cheap methods, but performance is insufficient to eliminate all pathogens. Besides it is a biocide and not a pesticide, which means by law, in the EU at least, it is legally forbidden to use it for the elimination of pathogens.

4.5.3 *Biofouling and Pretreatment*

Disinfection methods are not very selective between pathogens and other organic material in the solution. Therefore, pretreatment (rapid sand filters, or 50–80 um mechanical filters) of the solution before disinfection is recommended at heat treatment, UV radiation and ozone treatment. If after disinfection residuals of the chemical methods remain in the water, they may react with the biofilms which have been formed in the pipe lines of the watering systems. If the biofilm is released from the walls of the pipes, they will be transported to the drippers and cause clogging. Several oxidizing methods (sodium hypochlorite, hydrogen peroxide with activators, chlorine dioxide) are mainly used to clean pipe lines and equipment, and these create a special risk for clogging drippers over time.

References

Abad M, Noguera P, Puchades R, Maquieira A, Noguera V (2002) Physico-chemical and chemical properties of some coconut coir dusts for use as a peat substitute for containerised ornamental plants. Bioresour Technol 82:241–245

Adams P (1991) Effect of increasing the salinity of the nutrient solution with major nutrients or sodium chloride on the yield, quality and composition of tomatoes grown in rockwool. J Hortic Sci 66:201–207

Adams P, Ho LC (1992) The susceptibility of modern tomato cultivars to blossom-end rot in relation to salinity. J Hortic Sci 67:827–839

Baxter I (2015) Should we treat the ionome as a combination of individual elements, or should we be deriving novel combined traits? J Exp Bot 66:2127–2131

Bittsanszky A, Uzinger N, Gyulai G, Mathis A, Junge R, Villarroel M, Kotzen B, Komives T (2016) Nutrient supply of plants in aquaponic systems. Ecocycles 2:17–20

Blok C, de Kreij C, Baas R, Wever G (2008) Analytical methods used in soilless cultivation. In: Raviv, Lieth (eds) Soilless culture, theory and practice. Elsevier, Amsterdam, pp 245–290

Briat JF, Dubos C, Gaymard F (2015) Iron nutrition, biomass production, and plant product quality. Trends Plant Sci 20:33–40

Bunt BR (2012) Media and mixes for container-grown plants: a manual on the preparation and use of growing media for pot plants. Springer, Dordrecht

Cooper A (1979) The ABC of NFT. Grower Books, London

Cortella G, Saro O, De Angelis A, Ceccotti L, Tomasi N, Dalla Costa L, Manzocco L, Pinton R, Mimmo T, Cesco S (2014) Temperature control of nutrient solution in floating system cultivation. Appl Therm Eng 73:1055–1065

De Kreij C, Voogt W, Baas R (1999) Nutrient solutions and water quality for soilless cultures, report 196. Research station for Floriculture and Glasshouse Vegetables, Naaldwijk, p 36

De Rijck G, Schrevens E (1997) pH influenced by the elemental composition of nutrient solutions. J Plant Nutr 20:911–923

De Rijck G, Schrevens E (1998) Elemental bioavailability in nutrient solutions in relation to complexation reactions. J Plant Nutr 21:2103–2113

De Rijck G, Schrevens E (1999) Anion speciation in nutrient solutions as a function of pH. J Plant Nutr 22:269–279

Dong S, Scagel CF, Cheng L, Fuchigami LH, Rygiewicz PT (2001) Soil temperature and plant growth stage influence nitrogen uptake and amino acid concentration of apple during early spring growth. Tree Physiol 21:541–547

Dorais M, Menard C, Begin G (2006) Risk of phytotoxicity of sawdust substrates for greenhouse vegetables. In: XXVII International Horticultural Congress-IHC2006: International symposium on advances in environmental control, automation 761, pp 589–595

Enzo M, Gianquinto G, Lazzarin R, Pimpini F, Sambo P (2001) Principi tecnico-agronomici della fertirrigazione e del fuori suolo. In: Tipografia-Garbin. Padova, Italy

EU Aquaponics Hub. Plant Protection Fact Sheet. http://euaquaponicshub.com/hub/wp-content/uploads/2016/05/Plant-protection-factsheet.pdf

European Parliament and European Council (2000) Directive 2000/60/EC of the European Parliament and the Council of 23 October 2000 establishing a framework for Community action in the field of water policy (short: Water Framework Directive) Off J Eur Communities, p 327/1

Fageria NK, Baligar VC, Clark RB (2002) Micronutrients in crop production. Adv Agron 77:185–268

Fanasca S, Colla G, Maiani G, Venneria E, Rouphael Y, Azzini E, Saccardo F (2006) Changes in antioxidant content of tomato fruits in response to cultivar and nutrient solution composition. J Agric Food Chem 54:4319–4325

Fernandez D (2016) HydroBuddy v1.50: The first free Open source hydroponic nutrient calculator program Available Online [WWW document]. Sci Hydroponics. http://scienceinhydroponics.com/

Fornes F, Belda RM, Abad M, Noguera P, Puchades R, Maquieira A, Noguera V (2003) The microstructure of coconut coir dusts for use as alternatives to peat in soilless growing media. Aust J Exp Agr 43:1171–1179

Gibson JL (2007) Nutrient deficiencies in bedding plants. Ball Publishing, Batavia

Goddek S, Keesman KJ (2018) The necessity of desalination technology for designing and sizing multi-loop aquaponics systems. Desalination 428:76–85

Haynes RJ (1990) Active ion uptake and maintenance of cation-anion balance: a critical examination of their role in regulating rhizosphere pH. Plant Soil 126:247–264

Heuvelink E, Kierkels T (2016) Iron is essential for photosynthesis and respiration: iron deficiency. In Greenhouses: the international magazine for greenhouse growers 5, pp 48–49

Ho LC, Belda R, Brown M (1993) Uptake and transport of calcium and the possible causes of blossom-end rot in tomato. J Exp Bot 44:509–518

Hoagland DR, Arnon DI (1950) The water culture method for growing plants without soil, Circular 347, Agricultural Experimental station, University of California, Berkeley, CA, USA

Huett DO (1994) Growth, nutrient uptake and tipburn severity of hydroponic lettuce in response to electrical conductivity and K:Ca ratio in solution. Aust J Agric Res 45:251–267

Hussain A, Iqbal K, Aziem S, Mahato P, Negi AK (2014) A review on the science of growing crops without soil (soilless culture)-a novel alternative for growing crops. Int J Agric Crop Sci 7:833–842

Jensen MH, Collins WL (1985) Hydroponic vegetable production. Hortic Rev 7:483–558

Kipp JA, Wever G, de Kreij C (2001) International substrate manual. Elsevier, Amsterdam

Kuhry P, Vitt DH (1996) Fossil carbon/nitrogen ratios as a measure of peat decomposition. Ecology 77:271–275

Lamanna D, Castelnuovo M, D'Angelo G (1990) Compost-based media as alternative to peat on ten pot ornamentals. Acta Hortic 294:125–130

Le Bot J, Adamowicz S, Robin P (1998) Modelling plant nutrition of horticultural crops: a review. Sci Hortic 74:47–82

Le Quillec S, Fabre R, Lesourd D (2003) Phytotoxicité sur tomate et chlorate de sodium. Infos-ctifl, pp 40–43

Liu S, Hou Y, Chen X, Gao Y, Li H, Sun S (2014) Combination of fluconazole with non-antifungal agents: a promising approach to cope with resistant *Candida albicans* infections and insight into new antifungal agent discovery. Int J Antimicrob Agents 43:395–402

Maher MJ, Prasad M, Raviv M (2008) Organic soilless media components. In: Raviv, Lieth (eds) Soilless culture, theory and practice. Elsevier, Amsterdam, pp 459–504

Marschner P (2012) Rhizosphere biology. In: Marschner's mineral nutrition of higher plants, 3rd edn. Academic, Amsterdam, pp 369–388

Masclaux-Daubresse C, Daniel-Vedele F, Dechorgnat J, Chardon F, Gaufichon L, Suzuki A (2010) Nitrogen uptake, assimilation and remobilization in plants: challenges for sustainable and productive agriculture. Ann Bot 105:1141–1157

Maucieri C, Nicoletto C, Junge R, Schmautz Z, Sambo P, Borin M (2018) Hydroponic systems and water management in aquaponics: a review. Ital J Agron 13:1–11

McCutchan JH, Lewis WM, Kendall C, McGrath CC (2003) Variation in trophic shift for stable isotope ratios of carbon, nitrogen, and sulfur. Oikos 102:378–390

Moriarty MJ, Semmens K, Bissonnette GK, Jaczynski J (2018) Inactivation with UV-radiation and internalization assessment of coliforms and *Escherichia coli* in aquaponically grown lettuce. LWT 89:624–630

Muneer S, Lee BR, Kim KY, Park SH, Zhang Q, Kim TH (2014) Involvement of Sulphur nutrition in modulating iron deficiency responses in photosynthetic organelles of oilseed rape (*Brassica napus* L.). Photosynth Res 119:319–329

NGS, Multilayer. http://ngsystem.com/en/ngs/multibanda

Nicoletto C, Maucieri C, Sambo P (2017) Effects on water management and quality characteristics of ozone application in chicory forcing process: a pilot system. Agronomy 7:29

Nicoletto C, Maucieri C, Mathis A, Schmautz Z, Komives T, Sambo P, Junge R (2018) Extension of aquaponic water use for NFT baby-leaf production: mizuna and rocket salad. Agronomy 8:75

Nozzi V, Graber A, Schmautz Z, Mathis A, Junge R (2018) Nutrient management in aquaponics: comparison of three approaches for cultivating lettuce, mint and mushroom herb. Agronomy 8:27

Olle M, Ngouajio M, Siomos A (2012) Vegetable quality and productivity as influenced by growing medium: a review. Agriculture 99:399–408

Parent S-É, Parent LE, Egozcue JJ, Rozane D-E, Hernandes A, Lapointe L, Hébert-Gentile V, Naess K, Marchand S, Lafond J, Mattos D, Barlow P, Natale W (2013) The plant ionome revisited by the nutrient balance concept. Front Plant Sci 4:39

Perelli M, Graziano PL, Calzavara R (2009) Nutrire le piante. Arvan Ed., Venice.

Pickering HW, Menzies NW, Hunter MN (2002) Zeolite/rock phosphate – a novel slow release phosphorus fertiliser for potted plant production. Sci Hortic 94:333–343

Postma J, van Os EA, Bonants PJM (2008) Pathogen detection and management strategies in soilless plant growing systems. In: Raviv, Lieth (eds) Soilless culture, theory and practice. Elsevier, Amsterdam, pp 425–458

Pregitzer KS, King JS (2005) Effects of soil temperature on nutrient uptake. In: Nutrient acquisition by plants. Springer, Berlin/Heidelberg, pp 277–310

Rooney CP, Zhao FJ, McGrath SP (2006) Soil factors controlling the expression of copper toxicity to plants in a wide range of European soils. Environ Toxicol Chem 25:726–732

Ruijs MNA (1994) Economic evaluation of closed production systems in glasshouse horticulture. Acta Hortic 340:87–94

Runia WT (1995) A review of possibilities for disinfection of recirculation water from soilless culture. Acta Hortic 382:221–229

Runia WT, van Os EA, Bollen GJ (1988) Disinfection of drain water from soilless cultures by heat treatment. Neth J Agric Sci 36:231–238

Schmautz Z, Graber A, Jaenicke S, Goesmann A, Junge R, Smits TH (2017) Microbial diversity in different compartments of an aquaponics system. Arch Microbiol 199:613–620

Schnug E, Haneklaus S (2005) Sulphur deficiency symptoms in oilseed rape (*Brassica napus* L.)- the aesthetics of starvation. Phyton 45:79–95

Silber A, Bar-Tal A (2008) Nutrition of substrate-grown plants. In: Raviv, Lieth (eds) Soilless culture, theory and practice. Elsevier, Amsterdam, pp 292–342

Sonneveld C, Voogt W (2009) Plant nutrition of greenhouse crops. Springer, Dordrecht, p 403

Steiner AA (1961) A universal method for preparing nutrient solutions of a certain desired composition. Plant Soil 15:134–154

Steiner AA (1984) The universal nutrient solution. In: Proceeding 6th International congress on soilless culture. Lunteren, Netherlands, ISOSC, pp 633–649

Tindall JA, Mills HA, Radcliffe DE (1990) The effect of root zone temperature on nutrient uptake of tomato. J Plant Nutr 13:939–956

Trang NTD, Brix H (2014) Use of planted biofilters in integrated recirculating aquaculture-hydroponics systems in the Mekong Delta, Vietnam. Aquac Res 45:460–469

Uchida R (2000) Essential nutrients for plant growth: nutrient functions and deficiency symptoms. In: Silva JA, Uchida R (eds) Plant nutrient management in Hawaii's soils, approaches for tropical and subtropical agriculture. College of Tropical Agriculture and Human Resources, University of Hawaii at Manoa, Honolulu, pp 31–55

Van Os EA (1994) Closed growing systems for more efficient and environmental friendly production. Acta Hortic 361:194–200

Van Os EA (1999) Closed soilless growing systems: a sustainable solution for Dutch greenhouse horticulture. Water Sci Technol 39:105–112

Van Os EA (2009) Comparison of some chemical and non-chemical treatments to disinfect a recirculating nutrient solution. Acta Hortic 843:229–234

Van Os EA (2017) Recent advances in soilless culture in Europe. Acta Hortic 1176:1–8

Van Os EA, Gieling TH, Lieth JH (2008) Technical equipment in soilless production systems. In: Raviv, Lieth (eds) Soilless culture, theory and practice. Elsevier, Amsterdam, pp 157–207

Vance CP, Uhde-Stone C, Allan DL (2003) Phosphorus acquisition and use: critical adaptations by plants for securing a nonrenewable resource. New Phytol 157:423–447

Verdonck O, Penninck RD, De Boodt M (1983) The physical properties of different horticultural substrates. Acta Hortic 150:155–160

Verwer FLJA (1978) Research and results with horticultural crops grown in rockwool and nutrient film. Acta Hortic 82:141–148

Wallach J (2008) Physical characteristics of soilless media. In: Raviv, Lieth (eds) Soilless culture, theory and practice. Elsevier, Amsterdam, pp 41–116

Wang M, Zheng Q, Shen Q, Guo S (2013) The critical role of potassium in plant stress response. Int J Mol Sci 14:7370–7390

Wu M, Kubota C (2008) Effects of high electrical conductivity of nutrient solution and its application timing on lycopene, chlorophyll and sugar concentrations of hydroponic tomatoes during ripening. Sci Hortic 116:122–129

Zekki H, Gauthier L, Gosselin A (1996) Growth, productivity, and mineral composition of hydroponically cultivated greenhouse tomatoes, with or without nutrient solution recycling. J Am Soc Hortic Sci 121:1082–1088

Zoschke K, Börnick H, Worch E (2014) Vacuum-UV radiation at 185 nm in water treatment–a review. Water Res 52:131–145

Part II
Diverse Technologies in Aquaponics

5

Fundamentals of Aquaponics

Wilson Lennard and Simon Goddek

Abstract Aquaponics is a technology that is part of the broader integrated agri-aquaculture systems discipline which seeks to combine animal and plant culture technologies to confer advantages and conserve nutrients and other biological and economic resources. It emerged in the USA in the early 1970s and has recently seen a resurgence, especially in Europe. Whilst aquaponics broadly combines recirculating fish culture with hydroponic plant production, the application of the term aquaponic is broad and many technologies claim use of the name. Combining fish culture with aquatic-based, terrestrial plant culture via aquaponics may be better defined via its nutrient resource sharing credentials. Aquaponics applies several principles including, but not limited to, efficient water use, efficient nutrient use, lowered or negated environmental impact and the application of biological and ecological approaches to agricultural fish and plant production. Water sources are important so that the nutrients required for fish and plant production are available and balanced, and system water chemistry is paramount to optimised fish and plant production. Systems may be configured in several ways, including those that are fully recirculating and those that are decoupled. Aquaponics importantly seeks to apply methods that provide technical, biological, chemical, environmental and economic advantages.

Keywords Aquaponics · Agri-aquaculture · Aquaculture · Hydroponics · Agriculture · Fish · Plants · Nutrients · Ecology

---

W. Lennard (✉)
Aquaponic Solutions, Blackrock, VIC, Australia
e-mail: willennard@gmail.com

S. Goddek
Mathematical and Statistical Methods (Biometris), Wageningen University, Wageningen, The Netherlands
e-mail: simon.goddek@wur.nl; simon@goddek.nl

5.1 Introduction

Aquaponics is a technology that is a subset of a broader agricultural approach known as integrated agri-aquaculture systems (IAAS) (Gooley and Gavine 2003). This discipline consists of integrating aquaculture practices of various forms and styles (mostly fin fish farming) with plant-based agricultural production. The rationale of integrated agri-aquaculture systems is to take advantage of the resources shared between aquaculture and plant production, such as water and nutrients, to develop and achieve economically viable and environmentally more sustainable primary production practices (Gooley and Gavine 2003). In essence, both terrestrial plant and aquatic animal production systems share a common resource: water. Plants are generally consumptive of water via transpiration and release it to the surrounding gaseous environment, whereas fish are generally less consumptive of water, but their contained culture produces substantial waste water streams due to accumulated metabolic wastes. Therefore, aquaculture may be integrated within the water supply pathway of plant production in non-consumptive ways so that two crops (fish and plants) may be produced from a water source that is generally used to produce one crop (plants).

An interesting additional advantage of integrating aquaculture with the irrigation supply pathway for plant production is that aquaculture also produces waste nutrients via the dissolved and undissolved wastes produced from fish (and other aquatic animal) metabolism. Therefore, aquaculture may also produce waste nutrient streams that are suitable for, and assist, plant production by contributing to the plants nutrient requirements.

The advantages produced by integrating aquaculture with conventional terrestrial and aquatic plant production systems have been summarised by Gooley and Gavine (2003) as:

1. An increase in farm productivity and profitability without any net increase in water consumption (Chap. 2).
2. Farm diversification into higher-value crops, including high-value aquatic species.
3. Reuse of otherwise wasted on-farm resources (e.g. capture and reuse of nutrients and water).
4. Reduction of net environmental impacts of semi-intensive and intensive farming practices.
5. Net economic benefits by offsetting existing farm capital and operating expenses (Chap. 18).

Aquaponics has been said to have evolved from relatively ancient agriculture practices associated with integrating fish culture with plant production, especially those developed within the South East Asian, flooded rice paddy farming context and South American Chinampa, floating island, agriculture practices (Komives and Junge 2015). In reality, historically, fish were rarely actively added to rice paddy fields until the nineteenth century (Halwart and Gupta 2004) and were present in very low densities which would not contribute to any substantial nutritive assistance to the plants. Chinampas were traditionally built on lakes in Mexico where nutrient advantages may have been supplied via the eutrophic or semi-eutrophic lake sediments rather than directly from any designed or actively integrated fish production system (Morehart 2016; Baquedano 1993).

Modern aquaponics started in the USA in the 1970s and was co-evolved by several institutions with an interest in more sustainable farming practices. Early important work was performed by several researchers, but ultimately, the progenitor of nearly all modern aquaponics is thought to be the work performed by, and the systems produced by, James Rakocy and his team at the University of the Virgin Islands (UVI) starting in the early 1980s (Lennard 2017).

Aquaponics is now considered a new and emerging industry with a relevant place in the broader, global agricultural production context and there are a number of variations of the technology of integrating fish culture with aquatic plant culture that are collectively defined under the aquaponics banner or name (Knaus and Palm 2017). Therefore, aquaponics seeks to integrate *aqua*culture animal production with hydro*ponic* plant production using various methods to share water and nutrient resources between the major production components to produce commercial and saleable fish and plant products.

5.2 A Definition of Aquaponics

Aquaponics fits into the broader definition of integrated agri-aquaculture systems (IAAS). However, IAAS applies many different aquatic animal and plant production technologies in many contexts, whereas aquaponics is far more tightly associated with integrating tank-based fish culture technologies (e.g. recirculating aquaculture systems; RAS) with aquatic or hydroponic plant culture technologies (Lennard 2017). RAS technologies apply conserved and standard methods for the culture of fish in tanks with applied filtration to control and alter the water chemistry to make it suitable for fish (i.e. fast and efficient solid fish wastes removal, efficient, bacteria-mediated conversion of potentially toxic dissolved fish waste ammonia to less toxic nitrate and oxygen maintenance via assisted aeration or directly injected oxygen gas) (Timmons et al. 2002). Hydroponics and substrate culture technologies apply conserved and standard methods for the culture of edible terrestrial plants within aquatic environments (i.e. the plants gain access to the nutrients required for growth via a water-based delivery method) (Resh 2013).

The association of aquaponics with standard RAS aquaculture and hydroponics/ substrate culture means that aquaponics is often defined simply as "... the combination of fish production (aquaculture) and soil-less plant cultivation hydroponics under coupled or decoupled water circulation" (Knaus and Palm 2017). This broad definition places an emphasis on the integration of hardware, equipment or technologies and places little, if any, emphasis on any other aspects of the method.

Because aquaponics is a relatively new industrial-scale technology that applies different methods and approaches, the applied definition appears very broad. Some define aquaponics within a recirculating context only (Cerozi and Fitzsimmons

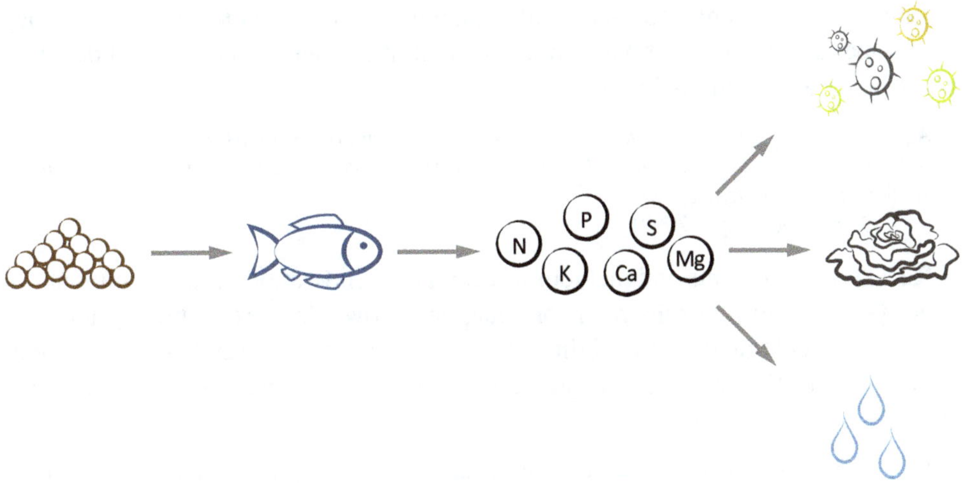

Fig. 5.1 Schematic representation of the nutrient flows within an aquaponic system. Fish feed is the major nutrient entry point. The fish eat the feed, use what nutrients they need, release the rest as waste and this waste is then partitioned between the microbes, plants and system water. (adapted from Lennard 2017)

2017), some concentrate on approaches that do not return the water from the plants to the fish (Delaide et al. 2016) and others include both recirculating and decoupled methods (Knaus and Palm 2017). Further still, some researchers are including the use of aquaculture effluents irrigated to soil-based crop production under the aquaponic title (Palm et al. 2018). Historically, aquaponics, as the breakdown of the word (*aqua*culture and hydro*ponics*) suggests, was defined as only concerning aquaculture and hydroponic plant production (Rakocy and Hargreaves 1993), so current attempts at associations with soil-based culture seem incongruous.

Whilst aquaponic systems do integrate tank-based aquaculture technologies with hydroponic plant culture technologies, aquaponic systems work by supplying nutrients to, and partitioning nutrients between, the production inhabitants (fish and plants) and the inhabitants that perform biological and chemical services that assist the production inhabitant outcome (microflora) (Fig. 5.1) (Lennard 2017). Therefore, is aquaponics more a system associated with nutrient supply, dynamics and partitioning rather than one associated with the technology, equipment or hardware applied?

Over the past decades, the definition of aquaponics has included a similar theme, with subtle variations. The broadest definition has generally been provided in the scientific publications of Rakocy and his UVI team, for example:

> *Aquaponics is the combined culture of fish and plants in closed recirculating systems.*
> – Rakocy et al. (2004a, b)

This early definition was based on the assumption that one-loop, fully recirculating systems, consisting of a recirculating aquaculture component and a hydroponic component, represented all aquaponic systems, which at the time, they did. Graber and Junge (2009) expanded the definition, due to changes and developments in the approach, as follows:

> *Aquaponic is a special form of recirculating aquaculture systems (RAS), namely a polyculture consisting of fish tanks (aquaculture) and plants that are cultivated in the same water circle (hydroponic).*
>
> – Graber and Junge (2009)

Recent developments and methods ask for a reconsideration of this standpoint. In recent years a shifting of the focus of aquaponics towards a production system that tackles both ecological responsibility and economic sustainability has been present. Kloas et al. (2015) and Suhl et al. (2016) were one of the first to address this economic consideration:

> *[…] a unique and innovative double recirculating aquaponic system was developed as a prerequisite for a high productivity comparable to professional stand-alone fish/plant facilities.*
>
> – Suhl et al. (2016)

The definition issue, or clarifying "what can be defined as aquaponics", has been a point of discussion over the past years. One of the main areas of development has been that of multi-loop (or decoupled) aquaponic systems that aim at providing additional fertilisers to the plants in order to expose them to an optimal nutrient concentration (Goddek 2017). There should be no opposition between the ideologies of fully recirculating and multi-loop aquaponic methodologies, both have their respective places and applications within the appropriate industrial context and a single common driving force of both should be that the technology, whilst being nutrient and water efficient, also needs to be economically competitive to establish itself in the market. In order to replace conventional practices, more than an ideology needs to be offered to potential clients/users – i.e. technical and economic feasibility.

The European COST sponsored Aquaponics Hub (COST FA1305 2017) applies the definition "*…a production system of aquatic organisms and plants where the majority (> 50%) of nutrients sustaining the optimal plant growth derives from waste originating from feeding the aquatic organisms*", which clearly places an emphasis on the nutrient sharing aspect of the technology.

It must also be stated that the proportion of fish to plants should remain at a level that supports a core prospect of aquaponics; that plants are grown using fish wastes. A system containing one fish and several hectares of hydroponic plant cultivation, for example, should not be considered as aquaponics, simply because that one fish effectively contributes nothing to the nutrient requirements of the plants. Since the labelling of aquaponic products plays an increasingly important role in consumer choice, we want to encourage a discussion by redefining aquaponics based on these multiple developments of the technology. Even though we advocate closing the nutrient cycle to the highest possible degree in the context of best practicable means, a potential definition should also take all developments into consideration.

Therefore, the definition should contain as a minimum, the requirement for a majority of aquaculture-derived nutrients for the plants. A new definition may therefore be represented as:

> *Aquaponics is defined as an integrated multi-trophic, aquatic food production approach comprising at least a recirculating aquaculture system (RAS) and a connected hydroponic unit, whereby the water for culture is shared in some configuration between the two units. Not less than 50% of the nutrients provided to the plants should be fish waste derived.*

Nutrient-based definitions are open and non-judgemental of the applied technology choice, or even the proportions of each component (fish and plants), as long as fish culture and some form of aquatic (hydroponic or substrate culture) plant production technology is utilised. However, it also focuses the definition on the nutrient dynamics and nutrient sharing aspects of the methods applied and therefore ensures, to at least some extent, that the advantages often associated with aquaponics (water saving, nutrient efficiency, lowered environmental impact, sustainability) are present in some proportion.

The nutrient association definition applied to aquaponics will always be a source of further contention among those who practise it. This is supported by the fact that the name aquaponics is applied to a vast array of different technologies with different nutrient supply motivations and usage outcomes: from system designs and methods that expect, if not demand, that the vast majority of the nutrients required to grow the plants arise from the fish wastes (in some cases, greater than 90%; Lennard 2017) to designs that share plant nutrient supplies between fish wastes and more substantial external additions (e.g. approximately 50:50 fish waste to external supplementation – as many modern, European decoupled aquaponic system designs do; COST FA1305 2017) to those designs that add so few fish that no discernible nutrient supply from the fish wastes to the plants is present (Lennard 2017).

The name aquaponics, until relatively recently (i.e. the last 3–5 years), has been universally applied to coupled and fully recirculating system designs that seek to supply as much of the required plant nutrition from the fish wastes as possible (Rakocy and Hargreaves 1993; Lennard 2017) (Fig. 5.2).

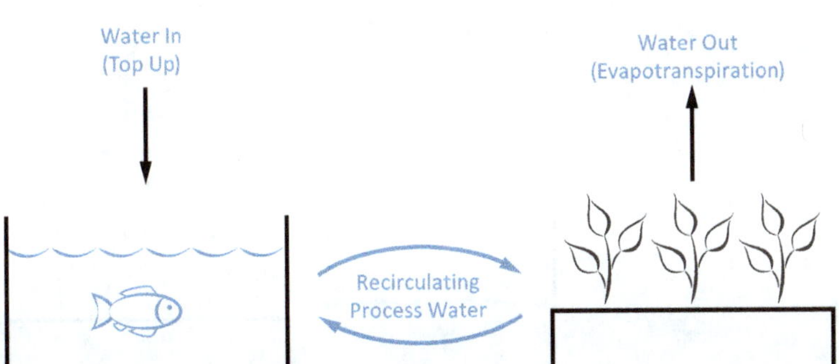

Fig. 5.2 Simplified scheme of the main water flows within a coupled aquaponic system. The nutrient concentrations in the process water are equally distributed throughout the whole system

However, decoupled approaches now represent a proportion of the systems being researched or commercially applied, especially in Europe, and in current practice do not supply plant nutrient requirements from the fish wastes to the same extent as fully recirculating systems do (Lennard 2017; Goddek and Keesman 2018). For example, Goddek and Keesman (2018) state that for 3 examples of current European decoupled aquaponic system designs, the relative addition requirements for external hydroponic-derived nutrients are 40–60% (NerBreen), 60% (Tilamur) and 38.1% (IGB Berlin). Because these decoupled designs are based on integrating existing RAS and hydroponic/substrate culture technologies, they are regarded as aquaponic in nature (Delaide et al. 2016) (Fig. 5.3) (see Chap. 8).

The definition of aquaponics is now being expanded beyond ecological, water and nutrient efficiency drivers and optimisation to also include economic drivers (Goddek and Körner 2019; Goddek and Keesman 2018; Goddek 2017; Kloas et al. 2015; Reyes Lastiri et al. 2016; Yogev et al. 2016) (Chap. 8). The benefits of such an approach are that a positive economic outcome from aquaponics technology is as important as its biological, chemical, engineering, ecological and sustainable credentials and therefore, the economic outcome should play a role within the overall definition (Chap. 8).

Many advantages are often associated with aquaponics, especially in terms of its water-use efficiency, its nutrient use efficiency, its sustainable nature, its ability to produce two crops from the one input source (fish feed) and its lowered environmental impact (Timmons, et al., 2002; Buzby and Lian-shin 2014; Wongkiew et al. 2017; Roosta and Hamidpour 2011; Suhl et al. 2016). These advantages are regularly quoted and applied by commercial aquaponic operators and are used as a marketing and price regulation pathway for the products (fish and plants), and therefore, the use of the name "aquaponics" directly and immediately associates that the products labelled as such have been produced with methods that contain or utilise the advantages listed. However, there is no formal regulation of the industry that dictates that the use of the word (aquaponics) only occurs when the advantages are apparent and present within the technology and methods applied. If the above advantages are assigned to aquaponics as a technology, then surely the technology

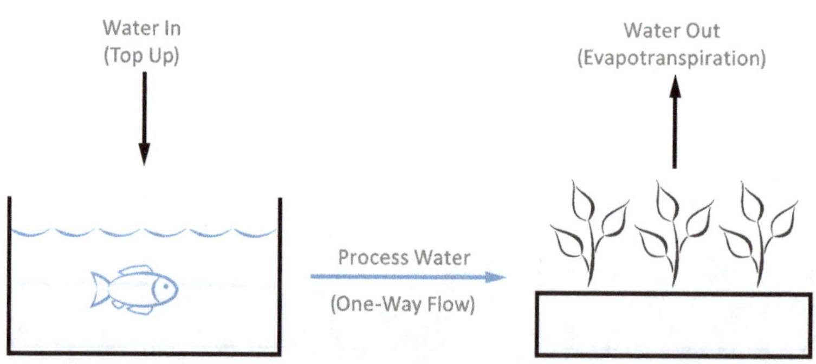

Fig. 5.3 Simplified scheme of the main water flows within a decoupled aquaponic system. The nutrient concentrations in each component may be separately tailored to the individual component requirement

should provide the prescribed advantages, and if the technology does not provide the advantages, then the word should not be applied (Lennard 2017).

Because aquaponics may be defined either in terms of its hardware equipment integration aspect (RAS with hydroponics), its nutrient sharing or partitioning properties or its ability to provide important advantages, there is still a wide spectrum of possible applications of the name to many different technical approaches that utilise different methods and demand different outcomes. Therefore, it appears that the actual definition of aquaponics is still unresolved.

It appears therefore that very important questions are yet to be answered: what is aquaponics and how is it defined?

This would suggest that one very important aspect for the aquaponic industry to consider is the development of a truthful and agreed-upon definition. The broader aquaponics industry will continue to be full of disagreement if a definition is not agreed upon, and more importantly, consumers of the products produced within aquaponic systems will become more and more confused about what aquaponics actually is – a state of affairs that will not assist the growth and evolution of the industry.

5.3 General Principles

Even though the definition of aquaponics has not been entirely resolved, there are some general principles that are associated with the broad range of aquaponic methods and technologies.

Using the nutrients added to the aquaponic system as optimally and efficiently as possible to produce the two main products of the enterprise (i.e. fish and plant biomass) is an important and shared first principle associated with the technology (Rakocy and Hargreaves 1993; Delaide et al. 2016; Knaus and Palm 2017). There is no use in adding nutrients (which possess an inherent cost in terms of money, time and value) to a system to watch a high percentage of those nutrients are partitioned into processes, requirements or outcomes that are not directly associated with the fish and plants produced, or any intermediary life forms that may assist nutrient access by the fish and plants (i.e. microorganisms – bacteria, fungi, etc.) (Lennard 2017). Therefore, probably the most important general principle associated with aquaponics is to use the applied nutrients as efficiently as possible to achieve the optimised production of both fish and plants.

This same argument may also be applied to the water requirement of the aquaponic system in question; again, the water added to the system should be utilised principally by the fish and plants and used as efficiently as possible and not allowed to leak to processes, life forms or outcomes that are not directly associated with fish and plant production or may impact on the surrounding environment (Lennard 2017).

In real terms, the efficient use of nutrients and water leads to several design principles that are broadly applied to the aquaponic method:

1. The most important principle of aquaponics is to use the wastes produced by the fish as a principle nutrient source for the plants. In fact, this is the entire idea of aquaponics and so should be a first order driver for the method. Aquaponics was historically envisaged as a system to grow plants using fish aquaculture wastes so that those aquaculture wastes had less environmental impact and were seen as a positive and profitable commodity, rather than a troublesome waste product with an associated cost to meet environmental legislative requirements (Rakocy and Hargreaves 1993; Love et al. 2015a, b).
2. The system design should encourage the use of fish keeping and plant culturing technologies that do not inherently uptake or destructively utilise the water or nutrient resources added. For example, fish keeping components based on using earthen ponds are discouraged, because the earthen pond has the ability to use and make unavailable water and nutrient resources to the associated fish and plants, thus lowering the water and nutrient use efficiency of the system. Similarly, hydroponic plant culturing methods should not use media that uptakes excessive amounts of nutrients or water and renders them unavailable to the plants (Lennard 2017).
3. The system design should not waste nutrients or water via the production of external waste streams. Principally, if water and nutrients leave the system via a waste stream, then that water and those nutrients are not being used for fish or plant production, and therefore, that water and those nutrients are being wasted, and the system is not as efficient as possible. In addition, the production of a waste streams can have a potential environmental impact. If waste water and nutrients do leave the aquaponic system, they should be used in alternate, exterior-to-system plant production technologies so water and nutrients are not wasted, contribute to the overall production of edible or saleable biomass and do not present a broader environmental impact potential (Tyson et al. 2011).
4. The system should be designed to lower or ideally, completely negate, direct environmental impact from water or nutrients. A first order goal of aquaponics is to use the wastes produced by the fish as a nutrient source for the plants so as to negate the release of those nutrients directly to the surrounding environment where they may cause impacts (Tyson et al. 2011).
5. Aquaponic system designs should ideally lend themselves to being located within environmentally controlled structures and situations (e.g. greenhouses, fish rooms). This allows the potential to achieve the best productive rates of fish and plants from the system. Most aquaponic designs are relatively high in terms of capital costs and ongoing costs of production, and therefore, the ability to house the system in the perfect environment enhances profit potentials that financially justify the high capital and costs of production (Lennard 2017).

The above outlined principles of design directly associate with a set of general principles that are often, but not always, applied to the aquaponic production environment. These general principles relate to how the system operates and how nutrients are portioned among the system and its inhabitants.

The basic premise of aquaponics, in a nutrient dynamic context, is that fish are fed fish feed, fish metabolise and utilise the nutrients in the fish feed, fish release wastes based on the substances in the fish feed they do not utilise (including elements), microflora access those fish metabolic wastes and use small amounts of them, but transform the rest, and the plants then access and remove those microflora transformed, fish metabolic wastes as nutrient sources and, to some extent, clean the water medium of those wastes and counteract any associated accumulation (Rakocy and Hargreaves 1993; Love et al. 2015a, b).

Because earthen-based fish production systems remove nutrients themselves, aquaponics generally utilises what are known as recirculating aquaculture system (RAS) principles for the fish production component (Rakocy and Hargreaves 1993; Timmons et al. 2002). Fish are kept in tanks made of materials that do not remove nutrients from the water (plastic, fibreglass, concrete, etc.), the water is filtered to treat or remove the metabolic waste products of the fish (solids and dissolved ammonia gases) and the water (and associated nutrients) is then directed to a plant culturing component whereby the plants use the fish wastes as part of their nutrient resource (Timmons et al. 2002). As for the fish, earthen-based plant culturing components are not used because the soils involved remove nutrients and may not necessarily make them fully available to the plants. In addition, hydroponic plant culturing techniques do not use soil and are cleaner than soil-based systems and allow some passive control of the microorganism mixtures present.

Plants cultured in conventional hydroponics require the addition of what are known as mineral fertilisers: nutrients that are present in their basal, ionic forms (e.g. nitrate, phosphate, potassium, calcium, etc. as ions) (Resh 2013). Conversely, recirculating aquaculture systems must apply regular (daily) water exchanges to control the accumulation of fish waste metabolites (Timmons et al. 2002). Aquaponics seeks to combine the two separate enterprises to produce an outcome that achieves the best of the two technologies while negating the worst (Goddek et al. 2015).

Plants require a suite of macro and micro elements for optimal and efficient growth. In aquaponics, the majority of these nutrients arise from the fish wastes (Rakocy and Hargreaves 1993; Lennard 2017; COST FA1305 2017). However, fish feeds (the major source of aquaponic system nutrients) do not contain all the nutrients required for optimised plant growth, and therefore, external nutrition, to varying extents, is required.

Standard hydroponics and substrate culture add nutrients to the water in forms that are directly plant-available (i.e. ionic, inorganic forms produced via designed salt variety additions) (Resh 2013). A proportion of the wastes released by fish are in forms that are directly plant-available (e.g. ammonia) but potentially toxic to the fish (Timmons et al. 2002). These dissolved, ionic fish waste metabolites, like ammonia, are transformed by ubiquitous bacterial species that replace hydrogen ions with oxygen ions, the product from ammonia being nitrate, which is far less toxic to the fish and the preferred nitrogen source for the plants (Lennard 2017). Other nutrients appropriate to plant uptake are bound in the solid fraction of the fish waste as organic compounds and require further treatment via microbial interaction to render the nutrients available to plant uptake (Goddek et al. 2015). Therefore, aquaponic systems require a suite of microflora to be present to perform these transformations.

The key to optimised aquaponic integration is determining the ratio between fish waste output (as directly influenced by fish feed addition) and plant nutrient utilisation (Rakocy and Hargreaves 1993; Lennard and Leonard 2006; Goddek et al. 2015). Various rules of thumb and models have been developed in an attempt to define this balance. Rakocy et al. (2006) developed an approach that matches the plant growing area requirement with the daily fish feed input and called it "The Aquaponic Feeding Rate Ratio". The feeding rate ratio is set between 60 and 100 grams of fish feed added per day, per square meter of plant growing area (60–100 g/m^2/day). This feeding rate ratio was developed using *Tilapia* spp. fish eating a standard, 32% protein commercial diet (Rakocy and Hargreaves 1993). In addition, the aquaponic system this ratio is particular to (known as the University of the Virgin Islands Aquaponic System – UVI System) does not utilise the solid fish waste fraction, is over-supplied with nitrogen and requires in-system, passive de-nitrification to control the nitrogen accumulation rate (Lennard 2017). Others have determined alternate ratios based on different fish and plant combinations, tested in different specific conditions (e.g. Endut et al. 2010 – 15–42 g/m^2/day for African catfish, *Clarias gariepinus* and water spinach plants, *Ipomoea aquatica*).

The UVI feeding rate ratio was developed by Rakocy and his team as an approximate approach; hence why it is stated as a range (Rakocy and Hargreaves 1993). The UVI ratio tries to account for the fact that different plants require different nutrient amounts and mixtures and therefore a "generic" aquaponic design approach is a difficult prospect. Lennard (2017) has developed an alternate approach that seeks to directly match individual fish waste nutrient production rates (based on the fish feed utilised and the fish conversion and utilisation of that feed) with specific plant nutrient uptake rates so that exacting fish to plant ratio matching for any fish or plant species chosen may be realised and accounted for in the aquaponic system design. He matches this design approach with a specific management approach that also utilises all the nutrients available within the fish solid waste fraction (via aerobic remineralisation of the fish solid wastes) and only adds the nutrients required by the chosen plant species for culture that are missing from the fish waste production fractions. Therefore, this substantially lowers the associated feeding rate ratio (e.g. less than 11 g/m^2/day for some leafy green varieties as a UVI equivalent) and allows any fish species to be specifically and exactly matched to any plant species chosen (Lennard 2017). Similarly, Goddek et al. (2016) have proposed models that allow more exacting fish to plant component ratio determination for decoupled aquaponic systems.

The general principles of efficient nutrient use, low and efficient water use, low or negated environmental impact, ability to be located away from traditional soil resources and sustainability of resource use are the general principles applied to aquaponic system design and configuration and their ongoing application should be encouraged within the field and industry.

5.4 Water Sources

Water is the key medium used in aquaponic systems because it is shared between the two major components of the system (fish and plant components), it is the major carrier of the nutrient resources within the system and it sets the overall chemical environment the fish and plants are cultured within. Therefore, it is a vital ingredient that may have a substantial influence over the system.

In an aquaponic system, water-based environment context, the source of water and what that source water contains chemically, physically and biologically are a major influence over the system because it sets a baseline for what is required to be added to the system by the various inputs of the system. These inputs, in turn, effect and set the environment that the fish and plants are cultured within. For example, some of the major inputs in terms of nutrients to any aquaponic system include, but are not limited to, the fish feed (a primary nutrient resource for the system), the buffers applied (which assist to control and set the pH values associated with both the fish and plant components) and any external nutrient additions or supplementations required to meet the nutrient needs of the fish and plants (Lennard 2017).

Fish feeds are designed to provide the nutrition required for fish growth and health and therefore contain nutrient mixtures and quantities primarily to aid the fish being cultured (Timmons et al. 2002; Rakocy et al. 2006). Plants, on the other hand, have different nutrient requirements to fish, and fish feeds rarely, if ever, meet the total nutrient requirements of the plants (Rakocy et al. 2006). Because of this, aquaponic systems that culture fish and plants solely using fish feed-derived nutrient resources may efficiently and optimally produce fish, but they rarely do so for the plants. The best aquaponic system designs recognise that the ultimate outcome is to produce both fish and plants at optimal and efficient growth rates and therefore, also recognise that some form of additional nutrition is required to meet the total plant nutrient requirement (Rakocy et al. 2006; Suhl et al. 2016).

Classical, fully recirculating aquaponic systems generally rely on fish feeds (after the fish have consumed that feed, metabolised it and utilised the nutrients within it) as the major nutrient source for the plants and supplement any missing nutrients required by the plants via some form of buffering regime (Rakocy et al. 2006) or via additional nutrient supplementation (e.g. adding chelated nutrient forms directly to the culture water or by adding nutrients via foliar sprays) (Roosta and Hamidpour 2011).

The best example of this classical recirculating aquaponic approach is the UVI (University of the Virgin Islands) aquaponic system developed by Dr. James Rakocy and his UVI team (Rakocy and Hargreaves 1993; Rakocy et al. 2006). The UVI design principally adds nutrients for both fish and plant culture via fish feed additions. However, fish feeds do not contain enough calcium (Ca^{+}) and potassium (K^{+}) for optimal plant culture. The bacteria-mediated conversion of fish waste-dissolved ammonia to nitrate causes system-wide production of hydrogen ions within the water column, and the proliferation of these hydrogen ions results in a constant fall in the system water pH towards acid. The buffering regime employed adds the missing calcium and potassium by adding basic salts (often salts based on carbonate, bicarbonate or hydroxyl ions paired with calcium or potassium) to the system that assist to control the system water pH at a level that meets both the shared pH environmental requirements of the fish and the plants, whilst providing the additional calcium and potassium the plants require (Rakocy et al. 2006). In addition, the UVI system adds another major nutrient for plant growth that is not available in standard fish feeds, iron (Fe), via regular and controlled iron chelate additions. Therefore, the potassium, calcium and iron the plants require that are not found in the fish feed are available via these two additional nutrient supply mechanisms (Rakocy et al. 2006).

Decoupled aquaponic designs adopt an approach to culture the fish and plants in a way whereby the water is used by the fish and the fish waste nutrients are supplied to the plants, without recirculation of the water back to the fish (Karimanzira et al. 2016). Decoupled designs therefore allow more flexibility in customising the water chemistry, after fish use, for optimised plant production because supplementation of the nutrients not present in the fish feed (and fish waste) may be achieved with no concerns of the water returning to the fish (Goddek et al. 2016). This means decoupled designs potentially may apply more exacting nutrient mixtures and strengths to the culture water, post fish use, for plant culture, and this may be achieved with more exacting and intense nutrient supplementation.

In both cases (recirculating and decoupled aquaponic system designs), an understanding of the chemical quality of the source water is vital so that as close to optimal nutrient concentrations for the plants may be achieved. If, for example, the source water contains calcium (a case often seen when ground water resources are utilised), this will affect and change the buffering regime applied to recirculating aquaponic designs and the extent of the nutrient supplementation applied to a decoupled design because the calcium present in the source water will offset any required supplementation required for plant calcium needs (Lennard 2017). Or, if the source water contains elevated sodium (Na^+) concentrations (again, often seen with ground water resources and a nutrient plants do not use and which can accumulate in system waters), it is important to know how much is present so management methods may be applied to avoid potential plant nutrient toxicity (Rakocy et al. 2006). The chemical nature of the source water, therefore, is vital to overall aquaponic system health and management.

Ultimately, because source water chemistry can affect aquaponic system nutrient management and because aquaponic operators like to have the ability to manipulate aquaponic water and nutrient chemistry to a high degree, a water source with little, if any, associated water chemistry is highly desirable (Lennard 2017). In this sense, rainwater or water treated for chemical removal (e.g. reverse osmosis) is the best source water for aquaponics in a water chemistry context (Rakocy et al. 2004a, b; Lennard 2017). Ground waters are also suitable, but it must be ensured that they do not contain chemicals or salts in concentrations that are too high to be practical (e.g. high magnesium or iron concentrations) or contain chemical species that are not used by the fish or plants (e.g. high sodium concentrations) (Lennard 2017). River waters may also be suitable as aquaponic source water, but as for other water sources, they should be tested for chemical presence and concentrations. Town water sources (i.e. water reticulated and supplied for domestic and consumptive purposes) are broadly applied in aquaponics (Love et al. 2015a, b) and are also acceptable if they contain acceptable nutrient, salt or chemical concentrations. In the case of town- or municipal-supplied water resources, it should be noted that many supplies have some form of sterilisation applied to make the water drinkable for humans. If this source of water is to be used for aquaponics, then it is important to ensure that any chemicals that may be applied to achieve sterilisation (e.g. chlorine, chloramine, etc.) are not present in concentrations that could harm the fish, plants or microorganisms within the aquaponic system (Lennard 2017).

The chemistry associated with source water is not the only factor that needs consideration when supplying source water for aquaponic use. Many natural waters may also contain microbial and other microorganisms that may affect the overall ecological health of the aquaponic system or present a discernible human health risk. Rainwaters rarely contain microbes themselves; however, the vessels or tanks the rainwater may be stored within may contain or allow microbial proliferation. Ground waters are usually good in terms of microbial presence but may also contain high microbial loads, especially if sourced from areas associated with animal farming or human waste treatment. River waters may also contain high microbial loads due to farming or human waste treatment outflows and again should be checked via detailed microbial analysis (Lennard 2017).

Because the chemical and microbial nature of the source water used in aquaponic systems can have potential effects on system water chemistry and microbiology, it is recommended that any applied water source be sterilised and treated for chemical removal (e.g. reverse osmosis, distillation, etc.) before being used in an aquaponic system (Lennard 2017). If sterilisation is universally applied, the chance of introducing any foreign and unwanted microbes to the system is substantially lowered. If water treatment and filtration is applied, any chemicals, salts, unwanted nutrients, pesticides, herbicides, etc. will be removed and therefore cannot contribute negatively to the system.

A clean water source, free of microbes, salts, nutrients and other chemicals allows the aquaponic operator to manipulate the system water to contain the nutrient mixture and strength they require without the fear that any external influences may affect the operation of the system or the health and strength of the fish and plants and is a vital requirement for any commercial aquaponic operation.

5.5 Water Quality Requirements

Aquaponics represents an effort to control water quality so that all the present life forms (fish, plants and microbes) are being cultured in as close to ideal water chemistry conditions as possible (Goddek et al. 2015). If water chemistry can be matched to the requirements of these three sets of important life forms, efficiency and optimisation of growth and health of all may be aspired to (Lennard 2017).

Optimisation is important to commercial aquaponic production because it is only through optimisation that commercial success (i.e. financial profitability) may be realised. Therefore, water chemistry and water quality requirements within the aquaponic system are pivotal to the ultimate commercial and economic success of the enterprise (Goddek et al. 2015).

There is currently disagreement within the broader aquaponic industry and community in terms of what represents good or acceptable water quality within aquaponic systems. It appears that it is universally accepted that the natural water chemistry requirements of the individual life form subsets (fish, plants and microbes) are broadly agreed upon (Rakocy and Hargreaves 1993; Rakocy et al. 2006; Goddek et al. 2015; Delaide et al. 2016; Lennard 2017). However, the presence of a broad range of approaches, methods and technology choices that are called aquaponics and the background or history of the associated, stand-alone technologies of recirculating aquaculture systems (RAS) and hydroponic plant culture (including substrate culture) appears to lead to disagreements among operators, scientists and designers. For example, taking only one single water chemistry parameter into consideration, pH, some argue that the pH requirements of hydroponically cultured plants are very different to the pH requirements of RAS-cultured freshwater fish species (Suhl et al. 2016). The hydroponic industry generally applies pH settings between 4.5 and 6.0 for water-based plant culture (Resh 2013), whereas the RAS industry typically applies pH settings between 7.0 and 8.0 (Timmons et al. 2002) to meet the requirements of the fish and the microbes present (which perform important transformations of potentially toxic fish waste metabolites to less toxic forms). The argument, therefore, is that any pH set point is a compromise between the requirements of the plants, the fish and the microbes and that therefore an optimal pH for all life forms is not achievable which leads to suboptimal plant production (Suhl et al. 2016). Others argue, however, that a closer scrutiny of the complexities of the nutrient dynamics of plant nutrient uptake may elucidate a different opinion (Lennard 2017).

Hydroponic (and substrate culture) systems feed nutrients to the plants in their basal, ionic forms by adding nutrient salts to the water that dissociate to release the available nutrient ions (Resh 2013). Research has demonstrated that these ionic nutrient forms exist in a window of availability to the plant, based on the available system water pH. Therefore, in a standard hydroponic context, with no present microbial flora (i.e. sterilised – as most hydroponic systems are), it is important to set the pH of the system water to a level that makes the mixture of ionic nutrients the plant requires as available as possible (Resh 2013). Within any hydroponic system, this is a compromise itself, because as any ionic nutrient availability chart demonstrates (see Fig. 5.4), different ionic nutrient forms are the most available at differing pHs (Resh 2013). It is this standard ionic nutrient availability association that the hydroponic industry uses as its primer for pH set points and explains why the desired hydroponic operational pH is somewhere between 4.5 and 6.0 (an acid environment) in sterilised hydroponic and substrate culture systems.

Alternatively, RAS applies a water pH set point based on what is natural for the fish being cultured and the microbes treating and converting the fish waste products

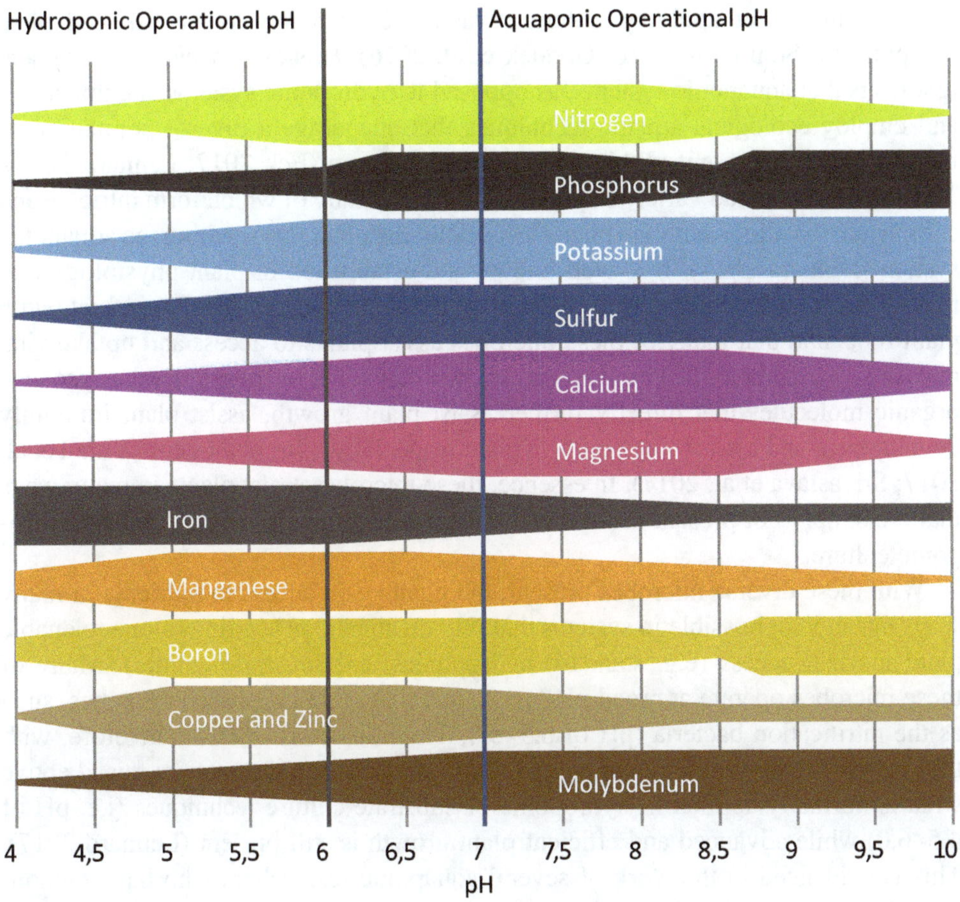

Fig. 5.4 Example of a standard pH mediated, nutrient availability chart for aquatically cultured plants. Red line represents a normal operating pH for a hydroponic system; the blue line that for an aquaponic system

(Timmons et al. 2002; Goddek et al. 2015; Suhl et al. 2016). In natural freshwater environments, most fish species require an environmental pH (i.e. water pH) that closely matches the internal pH of the fish, which is often close to a pH of 7.4 (Lennard 2017). In addition, the major microbes associated with dissolved metabolite transformation in RAS culture (the nitrification bacteria of several species) also require a pH around 7.5 for optimal ammonia transformation to nitrate (Goddek et al. 2015; Suhl et al. 2016). Therefore, RAS operators apply a pH set point of approximately 7.5 to RAS freshwater fish culture.

There is an obvious difference between a pH of 5.5 (an average for standard, sterilised, hydroponic plant culture) and a pH of 7.5 (an average standard for RAS fish culture). Therefore, it is argued broadly that pH represents one of the largest water quality compromises present in aquaponic science (Goddek et al. 2015; Suhl et al. 2016). Advocates of decoupled aquaponic designs often cite this difference in optimal pH requirement as an argument for the decoupled design approach, stating that fully recirculating designs must find a pH compromise when decoupled designs have the luxury of applying different water pH set points to the fish and plant components (Suhl et al. 2016; Goddek et al. 2016). However, what this argument ignores is that aquaponic systems, as opposed to hydroponic systems, are not sterile and employ ecological aquatic techniques that encourage a diverse population of microflora to be present within the aquaponic system (Eck 2017; Lennard 2017). This results in a broad variety of present microbes, many of which form intricate and complex associations with the plants, especially the plant roots, within the aquaponic system (Lennard 2017). It is well known and established in plant physiology that many microbes, associated with the soil medium and matrix, closely associate with plant roots and that many of these microbes assist plants to access and uptake vital nutrients (Vimal et al. 2017). It is also known that some of these microbes produce organic molecules that directly further assist plant growth, assist plant immunity development and assist to outcompete plant (especially root) pathogens (Vimal et al. 2017; Srivastava et al. 2017). In essence, these microbes assist plants in many ways that are simply not present in the sterilised environment applied in standard hydroponic culture.

With these diverse microbes present, the plants gain access to nutrients in many ways that are not possible in systems that rely on aquatic pH settings alone to enable plant nutrient access (e.g. standard hydroponics and substrate culture). Many of these microbes operate at broad pH levels, just like other soil-based microbes, such as the nitrification bacteria (pH of 6.5–8.0, Timmons et al. 2002). Therefore, with these microbes present in aquaponic systems, the pH set point may be raised above what is normally applied in hydroponic or substrate culture techniques (i.e. pH of 4.5–6.0) while advanced and efficient plant growth is still present (Lennard 2017). This is evidenced in the work of several aquaponic researchers who have demonstrated better plant growth rates in aquaponics than in standard hydroponics (Nichols and Lennard 2010).

Other water quality requirements in aquaponic systems relate to physical/chemical parameters and more specifically, plant nutrient requirement parameters. In terms of physical/chemical requirements, plants, fish and microbes share many commonalities. Dissolved oxygen (DO) is vital to fish, plant roots and microflora and must be maintained in aquaponic systems (Rakocy and Hargreaves 1993; Rakocy et al. 2006). Plant roots and microflora generally require relatively lower DO concentrations than most fish; plant roots and microbes can survive with DO below 3 mg/L (Goto et al. 1996), whereas most fish require above 5 mg/L (Timmons et al. 2002). Therefore, if the DO concentration within the aquaponic system is set and maintained for the fish requirement, the plant and microbe requirement is also met (Lennard 2017). Different fish species require different DO concentrations: warm water fish (e.g. *Tilapia* spp., barramundi) can generally tolerate lower DO concentrations than cool water fish species (e.g. salmonids like rainbow trout and arctic char); because the fish DO requirement is almost always greater than the plant roots and microfloral requirement, DO should be set for the specific fish species being cultured (Lennard 2017).

Water-carbon dioxide (CO_2) concentrations, like that for DO, are generally set by the fish because the plant roots and microbes can tolerate higher concentrations than the fish. Carbon dioxide concentrations are important to optimal fish health and growth and are often ignored in aquaponic designs. Parameters and set points for CO_2 concentrations should be the same as for the same fish species cultured in fish-only, RAS systems and in general, should be kept below 20 mg/L (Masser et al. 1992).

Water temperature is important to all the present life forms within an aquaponic system. Fish and plant species should be matched as closely as possible for water temperature requirements (e.g. *Tilapia* spp. of fish like 25° C plus, and plants like basil thrive in this relatively high water temperature; lettuce varieties like cooler water, and therefore, a better matched fish candidate is rainbow trout) (Lennard 2017). However, as for other water physical and chemistry parameters, meeting the fish's requirement for water temperature is paramount because the microbes have the ability to undergo specific species selection based on the ambient conditions (e.g. nitrification bacterial species differentiation occurs at different water temperatures and the species that matches best to the particular water temperature will dominate the nitrification bacterial biomass of the system) and many plants can grow very well at a broader range of water temperatures (Lennard 2017). Matching the water temperature, and maintaining it within plus or minus 2° C (i.e. a high-level temperature control) to the fish, is an important requirement in aquaponics because when water temperature is correct and does not deviate from the ideal average, the fish achieve efficient and optimised metabolism and eat and convert feed efficiently, leading to better fish growth rates and stable and predictable waste load releases, which assists plant culture (Timmons et al. 2002).

Maintaining water clarity (low turbidity) is another important parameter in aquaponic culture (Rakocy et al. 2006). Most water turbidity is due to suspended solids loads that have not been adequately filtered, and these solids may affect fish by adhering to their gills, which may lower potential oxygen transfer rates and ammonia release rates (Timmons et al. 2002). Suspended solids loads less than 30 mg/L are recommended for aquaponically cultured fish (Masser et al. 1992; Timmons et al. 2002). High suspended solids loads also affect plant roots because they have the ability to adhere to the roots which may cause nutrient uptake inefficiency, but more commonly provides increased potential for pathogenic organism colonisation, which leads to poor root health and ultimate plant death (Rakocy et al. 2006). These suspended solids also encourage the prevalence of heterotrophic bacteria (species that break down and metabolise organic carbon) which, if allowed to dominate systems, may outcompete other required species, such as nitrification bacteria.

Electrical conductivity (EC) is a measure often applied in hydroponics to gain an understanding of the amount of total nutrient present in the water. It, however, cannot provide information on the nutrient mix, the presence or absence of individual nutrient species or the amount of individual nutrient species present (Resh 2013). It is not often applied in aquaponics because it only measures the presence of ionic (charged) nutrient forms, and it has been argued that aquaponics is an organic nutrient supply method, and therefore, EC is not a relevant measure (Hallam 2017). However, plants generally only source ionic forms of nutrients, and therefore,

EC can be used as a general tool or guide to the total amount of plant-available nutrient in an aquaponic system (Lennard 2017).

For fully recirculating aquaponic systems, in terms of physical and chemical parameters, it is the fish that are more exacting in their requirements, and therefore, if systems are managed to maintain the requirements of the fish, the plants and microbes are having their requirements more than satisfied (Lennard 2017). The difference when it comes to the plants, however, is their requirement for the correct mixture and strength of nutrients to be present to allow optimised nutrient access and uptake (whether stand-alone or microbial-assisted) which leads to efficient and fast growth. Decoupled aquaponic systems may therefore be more attractive because of the perception that they allow more exacting nutrient delivery to the plants (Goddek et al. 2016). Fish feed and, therefore, fish waste do not contain the correct mixture of nutrients to meet the plant requirements (Rakocy et al. 2006). Therefore, the aquaponic system design must account for those missing nutrients and supplement them. Fully recirculating aquaponic systems generally supplement nutrients by adding them in the salt species used to manage the daily pH buffering regime; the basic portion of the salt adjusts the pH and the positive portion of the salt allows the supplementation of missing plant nutrients (e.g. potassium, calcium, magnesium) (Rakocy et al. 2006). Decoupled aquaponic designs take the waste water and associated solid wastes from the fish component and adjust the water to contain the nutrients required for plant production by adding nutrients in different forms (Goddek et al. 2016). These nutrient additions are generally based on using standard hydroponic salt species that do not necessarily provide any pH adjustment outcome (e.g. calcium phosphate, calcium sulphate, potassium phosphate, etc.).

The pathway to efficient plant growth in aquaponic systems is to provide an aquatic nutrient profile that provides all the nutrients the plant requires (mixture) at the strengths required (concentration) (Lennard 2017). In fully recirculating aquaponic designs, or decoupled aquaponic designs that do not apply sterilisation methods, there appears to be less of a requirement to meet the nutrient concentrations or strengths applied in standard hydroponics, because the ecological nature of the system associates many diverse microflora with the plant roots and these microflora assist plant nutrient access (Lennard 2017). For decoupled, or other, aquaponic designs that apply sterilisation to the plant component and follow a standard hydroponic analogue approach, there appears to be a requirement to try and approach standard hydroponic nutrient concentrations (Suhl et al. 2016; Karimanzira et al. 2016). The compromise, however, with the decoupled approach is that it leads to external supplementation ratios far beyond those of fully recirculating aquaponic designs; European decoupled designs currently average 50% or greater external nutrient additions (COST FA1305 2017; Goddek and Keesman 2018), while the UVI method supplies less than 20%, and other systems may supply less than 10% external nutrient supplementation (Lennard 2017).

No matter the method, all aquaponic systems should strive to supply the plants with the nutrition required for optimised growth so as to provide the enterprise with the greatest chance of financial viability. In this context, the nutrient content and strength of the water being delivered to the plants is very important and regular nutrient testing of the water should be employed so that nutrient mixture and strength may be maintained and managed as a very important water quality requirement.

5.6 Applicable Fish Culture Technologies

In aquaponics, the aquaculture portion of the integration equation is broadly applied in a tank-based context, where the fish are kept in tanks, the water is filtered via mechanical (solids removal) and biological (ammonia transformation to nitrate) mechanisms and dissolved oxygen is maintained, either via aeration or direct oxygen injection (Rakocy et al. 2006; Lennard 2017).

As has been argued in Sect. 5.0 (Introduction) of this chapter, historical examples of chinampas (Somerville et al. 2014) and Asian rice paddy farming (Halwart and Gupta 2004) as early iterations of aquaponics are unfounded and inappropriate examples of aquaponic principles, because modern aquaponics relies on designed additions of fish and fish feeds to supply a designed level of nutrition to the plants, and therefore, these historical examples cannot be considered in any way similar (Lennard 2017).

The above historical examples, which rely on soil-based plant culturing systems, lead to the question of what aquaculture technologies are suitable for aquaponic integration. Soil-based, extensive, freshwater pond aquaculture of fin fish is the largest culturing method applied to produce freshwater fish for human consumption (Boyd and Tucker 2012). A pond approach relies on the earthen base of the pond, and the associated microflora present in that soil, to treat and remediate the wastes produced from fish culture so the fish are not living in water that has a potential to be toxic to them (Boyd and Tucker 2012). Because this system relies on the inherent treatment capacity of the earthen pond itself, fish densities are relatively low compared to other aquaculture methods. Because the fish densities are low (and therefore the associated feeding rates are low) and the pond itself treats and uptakes the waste nutrients produced by the fish, pond waters exhibit extremely low water nutrient concentrations. These pond system aquatic nutrient concentrations are so low that they are often inappropriate as nutrient sources for substantial, commercial aquatic plant production methods (Lennard 2017). Therefore, ponds are not an appropriate aquaculture method to be integrated with hydroponics in terms of acceptable plant production rates.

Similarly, raceway fin fish culture methods (as regularly applied for freshwater Salmonid production), which supply very large volumes of water at high turnover rates, or low residence times, through controlled raceway fish culture tanks, are not appropriate for aquaponic integration because the high water turnover rates do not allow adequate nutrient accumulations to meet plant nutrient requirements (Rakocy and Hargreaves 1993).

The most appropriate fish culture technologies to apply within an aquaponic integration context are those that culture fish in tanks and allow a level of fish waste accumulation (plant nutrient accumulation) that has the potential to lead to water nutrient concentrations that are applicable for efficient hydroponic plant production (Rakocy et al. 2006). Recirculating Aquaculture System (RAS) principles are broadly applied to aquaponics because they provide the methodologies to successfully keep and grow the fish, in controlled volumes of water, with low daily water replacement rates, that allow fish waste (plant nutrient) accumulations that approach those required to efficiently hydroponically culture the plants (Rakocy and Hargreaves 1993; Lennard 2017). The complexities and design requirements of RAS are discussed in Chap. 3 of this book. Suffice to say that RAS fish culture is the only real appropriate method to apply for fish culturing components in an aquaponic context and as discussed above, soil-based aquaculture systems, such as extensive pond systems and raceway culture systems, cannot provide the nutrient requirements of the plants and therefore should not be considered.

5.7 Nutrient Sources

The major input to any aquaponic system are the nutrients added because aquaponic systems are designed to efficiently partition the nutrients added to them to the three important forms of life present: the fish and plants (which are the main products of the system) and the microflora (which assist to make the added nutrients available to the fish and plants) (Lennard 2017).

In classical, fully recirculating aquaponic designs, one of the key design drivers is to use the main nutrient input source, the fish feed, as efficiently as possible and therefore fully recirculating designs strive to supply as many of the nutrients required for the plants from the fish feed (Lennard 2017). Decoupled designs, on the other hand, place an emphasis on optimised plant growth by directly comparing the nutrient mixtures and strengths applied in standard hydroponics and substrate culture and trying to replicate those within the aquaponic context and therefore do not strive to supply as many of the nutrients required for the plants from the fish feed and utilise substantial external nutrient supplementations to achieve the required plant growth rates (Delaide et al. 2016). This means that a different emphasis is placed on the origin of the nutrients added, based on the technical design approach, and this, therefore, affects the main nutrient supply source of the aquaponic system; for fully recirculating designs, the major plant nutrient source is fish feed (via fish waste production), and for decoupled designs the major nutrient supply source for the plants is external supplements (e.g. nutrient salts) (Lennard 2017).

Fully recirculating aquaponic designs, such as the UVI aquaponic system model, rely on the fish feed as the major nutrient source for the system (Rakocy et al. 2006). The fish feed is added to the fish, which eat it, metabolise it and use the nutrients from it as required and then produce a waste stream (both solids and dissolved). This waste stream from the fish becomes the major nutrient source for the plants, and hence, the fish feed is the major nutrient source for the plants. The UVI system provides approximately 80% or more of the nutrients required to grow the plants from the fish feed alone (Lennard 2017). The remaining nutrients required for plant growth, because the fish feed does not contain them in the amounts required, are added via a nutrient supplementation method that provides the dual role of supplementing the additional nutrients and controlling the system aquatic pH (Rakocy et al. 2006). This dual role approach is referred to as "buffering" and the supplement is referred to as "buffer". For the UVI model, the two important plant nutrients identified as lacking in fish feed and which require supplementation are potassium (K) and calcium (Ca) and are supplemented daily via the buffering regime. In addition, plant-required iron (Fe) is also lacking in the fish feed and is supplemented in a chelated form via direct addition to the system water at a frequency measured in weeks (i.e. every 2–4 weeks based on weekly aquatic iron analysis) (Rakocy et al. 2006).

Other fully recirculating aquaponic design approaches or methods place an even higher emphasis on providing nutrients via the fish feed. Lennard (2017) has developed a method for fully recirculating systems that supplies greater than 90% of the nutrients required for plant growth from the fish feed added. The increase in the efficiency of nutrients supplied via the fish feed of this method when compared to the UVI method is that this approach remineralises the solid fish wastes (via external, bacteria-mediated biodigestion) and adds these nutrients back into the aquaponic system for plant utilisation, whereas the UVI method sends the majority of the solid fish wastes to an external waste stream (Rakocy et al. 2006; Lennard 2017). This approach also adds nutrients deficient in the fish feed for plant growth via a buffering regime; however, this regime is far more exacting and allows greater manipulation of nutrient strengths and mixtures than the UVI approach (Lennard 2017).

Therefore, the major nutrient addition pathways for most fully recirculating aquaponic system designs are the fish feed (major route), buffer external supplementation for added potassium and calcium (minor route) and direct supplementation of iron chelate (minor route).

Decoupled aquaponic system designs, such as those being widely adopted currently in Europe, rely on a mixture of fish feed nutrients and active, external supplementation to provide the nutrients required for plant growth (Suhl et al. 2016). Because decoupled designs do not return water from the plant component to the fish component, it is possible to customise the nutrient profile within the water specifically to the plant requirement (Goddek et al. 2016). Therefore, decoupled aquaponic designs almost always rely on substantial external nutrient supplementation to meet the plant requirement and place far less emphasis on providing as much nutrition as possible for the plants from the fish wastes. In addition, the amount of external supplementation is substantial when compared to fully recirculating approaches (Lennard 2017) with external fractions regularly 50% or more of the total plant nutrient requirement or greater (Goddek 2017). The external nutrients supplemented to decoupled aquaponic systems are most often hydroponic nutrient salt analogues or derivatives (Delaide et al. 2016; Karimanzira et al. 2016). This reliance on a source of substantial additional nutrients other than those that arise from fish wastes (fish feeds) that are hydroponic salt in nature, for plant supply of the European decoupled approach, has even directly affected the definition of aquaponics that the European aquaponics community currently applies, with the EU COST, EU Aquaponics Hub, defining aquaponics as "...a production system of aquatic organisms and plants where the majority (> 50%) of nutrients sustaining the plants derives from wastes originating from feeding the aquatic organisms" (Goddek 2017; COST FA 1305, 2017) compared to Lennard (2017) who defines aquaponics as requiring at least 80% nutrient supply from fish wastes. Some also argue whether a method that relies on 50% of the nutrients required for plant growth originating from external sources other than fish feeds is actually aquaponic in nature or rather, a hydroponic method with some fish integrated or added (Lennard 2017)?

Another proposed supply source for nutrients to aquaponic systems is that of external nutrient supplementation via the application of foliar plant sprays (Tyson et al. 2008; Roosta and Hamidpour 2011; Roosta and Hamidpour 2013; Roosta 2014). These foliar sprays are again, an aquatic delivery of standard hydroponic nutrient salts or derivatives. The difference is that in the decoupled examples above, the nutrient salts are added directly to the culture water and are therefore accessed by the plants via root uptake (Resh 2013), whereas foliar sprays, as the name implies, add dissolved nutrient salts to the plant leaves and uptake is achieved via plant leaf stomatal or cuticle access (Fernandez et al. 2013).

There are therefore several major nutrient sources applied in aquaponics: fish feeds, buffering systems (via basic, pH-adjusting salt species added to the water column), nutrient salt additions (hydroponic nutrient salts added to the water column) and foliar sprays (hydroponic nutrient salts added to the leaf surface). All of which supply nutrients to the aquaponic system for the health and growth of the fish and plants that are cultured.

5.8 Aquaponics as an Ecological Approach

Aquaponics, until recently, has been dominated by fully recirculating (or coupled) design approaches that share and recirculate the water resource constantly between the two major components (fish and plant culture) (Rakocy et al. 2006; Lennard 2017). In addition, the low to medium technology approaches historically applied to aquaponics have driven a desire to remove costly components so as to increase the potential of a positive economic outcome. One of the filtration components almost always applied to standard RAS and hydroponics/substrate culture technologies, that of aquatic sterilisation, has regularly not been included by aquaponic designers.

Sterilisation in a RAS and hydroponic/substrate culture context is universally applied because the high densities of either the fish or plants cultured usually attract pressure from aquatic, pathogenic organisms that substantially lower overall production rates (Van Os 1999; Timmons et al. 2002). The major reason for this increased aquatic pest pressure in both technologies is that each concentrates on providing minimal biotic, ecological resources and therefore allows considerable "ecological space" within the system water for biotic colonisation. In these "open" biological conditions, pest and pathogenic species proliferate and tend to colonise quickly to take advantage of the species present (i.e. fish and plants) (Lennard 2017). In this context, sterilisation or disinfection of the culture water has historically be seen as an engineered approach to counteract the issue (Van Os 1999; Timmons et al. 2002). This means that both RAS and hydroponic/substrate culture industries adopt a sterilisation approach to control pathogenic organisms within the associated culture water.

Aquaponics has always placed an emphasis on the importance of the associated microbiology to perform important biological services. In all the coupled aquaponic designs of Rakocy and his UVI team, a biological filter was not included because they demonstrated that the raft culture, hydroponic component provided more than enough surface area to support the nitrifying bacteria colony size to treat all the ammonia produced by the fish as a dissolved waste product and convert it to nitrate (Rakocy et al. 2006, 2011). Rakocy and his team therefore did not advocate applied sterilisation of the system water because it may have affected the nitrifying bacterial colonies. This historical UVI/Rakocy perspective dictated aquaponic system design into the future. Other advantages of not including aquatic sterilisation to aquaponic systems were identified and discussed, especially in the context of assistive plant microbiota (Savidov 2005; Goddek et al. 2016).

The current thinking in aquaponic research and industry is that not applying any form of aquatic sterilisation or disinfection allows the system water to develop a complex aquatic ecology that consists of many different microbiological life forms (Goddek et al. 2016; Lennard 2017). This produces a situation similar to a natural ecosystem whereby a high diversity of microflora interacts with each other and the other associated life forms within the system (i.e. fish and plants). The proposed outcome is that this diversity leads to a situation in which no single pathogenic organism can dominate due to the presence of all of the other microflora and can therefore not cause devastating effects to fish or plant production. It has been demonstrated that aquaponic systems contain a high diversity of microflora (Eck 2017) and via the proposed ecological diversity mechanism outlined above, assistance to both fish and plant health and growth is provided via this microbial diversity (Lennard 2017).

The non-sterilised, ecologically diverse approach to aquaponics has been historically applied to coupled or fully recirculating aquaponic designs (Rakocy et al. 2006), whereas a sterilised, hydroponic analogy has been proposed for some decoupled aquaponic design approaches (Monsees et al. 2016; Priva 2009; Goddek 2017). However, it appears more decoupled designers are now applying principles that take an ecological, non-sterilised approach into consideration (Goddek et al. 2016; Suhl et al. 2016; Karimanzira et al. 2016) and therefore acknowledge there is a positive effect associated with a diverse aquaponic microflora (Goddek et al. 2016; Lennard 2017).

5.9 Advantages of Aquaponics

Because there are two separate, existing, analogous technologies that produce fish and plants at high rates (RAS fish culture and hydroponic/substrate culture plant production), a reason for their integration seems pertinent. RAS produces fish at productive rates in terms of individual biomass gain, for the feed weight added, that rivals, if not betters, other aquaculture methods (Lennard 2017). In addition, the high fish densities that RAS allows lead to higher collective biomass gains (Rakocy et al. 2006; Lennard 2017). Hydroponics and substrate culture possess, within a controlled environment context, advanced production rates of plants that better most other agriculture and horticulture methods (Resh 2013). Therefore, initially, there is a requirement for aquaponics to produce fish and plants at rates that equal these two separate productive technologies; if not, then any loss of productive effort counts against any integration argument. If the productive rate of the fish and plants in an aquaponic system can equal, or better, the RAS and hydroponic industries, then a further case may be made for other advantages that may occur due to the integration process.

Standard hydroponics or substrate culture has been directly compared with aquaponics in terms of the plant growth rates of the two technologies. Lennard (2005) compared aquaponic system lettuce production to a hydroponic control in several replicated laboratory experiments. He demonstrated that aquaponic lettuce production was statistically lower in aquaponics (4.10 kg/m^2) when compared to hydroponics (6.52 kg/m^2) when a standard approach to media bed aquaponic system design and management was applied. However, he then performed a series of experiments that isolated specific parameters of the design (e.g. reciprocal vs constant hydroponic subunit water delivery, applied water flow rate to the hydroponic subunit and comparing different hydroponic subunits) or comparing specific management drivers (e.g. buffering methodologies and species and the overall starting nutrient concentrations) to achieve optimisation and then demonstrated that aquaponics (5.77 kg/m^2) was statistically identical to hydroponic lettuce production (5.46 kg/m^2) after optimisation of the aquaponic system based on the improvements suggested by his earlier experiments, the result suggesting that improvements to coupled or fully recirculating aquaponic designs can equal standard hydroponic plant production rates. Lennard (2005) also demonstrated fish survival, SGR, FCR and growth rates equal to those exhibited in standard RAS and extensive pond aquaculture for the fish species tested (Australian Murray Cod).

Pantanella et al. (2010) also demonstrated statistically similar lettuce production results within high fish density (5.7 kg/m^2 lettuce production) and low fish density (5.6 kg/m^2 lettuce production) aquaponic systems compared to a standard hydroponic control (6.0 kg/m^2).

Lennard (Nichols and Lennard 2010) demonstrated statistically equal or better results for all lettuce varieties and almost all herb variety production tested in a nutrient film technique (NFT) aquaponic system when compared to a hydroponic NFT system within the same greenhouse.

Delaide et al. (2016) compared nutrient-complemented RAS production water (nutrients were added to match one water nutrient mixture and strength example by Rakocy – denoted as an aquaponic analogue), fully nutrient-complemented RAS production water (RAS production water with added hydroponic nutrient salts to meet a water nutrient mixture and strength as used for standard hydroponics – denoted as a decoupled analogue) and a hydroponic control (standard hydroponic nutrient solution) in terms of plant growth rate and showed the aquaponic water analogue equalled the hydroponic control and the decoupled analogue water bettered the hydroponic control. However, it must be noted that these were not fully operating aquaponic systems containing fish (and the associated, full and active microbial content) that were compared, but simply water removed from an operating RAS and complemented, then compared to a hydroponic control water.

Rakocy and his UVI team have demonstrated with several studies that *Tilapia* spp. fish growth rates equal industry standards set by standard aquaculture production practices (Rakocy and Hargreaves 1993; Rakocy et al. 2004a, b, 2006, 2011).

These and other studies have demonstrated that aquaponics, no matter the configuration (coupled and decoupled), has the potential to produce plant production rates equal to, or better than, standard hydroponics and fish production rates of a similar standard to RAS. Therefore, the above discussed requirement for aquaponics to equal its industry analogues (RAS and hydroponics) seems to have been adequately proven, and therefore, the other advantages of aquaponics should be considered.

Efficient water use is regularly attributed to aquaponics. Lennard (2005) stated that the water savings associated with an optimised aquaponic test system (laboratory) were 90% or greater when compared to a standard RAS aquaculture control system where water was exchanged to control nitrate accumulations, whereas the plants in the aquaponics performed the same requirement. Therefore, he demonstrated that aquaponics provides a substantial water-saving benefit compared to standard RAS aquaculture. Interestingly, this 90% water-saving figure has subsequently been stated broadly within the global aquaponics community in a plant use context (e.g. aquaponics uses 90% less water than soil-based plant production (Graber and Junge 2009)) – an example of how scientific argument can be incorrectly adopted by nonscientific industry participants.

McMurtry (1990) demonstrated a water consumption rate in his aquaponic system of approximately 1% of that required in a similar pond culture system. Rakocy (1989) has demonstrated similar 1% water consumption rates compared to pond-based aquaculture. Rakocy and Hargreaves (1993) stated that the daily water replacement rate for the UVI aquaponic system was approximately 1.5% of the total system volume and Love et al. (2015a, b) stated an approximate 1% of system volume water loss rate per day for their aquaponic research system.

The comparison of aquaponics to RAS can realise substantial water savings and aquaponics uses small amounts of replacement water on a daily basis. A well-designed aquaponic system will seek to use water as efficiently as possible and therefore only replace that water lost via plant evapotranspiration (Lennard 2017). In fact, it has been proposed that water may even be recovered from that lost due to plant evapotranspiration via employing some form of air water content harvesting scheme or technology (Kalantari et al. 2017). Coupled aquaponic systems appear to provide a greater potential to conserve and lower water use (Lennard 2017). If the nutrient dynamics between fish production and plant use can be balanced, the only water loss is via plant evapotranspiration, and because the water is integrally shared between the fish and plant components, daily makeup water volumes simply represent all the water lost from the system plants (Lennard 2017). Decoupled aquaponic designs present a more difficult proposition because the two components are not integrally linked and the daily water use of the fish component does not match the daily water use of the plant component (Goddek et al. 2016; Goddek and Keesman 2018). Therefore, water use and replacement rates for aquaponic systems are not completely resolved and probably never will be due to the broad differences in system design approaches.

Efficient nutrient utilisation is assigned to the aquaponic method and cited as an advantage of the aquaponic approach (Rakocy et al. 2006; Blidariu and Grozea 2011; Suhl et al. 2016; Goddek et al. 2015). This is generally because standard RAS aquaculture utilises the nutrients within the fish feed to grow the fish, with the remainder being sent to waste. Fish metabolise much of the feed they are fed, but only utilise approximately 25–35% of the nutrients added (Timmons et al. 2002; Lennard 2017). This means up to 75% of the nutrients added to fish-only RAS are wasted and not utilised. Aquaponics seeks to use the nutrients wasted in RAS for plant production, and therefore, aquaponics is said to use the nutrients added more efficiently because two crops are produced from the one input source (Rakocy and Hargreaves 1993; Timmons et al. 2002; Rakocy et al. 2006; Lennard 2017). The extent of fish waste nutrient use does differ between the various aquaponic methods. The fully recirculating UVI model does not utilise the majority of the solid fish wastes generated in the fish component and sends them to waste (Rakocy et al. 2006), the fully recirculating Lennard model takes this a step further by using all the wastes generated by the fish component (dissolved wastes directly and solids via external microbial remineralisation with main system replacement) (Lennard 2017). Many decoupled approaches also attempt to utilise all the wastes generated by the fish component, via direct use of dissolved wastes and again, via external microbial remineralisation with main system replacement (Goddek et al. 2016; Goddek and Keesman 2018). All of these methods and approaches demonstrate that a primary driver for the aquaponic method is to utilise as many of the added nutrients as possible and therefore attempt to use the added nutrients as efficiently as possible.

Independence from soil has been cited as an advantage of the aquaponic method (Blidariu and Grozea 2011; Love et al. 2015a, b). The advantage perceived is that because soil is not required, the aquaponic system or facility may be located where the operator chooses, rather than where suitable soil is present (Love et al. 2015a, b). Therefore, the aquaponic method is independent of location based on soil availability, which is an advantage over soil-based agriculture.

It has been argued that aquaponics provides an advantage by mimicking natural systems (Blidariu and Grozea 2011; Love et al. 2014). This is supported by the ecological nature of the aquaponic approach/method, as outlined in Sect. 5.7 above, with the associated advantages related to diverse and dense microfloral communities (Lennard 2017).

Aquaculture has a potential direct environmental impact due to the release of nutrient-rich waste waters to the surrounding environment – generally, aquatic environments (Boyd and Tucker 2012). Some hydroponic methods may also possess this potential. However, aquaponics can exhibit lowered or negated direct environmental impact from nutrient-rich waste streams because the main waste-generating component (i.e. the fish) is integrated with a nutrient use component (i.e. the plants) (Rakocy et al. 2006; Blidariu and Grozea 2011; Goddek et al. 2015; Lennard 2017). However, some aquaponic methods do produce wastes (e.g. the UVI model); but these are generally treated and reused for other agricultural practices at the aquaponic facility site (Timmons et al. 2002; Rakocy et al. 2006). Many aquaponic methods rely on the use of standard aquaculture feeds, which contain varying concentrations of sodium, usually via the use of fish meal or fish oil as an ingredient (Timmons et al. 2002). Sodium is not utilised by plants and therefore may accumulate over time in aquaponic systems, which may lead to a requirement for some form of water replacement so sodium does not accumulate to concentrations that affect the plants (Lennard 2017). However, it has been reported that some species of lettuce had the ability to take up sodium, when being exposed to aquaculture water (Goddek and Vermeulen 2018).

Coupled or fully recirculating aquaponic systems integrally share the water resource between the two main components (fish and plant). Because of this fully connected and recirculating aquatic nature, coupled aquaponic systems exhibit a self-controlling mechanism in terms of an inability to safely apply herbicides and pesticides to the plants; if they are applied, their presence may negatively affect the fish (Blidariu and Grozea 2011). Fully recirculating advocates see this inability to applying pesticides and herbicides as an advantage, the argument being that it guarantees a spray-free product (Blidariu and Grozea 2011). Decoupled aquaponics advocates also seek to not apply herbicides or pesticides; however, due to the fact that the water is not recirculated back to the fish from the plants, the ability to apply pesticides and herbicides to the plants is present (Goddek 2017). Therefore, the application or lack of application of pesticides and herbicides to the plant component of aquaponic designs is seen differently by groups who advocate for different design approaches.

There is a perception that the presence of both fish and plants in the same aquatic system provides positive synergistic effects to fish and plant health (Blidariu and Grozea 2011). This has been indirectly demonstrated by the ability of aquaponics in some studies to produce plant growth rates greater than those seen in standard hydroponics (Nichols and Lennard 2010; Delaide et al. 2016). However, no direct causal link has been established between the presence of both fish and plants and any positive outcome to fish or plant health.

References

Baquedano E (1993) Aztec Inca & Maya. A Dorling Kindersley Book, Singapore

Blidariu F, Grozea A (2011) Increasing the economic efficiency and sustainability of indoor fish farming by means of aquaponics – review. Anim Sci Biotechnol 44(2):1–8

Boyd CE, Tucker CS (2012) Pond aquaculture water quality management. Springer Science and Business Media

Buzby KM, Lian-shin L (2014) Scaling aquaponic systems: balancing plant uptake with fish output. Aquac Eng 63:39–44

Cerozi BS, Fitzsimmons K (2017) Phosphorous dynamics modelling and mass balance in an aquaponics system. Agric Syst 153:94–100

COST FA1305 (2017) The EU aquaponics hub – realising sustainable integrated fish and vegetable production for the EU. https://euaquaponicshub.com

Delaide B, Goddek S, Gott J, Soyeurt H, Haissam Jijakli M (2016) Lettuce (*Lactuca sativa* L. var. Sucrine) (2016). Growth performance in complemented aquaponic solution outperforms hydroponics. Water 8(10):467

Eck M (2017) Taxonomic characterisation of bacteria communities from water of diversified aquaponic systems. Thesis for the partial fulfillment of a Masters Degree. Université de Liège, Liège

Endut A, Jusoh A, Ali N, Wan Nik WB, Hassan A (2010) A study on the optimal hydraulic loading rate and plant ratios in recirculation aquaponic system. Technol 101:1511–1517

Fernandez V, Sotiropoulos T, Brown P (2013) Foliar fertilization: scientific principles and field practices, 1st edn. International Fertilizer Industry Association (IFA), Paris

Goddek S (2017) Opportunities and challenges of multi-loop aquaponic systems. Wageningen University, Wageningen

Goddek S, Keesman KJ (2018) The necessity of desalination technology for designing and sizing multi-loop aquaponics systems. Desalination 428:76–85

Goddek S, Körner O (2019) A fully integrated simulation model of multi-loop aquaponics: a case study for system sizing in different environments. Agric Syst 171:143–154

Goddek S, Vermeulen T (2018) Comparison of *Lactuca sativa* growth performance in conventional and RAS-based hydroponic systems. Aquac Int 26:1–10. https://doi.org/10.1007/s10499-018-0293-8

Goddek S, Delaide B, Mankasingh U, Vala Ragnarsdottir K, Jijakli H, Thorarinsdottir R (2015) Challenges of sustainable and commercial aquaponics. Sustainability 7(4):4199–4224

Goddek S, Espinal CA, Delaide B, Jikali MH, Schmautz Z, Wuertz S, Keesman J (2016) Navigating towards decoupled aquaponic systems: a system dynamics design approach. Water 8(7):303. 1–29

Gooley GJ, Gavine FM (2003) Integrated agri-aquaculture systems: a resource handbook for Australian industry development. RIRDC Publication No. 03/012

Goto E, Both AJ, Albright LD, Langhans RW, Leed AR (1996) Effect of dissolved oxygen concentration on lettuce growth in floating hydroponics. Proceedings of the International Symposium in Plant Production in Closed Systems. Acta Hortic 440:205–210

Graber A, Junge R (2009) Aquaponic systems: nutrient recycling from fish wastewater by vegetable production. Desalination 246:147–156

Hallam M (2017) EC. Murray Hallam's practical aquaponics. https://aquaponics.net.au/ec/

Halwart M, Gupta MV (eds) (2004) Culture of fish in rice fields. FAO and The WorldFish Center, Penang

Kalantari F, Tahir OM, Lahijani AM, Kalantari S (2017) A review of vertical farming technology: a guide for implementation. Adv Eng Forum 24:76–91

Karimanzira D, Keesman KJ, Kloas W, Baganz D, Raushenbach T (2016) Dynamic modelling of the INAPRO aquaponic system. Aquac Eng 75:29–45

Kloas W, Groß R, Baganz D, Graupner J, Monsees H, Schmidt U, Staaks G, Suhl J, Tschirner M, Wittstock B, Wuertz S, Zikova A, Rennert B (2015) A new concept for aquaponic systems to improve sustainability, increase productivity, and reduce environmental impacts. Aquac Environ Interact 7:179–192. https://doi.org/10.3354/aei00146

Knaus U, Palm HW (2017) Effects of the fish species choice on vegetables in aquaponics under spring-summer conditions in northern Germany (Mecklenburg Western Pomerania). Aquaculture 473:62–73

Komives T, Junge R (2015) Editorial: on the "aquaponic corner" section of the journal. Ecocycles 1 (2):1–2

Lennard WA (2005) Aquaponic integration of Murray Cod (*Maccullochella peelii peelii*) aquaculture and lettuce (*Lactuca sativa*) hydroponics. Thesis (Ph.D.). RMIT University, 2005

Lennard W (2017) Commercial aquaponic systems: integrating recirculating fish culture with hydroponic plant production. In press

Lennard WA, Leonard BV (2006) A comparison of three different hydroponic sub-systems (gravel bed, floating and nutrient film technique) in an aquaponic test system. Aquac Int 14:539–550

Love DC, Fry JP, Genello G, Hill ES, Frederick JA, Li X, Semmens K (2014) An international survey of aquaponics practitioners. PLoS One 9(7):E102662

Love DC, Fry JP, Li X, Hill ES, Genello L, Semmens K, Thompson RE (2015a) Commercial aquaponics production and profitability: findings from an international survey. Aquaculture 435:67–74

Love DC, Uhl MS, Genello L (2015b) Energy and water use of a small-scale raft aquaponics system in Baltimore, Maryland, United States. Aquac Eng 68:19–27

Masser MP, Rakocy J, Losordo TM (1992) Recirculating aquaculture tank production systems. SRAC Publication No. 452. Southern Regional Aquaculture Center. USA

McMurtry M (1990) Performance of an integrated aquaculture-olericulture system as influenced by component ratio. PhD. dissertation, North Carolina State University, Raleigh, North Carolina, USA

Monsees H, Kloas W, Wuertz S (2016) Comparison of coupled and decoupled aquaponics - Implications for future system design. Abstract from aquaculture Europe, 2016. Edinburgh, Scotland

Morehart CT (2016) Chinampa agriculture, surplus production and political change at Xaltocan, Mexico. Anc Mesoam 27(1):183–196

Nichols MA, Lennard W (2010) Aquaponics in New Zealand. Practical Hydroponics and Greenhouses, 115: 46–51

Palm HW et al (2018) Towards commercial aquaponics: a review of systems, designs, scales and nomenclature. Aquac Int 26(3):813–842

Pantanella E, Cardarelli M, Colla G, Rea A, Marcucci A (2010) Aquaponic vs hydroponic: production and quality of lettuce crop. Acta Hortic 927:887–893

Priva (2009) Eindrapport project EcoFutura, visteelt in de glastuinbouw. Priva B.V., Aqua-Terra Nova B.V., Green Q Group B.V., Groen Agro Control. www.ecofutura.nl

Rakocy JE (1989) Vegetable hydroponics and fish culture, a productive interface. World Aquacult 20:42–47

Rakocy JE, Hargreaves JA (1993) Integration of vegetable hydroponics with fish culture: a review. In: Wang J (ed) Techniques for modern aquaculture. American Society of Agricultural Engineers, St Joseph

Rakocy JE, Bailey DS, Shultz RC, Thoman ES (2004a) Update on tilapia and vegetable production in the UVI aquaponic system. In: New Dimensions on Farmed Tilapia: Proceedings of the Sixth International Symposium on Tilapia in Aquaculture, Manila, pp 676–690

Rakocy JE, Shultz RC, Bailey DS, Thoman ES (2004b) Aquaponic integration of Tilapia and Basil: Comparing a batch and staggered cropping system. Acta Hortic 648:63–69

Rakocy JE, Masser MP, Losordo TM (2006) Recirculating aquaculture tank production systems: aquaponics – integrating fish and plant culture. SRAC Publication No. 454. Southern Regional Aquaculture Center. USA

Rakocy JE, Bailey DS, Shultz RC, Danaher JJ (2011) A commercial scale aquaponic system developed at the University of the Virgin Islands. Proceedings of the 9th International Symposium on Tilapia in Aquaculture

Resh HM (2013) Hydroponic food production, 7th edn. CRC Press, Boca Raton

Reyes Lastiri D, Slinkert T, Cappon HJ, Baganz D, Staaks G, Keesman KJ (2016) Model of an aquaponic system for minimised water, energy and nitrogen requirements. Water Sci Technol 74:1. https://doi.org/10.2166/wst.2016.127

Roosta HR (2014) Effects of foliar spray of K on mint, radish, parsley and coriander plants in aquaponic system. J Plant Nutr 37(14):2236–2254

Roosta HR, Hamidpour M (2011) Effects of foliar application of some macro- and micro-nutrients on tomato plants in aquaponic and hydroponic systems. Sci Hortic 129:396–402

Roosta HR, Hamidpour M (2013) Mineral nutrient content of tomato plants in aquaponic and hydroponic systems: Effect of foliar application of some macro-and micro-nutrients. J Plant Nutr 36(13):2070–2083

Savidov N (2005) Comparative study of aquaponically and hydroponically grown plants in model system. In: Evaluation and development of aquaponics production and product market capabilities in Alberta. Chapter 3.2., Phase II, pp 21–31

Somerville C, Cohen M, Pantanella E, Stankus A, Lovatelli A (2014) Small-scale aquaponic food production: integrated fish and plant farming. FAO Fisheries and Aquaculture Technical Paper No. 589

Srivastava JK, Chandra H, Kalra SJ, Mishra P, Khan H, Yadav P (2017) Plant–microbe interaction in aquatic system and their role in the management of water quality: a review. Appl Water Sci 7:1079–1090

Suhl J, Dannehl D, Kloas W, Baganz D, Jobs S, Schiebe G, Schmidt U (2016) Advanced Aquaponics: evaluation of intensive tomato production in aquaponics vs conventional hydroponics. Agric Water Manag 178:335–344

Timmons MB, Ebeling JM, Wheaton FW, Summerfelt ST, Vinci BJ (2002) Recirculating aquaculture systems, 2nd edn. Cayuga Aqua Ventures, Ithaca

Tyson RV, Simonne EH, Treadwell DD, Davis M, White JM (2008) Effect of water pH on yield and nutritional status of greenhouse cucumber grown in recirculating hydroponics. J Plant Nutr 31 (11):2018–2030

Tyson RV, Treadwell DD, Simonne EH (2011) Opportunities and challenges to sustainability in aquaponic systems. Hort Technol 21(1):6–13

Van Os E (1999) Design of sustainable hydroponic systems in relation to environment-friendly disinfection methods. Acta Hortic 548:197–205

Vimal SR, Singh JS, Arora NK, Singh S (2017) Soil-plant-microbe interactions in stressed agriculture management: a review. Pedosphere 27(2):177–192

Wongkiew S, Zhen H, Chandran K, Woo Lee J, Khanal SK (2017) Nitrogen transformations in aquaponic systems: a review. Aquac Eng 76:9–19

Yogev U, Barnes A, Gross A (2016) Nutrients and energy balance analysis for a conceptual model of a three loops off grid, aquaponics. Water 8:589. https://doi.org/10.3390/W8120589

6

Aquaponics System and the Role of Microbial Communities

Alyssa Joyce, Mike Timmons, Simon Goddek, and Timea Pentz

Abstract The growth rates and welfare of fish and the quality of plant production in aquaponics system rely on the composition and health of the system's microbiota. The overall productivity depends on technical specifications for water quality and its movement amongst components of the system, including a wide range of parameters including factors such as pH and flow rates which ensure that microbial components can act effectively in nitrification and remineralization processes. In this chapter, we explore current research examining the role of microbial communities in three units of an aquaponics system: (1) the recirculating aquaculture system (RAS) for fish production which includes biofiltration systems for denitrification; (2) the hydroponics units for plant production; and (3) biofilters and bioreactors, including sludge digester systems (SDS) involved in microbial decomposition and recovery/remineralization of solid wastes. In the various sub-disciplines related to each of these components, there is existing literature about microbial communities and their importance within each system (e.g. recirculating aquaculture systems (RAS), hydroponics, biofilters and digesters), but there is currently limited work examining interactions between these components in aquaponics system, thus making it an important area for further research.

Keywords Microbiota · Aquaponics · Biofilters · Bioreactors · RAS · Hydroponics · Metagenomics

A. Joyce (✉)
Department of Marine Science, University of Gothenburg, Gothenburg, Sweden
e-mail: alyssa.joyce@gu.se

M. Timmons
Biological & Environmental Engineering, Cornell University, Ithaca, NY, USA
e-mail: mbt3@cornell.edu

S. Goddek
Mathematical and Statistical Methods (Biometris), Wageningen University, Wageningen, The Netherlands
e-mail: simon.goddek@wur.nl; simon@goddek.nl

T. Pentz
Eat & Shine VOF, Velp, The Netherlands
e-mail: timea.pentz@eatandshine.nl

6.1 Introduction

Recirculating water in the aquaculture portion of an aquaponics system contains both particulate and dissolved organic matter (POM, DOM) which enter the system primarily via fish feed; the portion of feed that is not eaten or metabolized by fish remains as waste in the recirculating aquaculture system (RAS) water, either in dissolved form (e.g. ammonia) or as suspended or settled solids (e.g. sludge). Once the majority of sludge is removed by mechanical separation, the remaining dissolved organic matter must still be removed from a RAS system. Such processes rely on microbiota in various biofilters in order to maintain water quality for the fish and to convert inorganic/organic wastes into forms of bioavailable nutrients for the plants. Microbial communities in aquaponics system include bacteria, archaea, fungi, viruses and protists in assemblages that fluctuate in composition based on an ebb and flow of nutrients and changes in environmental conditions such as pH, light and oxygen. Microbial communities play a significant role in denitrification and mineralization processes (see Chap. 10) and thus have key roles in the overall productivity of the system, including fish welfare and plant health.

The challenges within any aquaponics system are to control inputs – water, fingerlings, feed, plantlets – and their associated microbiota to maximize the benefits of organic matter and its breakdown into bioavailable forms for target organisms. Given that optimal environmental growth parameters and nutrients differ for fish and plants (see Chap. 8), various separation and aeration systems, and biofilters containing relevant microbial assemblages, must be situated at strategic points in the water supply in order to help maintain nutrient levels, pH and dissolved oxygen (DO) levels within desired ranges for both target fish and plant species. Indeed, water quality parameters, including temperature, DO, electrical conductivity, redox potential, nutrient levels, carbon dioxide, lighting, feed and flow rates, all affect the behaviour and composition of microbial communities within an aquaponics system (Junge et al. 2017). In this regard, it is important to refine setup and operation so that each unit contributes adequate quantities of bioavailable forms of nutrients to its successor, rather than enabling proliferation of pathogens or opportunistic microbes that can consume the bulk of macronutrients needed downstream.

Various techniques for the analysis of microbial communities can yield important information about changes in community structure and function over time in different aquaponic configurations. By correlating these changes with nutrient bioavailability and operational parameters, it is possible to reduce over- or under-production of essential nutrients or the production of noxious by-products. For instance, maximizing recovery of beneficial plant nutrients from waste organic matter in the fish component depends primarily on the ability of microbiota to facilitate breakdown of nutrients within a series of biofilters and sludge digesters, whose performance is based on a range of operational parameters such as flow rates, residence time and pH (Van Rijn 2013). Since not all aquaponics system include sludge digesters, we will address this aspect in more detail in the latter half of this review whilst referring the reader to Chap. 3 for more details on solid separation techniques and Chaps. 7 and 8 for discussions on coupled vs decoupled aquaponics system. If we consider here only dissolved and suspended particulates in the water (and not sludge), all aquaponics system employ a range of different biofilters that expose the attached microorganisms to organic matter passing through the filter and provide an appropriate substrate and sufficient surface area for microbial attachment and formation of biofilms. Degradation of this organic matter provides energy to the microbial communities, which in turn release macronutrients (e.g. nitrate, orthophosphate) and micronutrients (e.g. iron, zinc, copper) back to the system in usable forms (Blancheton et al. 2013; Schreier et al. 2010; Vilbergsson et al. 2016a).

There is considerable agricultural research on the role of microbiota in plant rooting, growth and health. The preponderance of this research focuses on soil-based systems; however, research on hydroponics has also increased in recent years (Bartelme et al. 2018). The microbiota in aquaculture have also been similarly well-characterized, where the role of microbes in fish health and digestion has received considerable attention as researchers attempt to better characterize the role of gut health on nutrient assimilation. Given the importance of biofiltration in RAS systems, bacteria involved in the nitrification process for RAS have also been comparatively well-studied and thus are not be addressed here (see Chaps. 10 and 12). However, there has been comparatively limited research on microbes in aquaponics system, especially the crucial interactions of microbiota amongst various compartments of the system. This lack of research currently limits the scope and productivity of such systems, where there is considerable potential for enhancement with pre- and probiotics, as well as other opportunities to improve the health of aquaponics system through a better understanding, and thus better ability to control, the vast set of uncharacterized microbiota that affect system health and performance.

As such, this chapter focuses primarily on recent studies that reveal how and where microbial communities determine productivity within compartments, whilst also highlighting the relatively small number of studies linking those microbial communities to interactions amongst components and overall system productivity. We attempt to identify gaps where further knowledge about microbial communities could address operational challenges and provide important insights for enhancing efficiency and reliability.

6.2 Tools for Studying Microbial Communities

New technologies for studying how microbial communities change over time, and which groups of organisms predominate under particular environmental conditions, have increasingly offered opportunities to anticipate adverse outcomes within system components and thus lead to the design of better sensors and tests for the effective monitoring of microbial communities in fish or plant cultures. For instance, various 'omics' technologies – metagenomics, metatranscriptomics, community proteomics, metabolomics – are increasingly enabling researchers to study the diversity of microbiota in RAS, biofilters, hydroponics and sludge digestor systems where sampling includes whole microbial assemblages instead of a given genome. Analysis of prokaryotic diversity in particular, has been helped enormously in recent decades by metagenomic and metatranscriptomic techniques. In particular, amplification and sequence analysis of the 16S rRNA gene, based on intraspecific conservation of neutral gene sequences flanking ribosomal operons in bacterial DNA, has been considered the 'gold standard' for taxonomic classification and identification of bacterial species. Such data is also used in microbiology to track epidemics and geographical distributions and study bacterial populations and phylogenies (Bouchet et al. 2008). The methodology can be labour-intensive and expensive, but recent automated systems, whilst not necessarily discriminatory at the species and strain level, offer opportunities for application in aquaponics settings (Schmautz et al. 2017). Recent reviews summarize applications of 16S rRNA as they pertain to RAS (Martínez-Porchas and Vargas-Albores 2017; Munguia-Fragozo et al. 2015; Rurangwa and Verdegem 2015). Advances in metagenomics of microbes other than bacteria found in RAS and hydroponics rely on similar methodologies but use 18S (eukaryotes), 26S (fungi) and 16S in combination with 26S (yeasts) rRNA clone libraries to characterize these microbiota (Martínez-Porchas and Vargas-Albores 2017). Detailed rRNA libraries, for instance, have also been used in hydroponics to characterize microbial communities in the rhizosphere (Oburger and Schmidt 2016). Such libraries can be particularly useful in aquaponics, given that they can examine assemblage of microorganisms such as bacteria, archaea, protozoans and fungi and provide feedback on changes within the system.

The development of automated next-generation sequencing (NGS) has also enabled data analysis of genomes from population samples (metagenomics) that can be used to characterize microbiota, reveal temporal-spatial phylogenetic changes and trace pathogens. Applications in RAS include tracking certain bacterial strains amongst cultured fish and eliminating populations that carry virulent strains, whilst preserving carriers of other strains (review: (Bayliss et al. 2017). Metagenomic approaches can be culture- and amplification-independent, which allows previously unculturable species to become known and investigated for their possible effects (Martínez-Porchas and Vargas-Albores 2017). Next-generation sequencing techniques are commonly used in plant microbiology along with follow-up metatranscriptomics analyses. An excellent example is the first whole-plant study of microbial communities in the rhizosphere, wherein root exudates were shown to correlate with developmental stages (Knief 2014).

Proteomics is most useful when studying a particular bacterial species or strain under specific environmental conditions in order to describe its pathogenicity or possible role in symbiosis. Nevertheless, there are advances in community proteomics that build on prior metagenomic studies and use various biochemical techniques to identify, for example, secreted proteins associated with commensal or symbiotic microbial communities, and further possibilities abound as the capability of NGS technologies advance rapidly (review: (Knief et al. 2011).

Metabolomics characterizes the functions of genes, but the techniques are not organism-specific or sequence-dependent and thus can reveal the wide range of metabolites that are end-products of cellular biochemistry in organisms, tissues, cells or cell compartment (depending on which samples are analysed). Nevertheless, knowledge about the metabolome of microbial communities under particular environmental conditions (microcosms) reveals a great deal about the biogeochemical cycling of nutrients and the effects of perturbations. Such knowledge characterizes various metabolic pathways and the range of metabolites present in samples. Subsequent biochemical and statistical analyses can point to physiological states that can in turn be correlated with environmental parameters which may not be evident from genomic or proteomic approaches. Nevertheless, combining metabolomics with gene function studies has tremendous potential in furthering aquaponics research; see review (van Dam and Bouwmeester 2016).

6.3 Biosecurity Considerations for Food Safety and Pathogen Control

6.3.1 Food Safety

Good food safety and ensuring animal welfare are high priorities in gaining public support for aquaponics. One of the most frequent issues raised by food safety experts in relation to aquaponics is the potential risk of contamination with human pathogens when using fish effluent as fertilizer for plants (Chalmers 2004; Schmautz et al. 2017). A recent literature search to determine zoonotic risks in aquaponics concluded that pathogens in contaminated intake water, or pathogens in components of feeds originating with warm-blooded animals, can become associated with fish gut microbiota, which, even if not detrimental to the fish themselves, can potentially be passed up the food chain to humans (Antaki and Jay-Russell 2015). The mechanisms of introduction of pathogens to an aquaponics system are thus of concern, with the likeliest source of faecal coliforms or other pathogenic bacteria stemming from feed inputs to fish. From a biological perspective, there are potential risks of these pathogens proliferating either in biofilters, or, in one-loop systems by introducing airborne pathogens from open plant components back to the fish tanks. Although biosecurity risks are low in the relatively closed environmental space of an aquaponics system – as compared for instance to open pond aquaculture – and are even lower in decoupled aquaponics system wherein portions of the system can be isolated, there is still a perception that fish sludge could be potentially dangerous when applied to plants for human consumption. *Escherichia coli* (*E. coli*) is a human enteric pathogen causing foodborne illnesses that has been a key concern regarding the use of animal waste as fertilizer in agriculture or aquaculture, e.g. integrated pig-fish systems (Dang and Dalsgaard 2012). However, it is generally not considered to present a risk in fish-plant aquaponics. For instance, Moriarty et al. (2018)

previously demonstrated that UV-radiation treatment can successfully reduce *E. coli* but also noted that the coliforms detected in the aquaponics system were at background levels and did not proliferate in the fish raceways or in the hydroponically grown lettuce within the experimental system, and thus did not present a health risk. There is limited research on these aspects, but a few preliminary studies have found very low risks of coliform contamination, for instance, by showing no difference in coliform levels from sterilized and non-sterilized RAS water treatments applied to plants (Pantanella et al. 2015). Even though there is a potential risk of internalization of microbes within plant leaves, and thus their transmission to the consumed portions of some edible leafy plants grown in aquaponics, other studies have come to similar conclusions that the risks are minimal of introducing potentially dangerous human pathogens (Elumalai et al. 2017).

However, managing risks, or more importantly managing the perceptions of those risks, remains a high priority for government authorities and aquaponics investors. It is assumed that the quality control of feed inputs and careful handling of fish/fish wastes can limit most of these potential concerns (Fox et al. 2012). Indeed, no known human health incidents have to our knowledge currently been reported in relation to aquaponics system, and this may be a function of the fact that RAS facilities and hydroponic greenhouses typically have good biosecurity measures, including hygiene and quarantine practices that are stringently observed. Recommended microbiological practices for biosecurity have been evaluated for different aquaculture production systems and recommendations formulated into Hazard Analysis Critical Control Points guidelines, an international system for controlling food safety (Orriss and Whitehead 2000). However, there is still a need for better scientific documentation of risks for pathogen transfers to humans, and direct research into management in this area of aquaponics production.

6.3.2 Fish and Plant Pathogens

There is existing discipline-specific literature in aquaculture, hydroponics and bioengineering that can help inform and enhance microbial performance in aquaponics. For instance, microbial communities serve a wide range of important functions in fish health, including playing a key role in the digestibility and assimilation of feed, as well as immunodulation, and these functions as well as the role of probiotics in enhancing aquaculture systems are well-reviewed (Akhter et al. 2015). The role of microbes in RAS systems specifically is also well covered, including microbial management of biofilters, as well as research into pathogen control, as well as various techniques to control off-flavours deriving from RAS systems (Rurangwa and Verdegem 2015). Likewise, the microbes in the rhizosphere of plants are important for rooting and plant growth (Dessaux et al. 2016) but also for controlling the spread of pathogens in hydroponic plant production; these areas are well explored in a recent review by Bartelme et al. (2018). However, there is still a very limited understanding of linkages in the microbiome amongst the compartments of aquaponics system, knowledge that is crucial for maximizing productivity and reducing pathogen transfer.

The proliferation of opportunistic pathogens that are dangerous to fish or plant health are important considerations in the economics of aquaponics operations, given that any use of antibiotics or disinfectants can have a potentially detrimental effect on biofilter function, as well as destabilizing microbial relationships in other compartments of the system. Disinfection protocols commonly used in RAS include treating water with ultraviolet light (Elumalai et al. 2017), which, combined with ozone (and usually a combination of both), comprises a first-line abiotic approach to maintaining water quality. Fish eggs/larvae are also often quarantined before being introduced, and any intake water treated, thus reducing direct potential sources of fish pathogen entry to the system.

Incoming water to RAS is also typically allowed to 'mature' in biofilters before being fed into the recirculating system. Experiments, for instance, have shown that inoculating a pre-biofilter with a mixture of nitrifying bacteria, and 'feeding' it with organic matter until bacterial populations match the carrying capacity of the fish tanks, means that the rearing tank water is less likely to be unstable and overtaken by opportunistic bacteria (Attramadal et al. 2016; Rurangwa and Verdegem 2015). However, should pathogens become problematic, the use of high-dose UV, ozone, chemical or antibiotic treatments can sometimes be necessary, although such use is generally disruptive to other compartments of the system, especially the biofilters (Blancheton et al. 2013). Indeed, depending on the dose and location within the system, non-selective treatments for pathogens can actually favour proliferation of opportunists. For instance, high levels of ozone treatment not only kills bacteria, protists and viruses but also oxidizes DOM and affects aggregation of POM, thereby exerting selection pressure on bacterial populations (ibid.).

A detailed discussion of plant pathogens in aquaponics system and their control is included in Chap. 14 and thus is not reiterated here. However, it is worth noting that *Bacillus* species are routinely used as commercial probiotics in aquaculture, and there is growing evidence that similar *Bacillus* species are also effective for plants, that are already available in some commercial hydroponics probiotics solutions (Shafi et al. 2017). A recent study has extended such studies on *Bacillus* to include experimentation in aquaponics system (Cerozi and Fitzsimmons 2016b). The location where the probiotics are introduced – in the fish, plant or biofilters – may be important, but it is not clear from existing work whether the addition of probiotics in the fish component, with potential benefits for the fish, also has better effects on plant growth and health relative to the addition of similar levels of probiotics directly to the hydroponics compartment.

In addition to standard application probiotics, there are a variety of innovative techniques for biocontrol that may in the future become increasingly valuable for reducing the presence and proliferation of harmful microbes. In one recent study, bacterial isolates were selected from an established aquaponics system based on their ability to exert inhibitory effects on both fish and plant fungal pathogens. The goal was to culture these isolates as inocula that could subsequently act as biological controls for diseases within that aquaponics system (Sirakov et al. 2016). For instance, Sirakov et al. demonstrated that a *Pseudomonas* sp. that they isolated was effective as a biocontrol for the pathogenic fungi *Saprolegnia parasitica* of fish and *Pythium ultimum* of plants. The researchers also reported in vitro inhibition of a variety of other bacterial isolates from the different aquaponics compartments, but without testing their in vivo effects. The potential for using such isolates as biological controls is not new, but applications of NGS techniques can now reveal more about interactions of such isolates with each other and with potential pathogens, thus making it possible to optimize the effectiveness of delivery. Use of other 'omics' techniques could help reveal overall community structure and associated metabolic functions, and begin elucidating which organisms and functions are most beneficial. In future, such techniques might allow selection for 'helper strains' within microbial communities, or the identification of exudates that have anti-microbial effects (Massart et al. 2015).

6.4 Microbial Equilibrium and Enhancement in Aquaponics Units

Productivity in aquaponics system involves monitoring and managing environmental parameters in order to provide each component, whether microbial, animal or plant, with optimal growth conditions. Whilst this is not always possible given trade-offs in requirements, one of the key goals of aquaponics revolves around the concept of homeostasis, wherein maintaining stability of the system involves adjusting operational parameters to minimize unnecessary perturbations that cause stress within a unit, or detrimental effects on other components. With ever-changing microbial assemblages, homeostasis never implies a permanent state of equilibrium, but rather a goal of achieving as much stability as possible, particularly within water quality parameters.

A RAS coupled to a hydroponics system will be ever-changing, but within this configuration, the RAS component remains relatively stable, particularly in decoupled systems (Goddek and Körner 2019). The hydroponics system, on the other hand, tends to be more erratic in water quality since the plant crops are often harvested in batch modes, and rarely in synchrony with fish production.

During the initial start-up phase of any aquaponics system, water quality – particularly with regard to microbial communities in biofilters – is a concern, and in order to minimize proliferation of opportunistic bacteria, a routine practice has been to allow microbial maturation of intake water before its introduction into the RAS, adding fish only after the capacity of the biofilters matches the carrying capacity of rearing tanks at a particular stocking density (Blancheton et al. 2013). A similar practice is observed in hydroponics where at least a portion of recycled water is used to inoculate a new crop, given that mature microbial communities take time to develop and introducing all new water results in long lag times. Such practices lead to greater stability in culture conditions and greater productivity. For instance, improved performance in RAS systems has been noted when the pre-intake filter is supplied with pulverized fish food to develop microbial communities more similar to those in the rearing tanks (Attramadal et al. 2014).

6.5 Bacterial Roles in Nutrient Cycling and Bioavailability

Considerable research has been conducted to characterize heterotrophic and autotrophic bacteria in RAS systems and to better understand their roles in maintaining water quality and cycling of nutrients (for reviews, see Blancheton et al. (2013); Schreier et al. (2010). Non-pathogenic heterotrophs, typically dominated by *Alphaproteobacteria* and *Gammaproteobacteria*, tend to thrive in biofilters, and their contributions to transformations of nitrogen are fairly well understood because nitrogen cycling (NC) has been of paramount importance in developing recirculating culture systems (Timmons and Ebeling 2013). It has long been recognized that the bacterial transformation of the ammonia excreted by fish in a RAS system must be matched with excretion rates, because excess ammonia quickly becomes toxic for fish (see Chap. 9). Therefore in freshwater and marine RAS, the functional roles of microbial communities in NC dynamics – nitrification, denitrification, ammonification, anaerobic ammonium oxidation and dissimilatory nitrate reduction – have received considerable research attention and are well described in recent reviews (Rurangwa and Verdegem 2015; Schreier et al. 2010). There are far fewer studies of nitrogen transformations in aquaponics, but a recent review (Wongkiew et al. 2017) provides a summary along with discussion of nitrogen utilization efficiency, which is a prime consideration for plant growth in hydroponics.

After nitrogen, the second most essential macronutrient in aquaponics is phosphorus, which is not a limiting factor for fish that acquire it from feed, but is crucial for plants in hydroponics. However, the forms of phosphate in fish wastes are not immediately bioavailable for plants. Plants must have adequate quantities of inorganic ionic orthophosphate ($H_2PO_4^-$ and HPO_4^{2-} = Pi) (Becquer et al. 2014), as this is the only bioavailable form for uptake and assimilation. Inorganic phosphate binds to calcium above pH 7.0, so aquaponics system must be careful to maintain pH conditions near pH 7.0. As pH values rise above 7.0, various insoluble forms of calcium phosphate can end up as precipitates in sludge (Becquer et al. 2014; Siebielec et al. 2014). Hence, RAS losses of P are primarily through removal of sludge from the system (Van Rijn 2013). However, somewhere in the aquaponics system, particulate matter must be captured and allowed to mineralize in order to provide sufficient supplies of usable nutrients for crops in the hydroponics unit. The mineralization step will also release other macro- and micronutrients so that there are fewer deficiencies, thus reducing the need for supplementation in the hydroponics compartment. Given that world supplies of phosphate-rich fertilizers are dwindling and supplementation with P is increasingly costly, efforts are being made to maximize the recovery of P from RAS sludge (Goddek et al. 2016b; Monsees et al. 2017).

The bioavailability of macro and micronutrients is currently poorly understood. Previous research (Cerozi and Fitzsimmons 2016a) suggests that the availability of nutrients becomes compromised as pH is reduced below 7.0 and has resulted in coupled hydroponics system for leafy greens being operated around pH 6.0. However, recent research comparing aquaponic conditions and pH 7.0 to hydroponic conditions of pH 5.8 showed no difference in productivity (Anderson et al. 2017a, b). In these studies, hydroponic conditions at pH 7.0 reduced productivity by ~ 22% compared to hydroponic pH 5.8. Initially, the hypothesis was that the differences in productivity could be ascribed to the microbiota of the aquaponic water, but subsequent research dismissed that theory (Wielgosz et al. 2017).

In RAS where C:N ratios increase due to availability of organic matter, denitrifying bacteria, especially *Pseudomonas* sp., use carbon as an electron donor in anoxic conditions, to produce N_2 at the expense of nitrate (Schreier et al. 2010; Wongkiew et al. 2017). Biofloc systems are sometimes used to augment feed for fish (Crab et al. 2012; Martínez-Córdova et al. 2015), and biofloc is increasingly being used in aquaponics system, especially in Asia (Feng et al. 2016; Kim et al. 2017; Li et al. 2018). When biofloc is used in aquaponics (da Rocha et al. 2017; Pinho et al. 2017), nutrient cycling becomes even more complex given that DO, temperature and pH influence whether heterotrophic (carbon-utilizing) microbial communities predominate over autotrophic denitrifiers that are capable of reducing sulphide to sulphate (Schreier et al. 2010). Heterotrophs tend to have a higher growth rate than autotrophs in the presence of adequate sources of carbon (Michaud et al. 2009); therefore, manipulating feed type or regimes, or adding an organic carbon source directly, whilst monitoring dissolved oxygen levels, can help keep populations equilibrated whilst still providing hydroponics with N in a usable form (Vilbergsson et al. 2016a).

In hydroponics system, nutrient cycling has been less well-studied since inorganic compounds containing the required balance of nutrients are typically added in order to ensure proper plant growth. However, high nutrient concentrations, especially in humid warm environments such as greenhouses, easily facilitate growth of microbial communities, especially phytopathogens such as fungi (*Fusarium*) and oomycota (*Phytophthora*, *Pythium* sp.), that can quickly spread in circulating water and may result in die-offs (Lee and Lee 2015). Recent efforts to better understand hydroponic rhizobacteria and their beneficial effects in promoting plant growth (but also for inhibiting pathogen proliferation) have utilized various 'omics' techniques to analyse microbial communities and their interactions with root systems (Lee and Lee 2015).

For instance, when probiotic bacteria such as *Bacillus*, are present, they were shown to enhance P availability and also appear to have an added plant growth-promoting effect in a tilapia-lettuce system (Cerozi and Fitzsimmons 2016b). In aquaponics system, the addition of probiotics to fish feed and RAS water, as well as to the hydroponic water supply, deserves further experimentation, since microbial communities can have multiple modulatory effects on plant physiology. For example, the microbial communities (bacteria, fungi, oomycetes) of four food crops were analysed by metagenomic sequencing when maintained in a constant nutrient film hydroponics system where pH and nutrient concentrations were allowed to fluctuate naturally throughout the plants' life cycles (Sheridan et al. 2017). The authors concluded that treatment with a commercial mixture of plant growth-promoting microbes (PGPMs), in this case bacteria, mycorrhizae and fungi, appeared to confer greater stability and similarities in community composition after 12–14 weeks than in controls. They suggest that this could be attributed to root exudates, which purportedly favour and even control the development of microbial communities appropriate to successive plant developmental stages. Given the known effects of PGPMs in soil-based crop production, and the few studies that are available for soilless systems, further investigation is warranted to determine how to enhance PGPMs and to improve their effects in aquaponics system (Bartelme et al. 2018). If hydroponic cultures are more stable and plant growth is more robust with PGPMs, then the goal should be to characterize microbial communities in aquaponics via metagenomics and correlate them with optimal macro- and micronutrient availability via metabolomics and proteomics.

6.6 Suspended Solids and Sludge

The parameters for operating aquaponics at a given scale – including water volume, temperature, feed and flow rates, pH, fish and crop ages and densities – all affect the temporal and spatial distribution of the microbial communities that develop within its compartments, for reviews: RAS (Blancheton et al. 2013); hydroponics (Lee and Lee 2015).

In addition to controlling dissolved oxygen, carbon dioxide levels and pH in aquaponics, it is also essential to control the accumulation of solids in the RAS system as fine suspended particles can adhere to gills, cause abrasion and respiratory distress and increase susceptibility to disease (Yildiz et al. 2017). More relevant, the particulate organic matter (POM) must be quickly and effectively removed from RAS systems, or else excessive heterotrophic growth will cause almost all unit processes to fail. RAS feeding rates must be carefully managed to minimize solids loading on the system (e.g. avoid over-feeding and minimize feeding costs). The biophysical properties of feed – particle size, nutrient content, digestibility, sensory appeal, density and settling rate – determine ingestion and assimilation rates, which in turn have an impact on solids build-up and thus water quality. Although water quality is frequently studied in the context of nutrient cycling (see Chap. 9), it is also important to obtain a better understanding of the composition of microbial communities and changes in these based on feed composition, particulate loading and how this influences the growth of heterotrophic and autotrophic bacterial communities.

Various features of RAS system designs have been developed specifically to deal with solids (Timmons and Ebeling 2013); see also review: (Vilbergsson et al. 2016b). For instance, some biofilters function to keep substantial portions of wastes suspended in order to facilitate degradation, whilst others mechanically filter through screens or granular media. Still others rely on sedimentation to simply collect and remove sludge. However, such methods are not particularly effective at recovering nutrients within the sludge and making it bioavailable for plant use. Historically, this sludge has been handled in bioreactors for its methanogenic value or dewatered to be used as fertilizer for soil-based crops, but various newer designs have attempted to improve recovery for use in the hydroponic component. Improving recovery of this sludge is an important area of investigation given that a significant portion of the essential macro- and micronutrients required for plant growth are bound to the particulate organic matter, which, if discarded, is lost from the system. By adding an additional sludge recycling loop to aquaponics system, solid wastes can be converted into dissolved nutrients for reuse by plants rather than being discarded (Goddek et al. 2018). Digesters or remineralizing bioreactors are one way of accomplishing this, however one of the key areas that is currently under-developed includes knowledge of how microbial communities within these sludge digesters can be enhanced (e.g. through addition of microbes) or better utilized (e.g. through better engineered design of linked reactors) to recover nutrients into bioavailable forms for plants. Even though the actual microbial communities within sludge digesters have not been well researched for aquaponics, there is considerable literature on the microbiota of sludge digesters for sewage and animal wastes in agriculture, including fish effluent, that can provide further insight into ideal designs for sludge recovery in aquaponics system. Current research on the incorporation of sludge into aquaponics system involves remineralization in digesters situated between the RAS and hydroponic unit (Goddek et al. 2016a, 2018). Within aerobic or anaerobic bioreactors, environmental conditions that are favourable for waste degradation can effectively break down this sludge into bioavailable nutrients, which can subsequently be delivered to hydroponics system without the presence of soil (Monsees et al. 2017). Many one-loop aquaponics system already include aerobic (Rakocy et al. 2004) and anaerobic (Yogev et al. 2016) digesters to transform nutrients that are trapped in the fish sludge and make them bioavailable for plants. The ability to decouple these has a number of advantages that are further discussed in Chap. 8 and appears to lead to higher growth rates (Goddek and Vermeulen 2018). However, despite the many advances, the actual technology to accomplish this remains challenging. For example, some heterotrophic denitrifying bacteria cultured in anoxic or even aerobic conditions with sludge from RAS will use nitrate as an electron receptor and oxidized carbon sources for energy, while storing excess P as polyphosphate along with divalent metal ions such as Ca^{+2} or Cu^{+2}. When stressed at alkaline pH, these bacteria degrade polyphosphate and release orthophosphate, which is the necessary form for assimilation of phosphate by plants (Van Rijn et al. 2006). Inserting remineralization bioreactor units, such as those in Goddek et al. (2018), could provide a way to better recover P for hydroponics. Similar methods have, for instance, been used with trout sludge from a RAS that were treated for nitrate and P content in excess of allowable disposal limits (Goddek et al. 2015). However, the microbial communities involved in these processes are sensitive to culture conditions such as C:N ratios, oxygenation, metal ions and pH, so nitrites and other noxious intermediates can accumulate. Despite a vast literature on digesters of various organic wastes, primarily anaerobic for biogas production (Ibrahim et al. 2016), there is far less research on treating RAS wastes (Van Rijn

2013), and in the case of aquaponics system, even less available research about the relationship between nutrient bioavailability and crop growth in hydroponics system (Möller and Müller 2012). At this time, more studies of RAS sludge bioreactors could provide important insights into culture conditions for microbial populations that produce favourable results, for instance, on P recovery, and its introduction into hydroponics units.

One of the current challenges in efforts to assess the recovery of P from sludge arises when comparing trials of anaerobic and aerobic digesters for their efficacy (Goddek et al. 2016b; Monsees et al. 2017). Although both studies used similar sludge composition initially, the results were quite different. In one study (Monsees et al. 2017), measures of various soluble nutrients in aerobic treatments resulted in a 330% increase in P concentration and a 16% decrease in nitrate concentration compared to minor increases in P and a 97% decrease in nitrate in anaerobic treatments. By contrast, results from a similar study (Goddek et al. 2016b) showed that growth of lettuce plants in a hydroponic unit was superior using anaerobic supernatant, even though both anaerobic and aerobic treatments only resulted in slightly better nitrate recovery from anaerobic conditions and almost complete loss of PO_4 from both treatments (Goddek et al. 2016b). Obviously, factors such as feed composition and rates, the suspension versus settling of solids, pH (maintained at 7 ± 1 with $CaOH_2$ in the former and variable 8.2–8.65 in the latter), sampling and fish strains differed in these two studies. Nevertheless, the contrasting results for PO_4 and NO_3 indicate the need for further research to optimize nutrient recovery, with the addition of a metagenomics approach to characterize microbial communities so as to better understand their role in these processes.

6.7 Conclusions

Formerly the domain of small-scale producers, technological advances are increasingly moving aquaponics into larger-scale commercial production by focusing on improved macro- and micronutrient recovery whilst providing technical innovations to reduce water and energy requirements. However, scaling up of aquaponics to an industrial scale requires a much better understanding and maintenance of microbial assemblages, and the implementation of strong biocontrol measures that favour the health and well-being of both fish and crops, whilst still meeting food safety standards for human consumption. Further research on biocontrol of microbial pathogens in aquaponics, including potential human, fish and plant pathogens are needed, in light of the sensitivity of such systems to perturbation, and the fact that the use of chemicals and antibiotics can have profound effects on microbial populations, fish and plant physiologies, as well as overall system operation. Elucidating microbial interactions can improve the productivity of aquaponics system given the crucial roles of microbes in converting organic matter into usable forms that can allow fish and plants to thrive.

References

Akhter N, Wu B, Memon AM, Mohsin M (2015) Probiotics and prebiotics associated with aquaculture: a review. Fish Shellfish Immunol 45:733–741

Anderson T, de Villiers D, Timmons M (2017a) Growth and tissue elemental composition response of butterhead lettuce (*Lactuca sativa*, cv. Flandria) to hydroponic and aquaponic conditions. Horticulturae 3:43

Anderson TS, Martini MR, de Villiers D, Timmons MB (2017b) Growth and tissue elemental composition response of Butterhead lettuce (*Lactuca sativa*, cv. Flandria) to hydroponic conditions at different pH and alkalinity. Horticulturae 3:41

Antaki ET, Jay-Russell M (2015) Potential zoonotic risks in aquaponics. IAFP, Portland

Attramadal KJK, Truong TMH, Bakke I, Skjermo J, Olsen Y, Vadstein O (2014) RAS and microbial maturation as tools for K-selection of microbial communities improve survival in cod larvae. Aquaculture 432:483–490

Attramadal KJ, Minniti G, Øie G, Kjørsvik E, Østensen M-A, Bakke I, Vadstein O (2016) Microbial maturation of intake water at different carrying capacities affects microbial control in rearing tanks for marine fish larvae. Aquaculture 457:68–72

Bartelme RP, Oyserman BO, Blom JE, Sepulveda-Villet OJ, Newton RJ (2018) Stripping away the soil: plant growth promoting microbiology opportunities in aquaponics. Front Microbiol 9:8

Bayliss SC, Verner-Jeffreys DW, Bartie KL, Aanensen DM, Sheppard SK, Adams A, Feil EJ (2017) The promise of whole genome pathogen sequencing for the molecular epidemiology of emerging aquaculture pathogens. Front Microbiol 8:121

Becquer A, Trap J, Irshad U, Ali MA, Claude P (2014) From soil to plant, the journey of P through trophic relationships and ectomycorrhizal association. Front Plant Sci 5:548

Blancheton J, Attramadal K, Michaud L, d'Orbcastel ER, Vadstein O (2013) Insight into bacterial population in aquaculture systems and its implication. Aquac Eng 53:30–39

Bouchet V, Huot H, Goldstein R (2008) Molecular genetic basis of ribotyping. Clin Microbiol Rev 21:262–273

Cerozi BD, Fitzsimmons K (2016a) The effect of pH on phosphorus availability and speciation in an aquaponics nutrient solution. Bioresour Technol 219:778–781

Cerozi BD, Fitzsimmons K (2016b) Use of Bacillus spp. to enhance phosphorus availability and serve as a plant growth promoter in aquaponics systems. Sci Hortic 211:277–282

Chalmers GA (2004) Aquaponics and food safety. Lethbridge, Alberta

Crab R, Defoirdt T, Bossier P, Verstraete W (2012) Biofloc technology in aquaculture: beneficial effects and future challenges. Aquaculture 356:351–356

da Rocha A, Biazzetti Filho M, Stech M, Paz da Silva R (2017) Lettuce production in aquaponic and biofloc systems with silver catfish *Rhamdia quelen*. Bol Inst Pesca 43:64

Dang STT, Dalsgaard A (2012) *Escherichia coli* contamination of fish raised in integrated pig-fish aquaculture systems in Vietnam. J Food Prot 75:1317–1319

Dessaux Y, Grandclément C, Faure D (2016) Engineering the rhizosphere. Trends Plant Sci 21:266–278

Elumalai SD, Shaw AM, Pattillo DA, Currey CJ, Rosentrater KA, Xie K (2017) Influence of UV treatment on the food safety status of a model aquaponics system. Water 9:27

Feng J, Li F, Zhou X, Xu C, Fang F (2016) Nutrient removal ability and economical benefit of a rice-fish co-culture system in aquaculture pond. Ecol Eng 94:315–319

Fox BK, Tamaru CS, Hollyer J, Castro LF, Fonseca JM, Jay-Russell M, Low T (2012) A preliminary study of microbial water quality related to food safety in recirculating aquaponic fish and vegetable production systems. College of Tropical Agriculture and Human Resources, Honolulu

Goddek S, Körner O (2019) A fully integrated simulation model of multi-loop aquaponics: a case study for system sizing in different environments. Agric Syst 171:143

Goddek S, Vermeulen T (2018) Comparison of *Lactuca sativa* growth performance in conventional and RAS-based hydroponics system. Aquac Int 26:1377. https://doi.org/10.1007/s10499-018-0293-8

Goddek S, Delaide B, Mankasingh U, Ragnarsdottir KV, Jijakli H, Thorarinsdottir R (2015) Challenges of sustainable and commercial aquaponics. Sustainability 7:4199–4224

Goddek S, Espinal CA, Delaide B, Jijakli MH, Schmautz Z, Wuertz S, Keesman KJ (2016a) Navigating towards decoupled aquaponics system: a system dynamics design approach. Water 8:303

Goddek S, Schmautz Z, Scott B, Delaide B, Keesman KJ, Wuertz S, Junge R (2016b) The effect of anaerobic and aerobic fish sludge supernatant on hydroponic lettuce. Agronomy-Basel 6:37

Goddek S, Delaide BP, Joyce A, Wuertz S, Jijakli MH, Gross A, Eding EH, Bläser I, Reuter M, Keizer LP (2018) Nutrient mineralization and organic matter reduction performance of RAS-based sludge in sequential UASB-EGSB reactors. Aquac Eng 83:10–19

Ibrahim MH, Quaik S, Ismail SA (2016) An introduction to anaerobic digestion of organic wastes, prospects of organic waste management and the significance of earthworms. Springer, Cham, pp 23–44

Junge R, König B, Villarroel M, Komives T, Haïssam Jijakli M (2017) Strategic points in aquaponics. Water 9:182

Kim SK, Jang IK, Lim HJ (2017) Inland aquaponics system using biofloc technology. Google Patents

Knief C (2014) Analysis of plant microbe interactions in the era of next generation sequencing technologies. Front Plant Sci 5:216

Knief C, Delmotte N, Vorholt JA (2011) Bacterial adaptation to life in association with plants–a proteomic perspective from culture to in situ conditions. Proteomics 11:3086–3105

Lee S, Lee J (2015) Beneficial bacteria and fungi in hydroponics system: types and characteristics of hydroponic food production methods. Sci Hortic 195:206–215

Li G, Tao L, Li X-l, Peng L, Song C-f, Dai L-l, Wu Y-z, Xie L (2018) Design and performance of a novel rice hydroponic biofilter in a pond-scale aquaponic recirculating system. Ecol Eng 125:1–10

Martínez-Córdova LR, Emerenciano M, Miranda-Baeza A, Martínez-Porchas M (2015) Microbial-based systems for aquaculture of fish and shrimp: an updated review. Rev Aquac 7:131–148

Martínez-Porchas M, Vargas-Albores F (2017) Microbial metagenomics in aquaculture: a potential tool for a deeper insight into the activity. Rev Aquac 9:42–56

Massart S, Martinez-Medina M, Jijakli MH (2015) Biological control in the microbiome era: challenges and opportunities. Biol Control 89:98–108

Michaud L, Lo Giudice A, Troussellier M, Smedile F, Bruni V, Blancheton J-P (2009) Phylogenetic characterization of the heterotrophic bacterial communities inhabiting a marine recirculating aquaculture system. J Appl Microbiol 107:1935–1946

Möller K, Müller T (2012) Effects of anaerobic digestion on digestate nutrient availability and crop growth: a review. Eng Life Sci 12:242–257

Monsees H, Keitel J, Paul M, Kloas W, Wuertz S (2017) Potential of aquacultural sludge treatment for aquaponics: evaluation of nutrient mobilization under aerobic and anaerobic conditions. Aquac Environ Interact 9:9–18

Moriarty MJ, Semmens K, Bissonnette GK, Jaczynski J (2018) Inactivation with UV-radiation and internalization assessment of coliforms and *Escherichia coli* in aquaponically grown lettuce. LWT 89:624–630

Munguia-Fragozo P, Alatorre-Jacome O, Rico-Garcia E, Torres-Pacheco I, Cruz-Hernandez A, Ocampo-Velazquez RV, Garcia-Trejo JF, Guevara-Gonzalez RG (2015) Perspective for aquaponics system: “omic” technologies for microbial community analysis. Biomed Res Int 2015:480386

Oburger E, Schmidt H (2016) New methods to unravel rhizosphere processes. Trends Plant Sci 21:243–255

Orriss GD, Whitehead AJ (2000) Hazard analysis and critical control point (HACCP) as a part of an overall quality assurance system in international food trade. Food Control 11:345–351

Pantanella E, Cardarelli M, Di Mattia E, Colla G (2015) Aquaponics and food safety: effects of UV sterilization on total coliforms and lettuce production. In: Carlile WR (ed) International conference and exhibition on soilless culture, pp 71–76

Pinho SM, Molinari D, de Mello GL, Fitzsimmons KM, Coelho Emerenciano MG (2017) Effluent from a biofloc technology (BFT) tilapia culture on the aquaponics production of different lettuce varieties. Ecol Eng 103:146–153

Rakocy JE, Bailey DS, Shultz RC, Thoman ES (2004) Update on tilapia and vegetable production in the UVI aquaponics system. New dimensions on farmed tilapia. Proceedings from the 6th international symposium on tilapia in aquaculture 000, 1–15

Rurangwa E, Verdegem MC (2015) Microorganisms in recirculating aquaculture systems and their management. Rev Aquac 7:117–130

Schmautz Z, Graber A, Jaenicke S, Goesmann A, Junge R, Smits THM (2017) Microbial diversity in different compartments of an aquaponics system. Arch Microbiol 199:613–620

Schreier HJ, Mirzoyan N, Saito K (2010) Microbial diversity of biological filters in recirculating aquaculture systems. Curr Opin Biotechnol 21:318–325

Shafi J, Tian H, Ji M (2017) Bacillus species as versatile weapons for plant pathogens: a review. Biotechnol Equip 31:446–459

Sheridan C, Depuydt P, De Ro M, Petit C, Van Gysegem E, Delaere P, Dixon M, Stasiak M, Aciksöz SB, Frossard E (2017) Microbial community dynamics and response to plant growth-promoting microorganisms in the rhizosphere of four common food crops cultivated in hydroponics. Microb Ecol 73:378–393

Siebielec G, Ukalska-Jaruga A, Kidd P (2014) Bioavailability of trace elements in soils amended with high-phosphate materials. In: Selim HM (ed) Phosphate in soils: interaction with micronutrients, radionuclides and heavy metals bioavailability of trace elements in soils amended with high-phosphate materials. CRC Press/Taylor & Francis Group, Boca Raton, pp 237–268

Sirakov I, Lutz M, Graber A, Mathis A, Staykov Y, Smits TH, Junge R (2016) Potential for combined biocontrol activity against fungal fish and plant pathogens by bacterial isolates from a model aquaponics system. Water 8:518

Timmons MB, Ebeling JM (2013) Recirculating aquaculture. Ithaca Publishing Company, Ithaca. 788 p

van Dam NM, Bouwmeester HJ (2016) Metabolomics in the rhizosphere: tapping into belowground chemical communication. Trends Plant Sci 21:256–265

Van Rijn J (2013) Waste treatment in recirculating aquaculture systems. Aquac Eng 53:49–56

Van Rijn J, Tal Y, Schreier HJ (2006) Denitrification in recirculating systems: theory and applications. Aquac Eng 34:364–376

Vilbergsson B, Oddsson GV, Unnthorsson R (2016a) Taxonomy of means and ends in aquaculture production-part 3: the technical solutions of controlling n compounds, organic matter, p compounds, metals, temperature and preventing disease. Water 8:506

Vilbergsson B, Oddsson GV, Unnthorsson R (2016b) Taxonomy of means and ends in aquaculture production – Part 2: The technical solutions of controlling solids, dissolved gasses and pH. Water 8:387

Wielgosz ZJ, Anderson TS, Timmons MB (2017) Microbial effects on the production of aquaponically grown lettuce. Horticulturae 3:46

Wongkiew S, Hu Z, Chandran K, Lee JW, Khanal SK (2017) Nitrogen transformations in aquaponics system: a review. Aquac Eng 76:9–19

Yildiz HY, Robaina L, Pirhonen J, Mente E, Domínguez D, Parisi G (2017) Fish welfare in aquaponics system: its relation to water quality with an emphasis on feed and faeces-a review. Water 9:13

Yogev U, Barnes A, Gross A (2016) Nutrients and energy balance analysis for a conceptual model of a three loops off grid, aquaponics. Water 8:589

7

System Design and Management Issues in Coupled Aquaponics

Harry W. Palm, Ulrich Knaus, Samuel Appelbaum, Sebastian M. Strauch, and Benz Kotzen

Abstract Coupled aquaponics is the archetype form of aquaponics. The technical complexity increases with the scale of production and required water treatment, e.g. filtration, UV light for microbial control, automatic controlled feeding, computerization and biosecurity. Upscaling is realized through multiunit systems that allow staggered fish production, parallel cultivation of different plants and application of several hydroponic subsystems. The main task of coupled aquaponics is the purification of aquaculture process water through integration of plants which add economic benefits when selecting suitable species like herbs, medicinal plants or ornamentals. Thus, coupled aquaponics with closed water recirculation systems has a particular role to fulfil.

Under fully closed recirculation of nutrient enriched water, the symbiotic community of fish, plants and bacteria can result in higher yields compared with standalone fish production and/or plant cultivation. Fish and plant choices are highly diverse and only limited by water quality parameters, strongly influenced by fish feed, the plant cultivation area and component ratios that are often not ideal. Carps, tilapia and catfish are most commonly used, though more sensitive fish species and crayfish have been applied. Polyponics and additional fertilizers are methods to improve plant quality in the case of growth deficiencies, boosting plant production and increasing total yield.

The main advantages of coupled aquaponics are in the most efficient use of resources such as feed for nutrient input, phosphorous, water and energy as well as in an increase of fish welfare. The multivariate system design approach allows

H. W. Palm (✉) · U. Knaus (✉) · S. M. Strauch
Faculty of Agricultural and Environmental Sciences, Department of Aquaculture and Sea-Ranching, University of Rostock, Rostock, Germany
e-mail: harry.palm@uni-rostock.de; ulrich.knaus@uni-rostock.de

S. Appelbaum
French Associates Institute for Agriculture and Biotechnology of Drylands, Jacob Blaustein Institutes for Desert Research, Ben-Gurion University of the Negev, Beersheba, Israel
e-mail: sappl@bgu.ac.il

B. Kotzen
School of Design, University of Greenwich, London, UK
e-mail: b.kotzen@greenwich.ac.uk

coupled aquaponics to be installed in all geographic regions, from the high latitudes to arid and desert regions, with specific adaptation to the local environmental conditions. This chapter provides an overview of the historical development, general system design, upscaling, saline and brackish water systems, fish and plant choices as well as management issues of coupled aquaponics especially in Europe.

Keywords Coupled aquaponics · Fish and plant choice · Nutrient cycles · Polyponic systems · Functions

7.1 Introduction

The combination of fish and plant cultivation in coupled aquaponics dates back to the first design by Naegel (1977) in Germany, using a 2000 L hobby scale system (Fig. 7.1) located in a controlled environment greenhouse. This system was developed in order to verify the use of nutrients from fish waste water under fully controlled water recirculating conditions intended for plant production including a dual sludge system (aerobic/anaerobic wastewater treatment). Naegel based his concept on the open pond aquaponic system of the South Carolina Agricultural Experiment Station, in the USA, where excess nutrients from the fishponds, stocked with channel catfish (*Ictalurus punctatus*), were eliminated by the hydroponic production of water chestnuts (*Eleocharis dulcis*) (Loyacano and Grosvenor 1973). By including nitrification and denitrification tanks to increase the nitrate concentration inside his system, Naegel (1977) attempted a complete oxidation of all nitrogenous compounds, reaching nitrate concentrations of 1200 mg/L, and demonstrating the effectiveness of the nitrification step. Although the system was stocked at a low density (20 kg/m^3 each) using tilapia (*Tilapia mossambica*) and carp (*Cyprinus carpio*), the tomatoes (*Lycopersicon esculentum*) and iceberg lettuce (*Lactuca scariola*) grew well and produced harvestable yield. These first research results led to the concept of coupled aquaponic systems, in which the plants eliminate the waste produced by the fish, creating adequate growth, demonstrating highly efficient water use in both units. The principle of coupled aquaponics was first described by Huy Tran at the World Aquaculture Conference in 2015 (Tran 2015).

Coupled aquaponic systems do not necessarily use mechanical particulate filtering in the classical sense and keep consistent nutrient flow between the aquaculture and hydroponic units. The main challenge is how to manage the faecal load in the coupled aquaponic system where the plants absorb the nutrients and particulate waste can be removed from the system by filter presses or geotextiles.

The development of modern agriculture, human population growth and shrinking resources worldwide, has promoted the development of coupled aquaponic systems. Since fish farming is considerably more efficient in protein production and water use compared with other farmed animals and since closed systems are largely site-independent, coupled aquaponic systems have been able to develop worldwide (Graber and Junge 2009), under arid conditions (Kotzen and Appelbaum 2010;

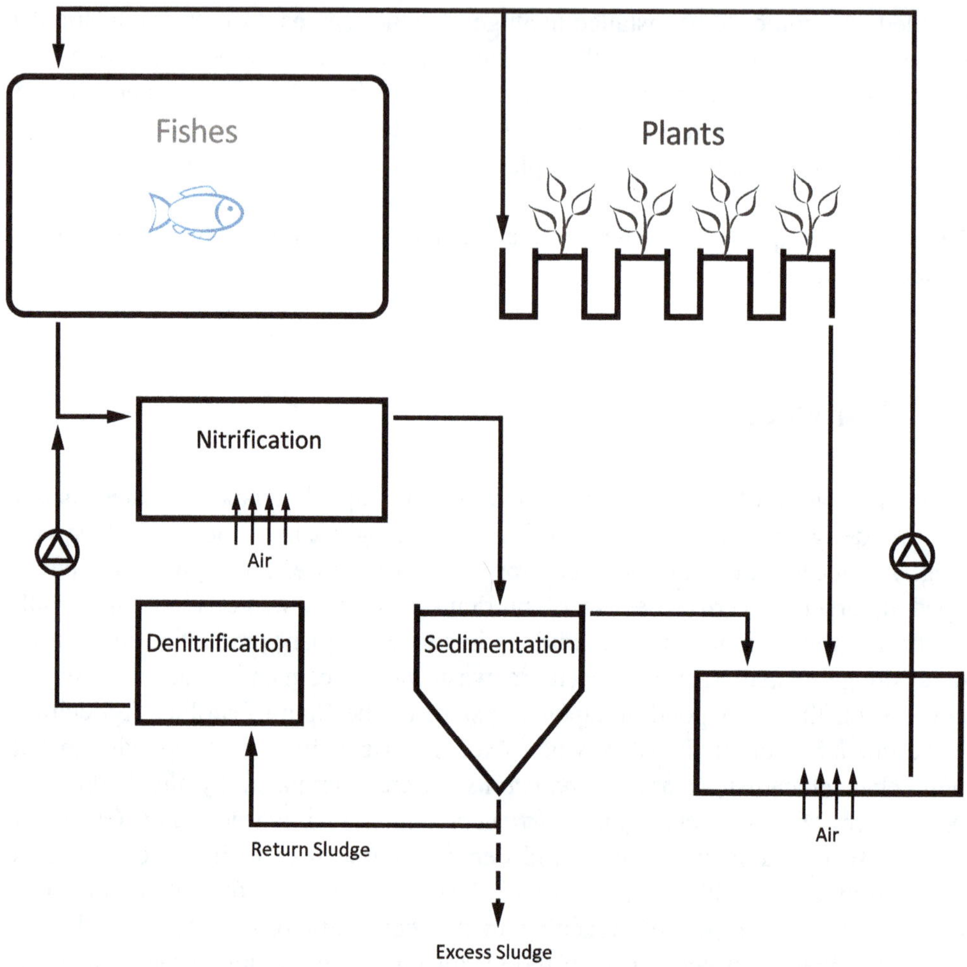

Fig. 7.1 Diagram of the first system by Naegel (1977) growing tilapia and common carp in combination with lettuce and tomatoes in a closed recirculation system

Appelbaum and Kotzen 2016) and even in urban settings (König et al. 2016). Most described systems belong to domestic, small-scale and semi-commercial installations (Palm et al. 2018) that are driven by hobby aquarists, enthusiasts or smaller start-up companies. New research results, summarized in this chapter, demonstrate both the potentials and constraints regarding the continued development of these systems into commercial aquaponics, being capable of making a significant contribution to future food production.

7.2 Historical Development of Coupled Aquaponics

Most original research efforts on coupled aquaponic systems took place in the USA with an increasing presence in the EU partly initiated by COST Action FA1305, The EU Aquaponics Hub and in other European research centres. Nowadays, fully recirculating aquaponic system designs almost completely dominate the American aquaponics industry, with estimates that over 90% of the existing aquaponic systems in the USA are of a fully recirculating design (Lennard, pers. comm.). The first American coupled aquaponics research was undertaken at the Illinois Fisheries and Aquaculture Center (formerly the SIU Cooperative Fisheries Research Laboratory) and the Department of Zoology, focusing on coupled aquaponic systems stocked with channel catfish (*Ictalurus punctatus*) in combination with tomatoes (*Lycopersicon esculentum*) (Lewis et al. 1978). The authors noted that an optimal plant growth is only possible when all the essential macro- and micronutrients are available in the process water, and thus nutrient supplementation is required in the event of nutrient deficiencies. The authors also demonstrated a deficiency in plant-available iron, constraining plant growth, which could be solved through iron-chelate supplementation. Other early studies in the USA focused on analysing technological functionality and the quality of the harvested channel catfish and tomatoes (Lewis et al. 1978; Sutton and Lewis 1982). Laboratory-scale aquaponic systems examined parameters, such as resource efficiencies with regard to materials, costs, water and energy consumption, and examined the use of other fish species such as *Tilapia* spp. in the US Virgin Islands (UVI) (Watten and Busch 1984). Dr. James Rakocy at the UVI developed the first commercial coupled aquaponic system, a raft system that combined the production of Nile tilapia (*Oreochromis niloticus*) and lettuce (*Lactuca sativa*), and later investigated the production of further plant species (Rakocy 1989, 2012; Rakocy et al. 2000, 2003, 2004, 2006, 2011). This medium scale commercial installation took advantage of the local climate where greenhouses were not necessary and the market conditions of the Virgin Islands to generate profit. The UVI aquaponic system was subsequently adopted in different countries with respect to the respective needs of different plants and the appropriateness of the technology, e.g. in Canada by Savidov (2005) and in Saudi Arabia by Al-Hafedh et al. (2008). This is the case in Europe as well, where coupled aquaponic systems have evolved from the original UVI design, e.g. the vertical aquaponic system at the Aquaponics Research Lab., University of Greenwich (Khandaker and Kotzen 2018). Several other research departments investigated the technological feasibility of closed – or 'coupled' – aquaponics production using various fish and plant species as well as hydroponic subsystems to increase yields and reducing different emission parameters (Graber and Junge 2009). For example, at Rostock University (Germany), the research focused on the stability of backyard systems (Palm et al. 2014a), combining different fish species, African catfish (*Clarias gariepinus*) and Nile tilapia (*Oreochromis niloticus*), with different plants (Palm et al. 2014b, 2015). In 2015, a modern experimental semi-commercial scale aquaponic system, the 'FishGlassHouse', was built on the campus of the University of Rostock (Palm et al. 2016). However, the system was designed allowing both coupled and decoupled operations. Other notable facilities were built at the Zürich University of Applied Sciences (ZHAW) at Waedenswil in Switzerland (Graber and Junge 2009; Graber et al. 2014), both coupled and decoupled research facility of the Icelandic company Svinna-verkfraedi Ltd. (Thorarinsdottir 2014; Thorarinsdottir et al. 2015), the cold water aquaponic system NIBIO Landvik at Grimstad (Skar et al. 2015; Thorarinsdottir et al. 2015), the PAFF Box (Plant And Fish Farming

Box) one loop aquaponic system at Gembloux Agro-Bio Tech – University of Liège, in Gembloux, Belgium (Delaide et al. 2017), the combined living wall and vertical farming aquaponic system at the University of Greenwich (Khandaker and Kotzen 2018), as well as the research-domestic coupled aquaponic system (changed from decoupled to coupled in 2018, Morgenstern and Dapprich 2018, pers. comm.) at the South Westphalia University of Applied Sciences, i.GREEN Institute for Green Technology & Rural Development.

7.3 Coupled Aquaponics: General System Design

The coupled aquaponics principle combines three classes of organisms: (1) aquatic organisms, (2) bacteria and (3) plants that benefit from each other in a closed recirculated water body. The water serves as a medium of nutrient transport, mainly from dissolved fish waste, which is converted into nutrients for plant growth by bacteria. These bacteria (e.g. *Nitrosomonas* spec., *Nitrobacter* spec.) oxidize ammonium to nitrite and finally to nitrate. Therefore, it is necessary for the bacteria to receive substantial amounts of ammonium and nitrite to stabilize colony growth and the quantity of nitrate production. Consequently, in a coupled aquaponic system, volumes are critically important, *i*) the aquaculture unit following the principles of recirculating aquaculture systems (RAS), *ii*) the bacterial growth substrate and *iii*) the space for the plant units and the amount of plants to be cultivated. Together, they form the aquaponics unit (Fig. 7.2).

The specific biological-chemical components of the process water have particular importance for coupled aquaponic systems. With food or uneaten feed particles, the organic fish waste and the bacteria inside the process water, an emulsion of nutrients combined with enzymes and digestive bacteria support the growth of fish and plants. There is evidence that compared to stand-alone systems such as aquaculture (fish) and hydroponics (plants), the growth of aquatic organisms and crops in a coupled aquaponics can be similar or even higher. Rakocy (1989) described a slightly higher

Fig. 7.2 Principle of coupled aquaponic system with fish, bacteria and plants in a fully closed water recirculation

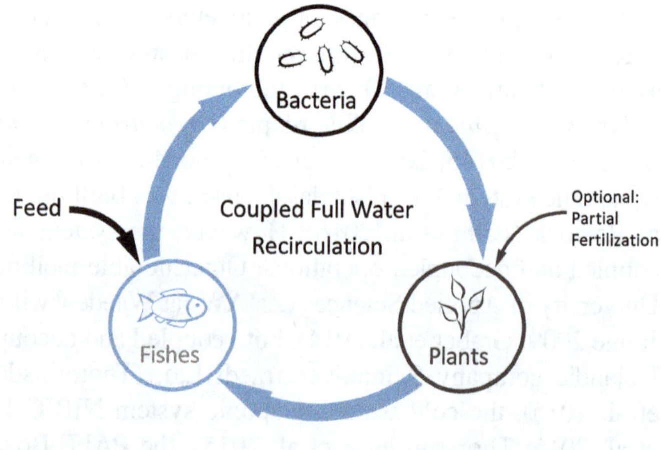

yield of tilapia (*Tilapia nilotica*, 46.8 kg) in coupled aquaponics in contrast to stand-alone fish culture (41.6 kg) and slight increases in Summer Bibb lettuce yield (385.1 kg) compared to vegetable hydroponic production (380.1 kg). Knaus et al. (2018b) recorded that aquaponics increased biomass growth of *O. basilicum*, apparently due to increased leaf generation of the plants (3550 leaves in aquaponics) compared to conventional hydroponics (2393 leaves). Delaide et al. (2016) demonstrated that aquaponic and hydroponic treatments of lettuce exhibited similar plant growth, whereas the shoot weight of the complemented aquaponic solution with nutrients performed best. Similar observations have been made by Goddek and Vermeulen (2018). Lehmonen and Sireeni (2017) observed an increased root weight, leaf area and leaf colour in Batavia salad (*Lactuca sativa* var. *capitata*) and iceberg lettuce (*L. sativa*) with aquaponics process water from *C. gariepinus* combined with additional fertilizer. Certain plants such as lettuce (*Lactuca sativa*), cucumbers (*Cucumis sativus*) or tomatoes (*Solanum lycopersicum*) can consume nutrients faster, and as a result flower earlier in aquaponics compared with hydroponics (Savidov 2005). Also, Saha et al. (2016) reported a higher plant biomass yield in *O. basilicum* in combination with crayfish *Procambarus* spp. and a low start-up fertilization of the aquaponic system.

The basic system design of coupled aquaponics consists of one or more fish tanks, a sedimentation unit or clarifier, substrates for the growth of bacteria or suitable biofilters and a hydroponic unit for plant growth (Fig. 7.3). These units are connected by pipes to form a closed water cycle. Often, after the mechanical filtration and the biofilter, a pump sump is used (one pump or one loop system) which, as the deepest point of the system, pumps the water back to the fish tanks from where it flows by gravity to the hydroponic unit.

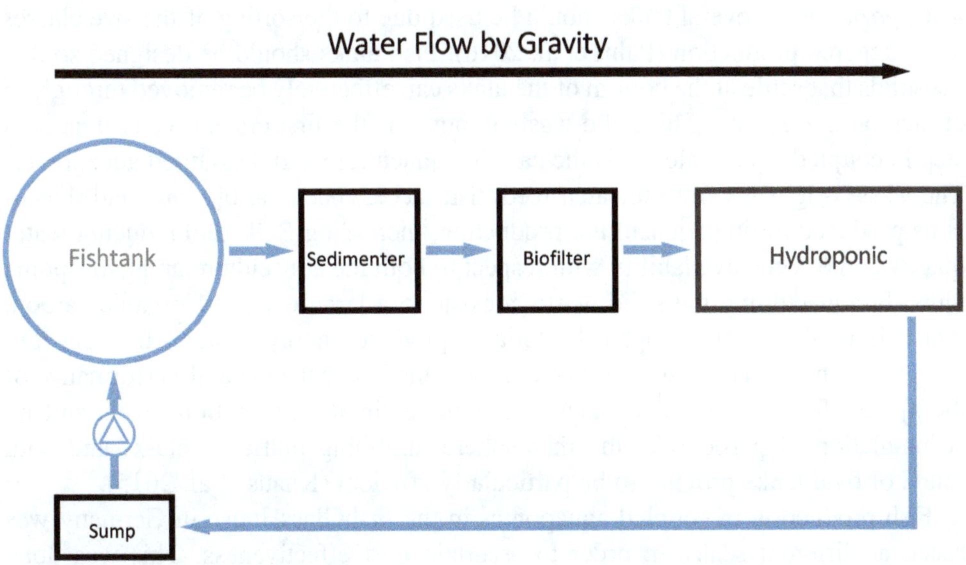

Fig. 7.3 Basic technical system design of a coupled aquaponic system with fish tank, sedimenter, biofilter, hydroponic unit and a sump where the water is pumped or airlifted back to the fish tanks and flows by gravity along the components

Coupled aquaponic systems are used in different scales. The closed-loop principle can be used in domestic systems (mini/hobby/backyard-coupled), demonstration units (e.g. living walls coupled), commercial aquaponics and aquaponics farming (with soil) ranging from small/semi-commercial to large-scale systems (Palm et al. 2018). A recent development in aquaponics has included partial fertilization, which is dependent on the tolerance of the fish species. This, however, can result in a short-term nutrient peak in the system but can be compensated through the nutrient retention by the plants. In coupled aquaponics, an optimal ratio of the production area (or fish volumes) of the aquaculture unit with the resulting feed demand as well as an adequate amount of plants to be cultured in the hydroponic unit (plant production area) must be achieved. (For discussions on the role of evapotranspiration and solar radiation within the systems, see Chaps. 8 and 11). For gravel aquaponics, Rakocy (2012) as a first attempt suggested 'component ratio principles', with a fish-rearing volume of 1 m^3 of fish tank volume to 2 m^3 hydroponic media of 3 to 6 cm pea gravel as a rule of thumb. Ultimately, the amount of fish determines the yield of crops in coupled aquaponics. Additionally, the technical conditions of the fish-rearing unit must be adapted according to the needs of the cultivated aquatic species.

7.4 Aquaculture Unit

The fish-rearing tanks (size, numbers and design) are selected depending on the scale of production and fish species in use. Rakocy et al. (2006) used four large fish-rearing tanks for the commercial production of *O. niloticus* in the UVI aquaponic system (USA). With the production of omnivorous or piscivorous fish species, such as *C. gariepinus*, several tanks should be used due to the sorting of the size classes and staggered production (Palm et al. 2016). Fish tanks should be designed so that the solids that settle at the bottom of the tanks can effectively be removed through an effluent at the bottom. This solid waste removal is the first crucial water treatment step in coupled aquaponics as is the case in aquaculture and decoupled aquaponics. The waste originates from uneaten feed, fish faeces, bacterial biomass and flocculants produced during aquaculture production, increasing BOD and reducing water quality and oxygen availability with respect to both the aquaculture and hydroponic units. In aquaculture, the solid waste consists to a large extent of organic carbon, which is used by heterotrophic bacteria to produce energy through oxygen consumption. The better the solid waste removal, the better the general performance of the system for both fish and plants, i.e. with optimal oxygenation levels and no accumulation of particles in the rhizosphere inhibiting nutrient uptake, and with round or oval tanks proving to be particularly efficient (Knaus et al. 2015).

Fish production in coupled aquaponics in the FishGlassHouse in Germany was tested at different scales in order to ascertain cost effectiveness. This was done effectively as extensive (max. 50 kg, 35 fish m^{-3}) or intensive (max. 200 kg, 140 fish m^{-3}) African catfish production. The semi-intensive production (max. 100 kg, 70 fish m^{-3}) cannot be recommended due to a negative cost benefit balance. In the semi-intensive production mode, system maintenance, labour and feed input were as much as under intensive production but with reduced fish and plant biomass output, and any economic gains in the aquaculture unit did not pay off (Palm et al. 2017). This resulted from the high biochemical oxygen demands (BOD), high denitrification because of the reduced oxygen availability, relatively high water exchange rates, predominantly anaerobic mineralization with distinct precipitation, low P and K-levels as well as a low pH-values with much less fish output compared with the intensive conditions. In contrast, the extensive fish production allowed higher oxygen availability with less water exchange rates and better nutrient availability for plant growth. Thus, under the above conditions, a RAS fish production unit for coupled aquaponics therefore either functions under extensive or intensive fish production conditions, and intermediate conditions should be avoided.

7.4.1 Filtration

Clarifiers, sometimes also called sedimenters or swirl separators (also see Chap. 3), are the most frequently used devices for the removal of solid waste in coupled aquaponics (Rakocy et al. 2006; Nelson and Pade 2007; Danaher et al. 2013, Fig. 7.4). Larger particulate matters must be removed from the system to avoid anoxic zones with denitrifying effects or the development of H_2S. Most clarifiers use lamella or plate inserts to assist in solids removal. Conical bottoms support sludge concentration at the bottom during operation and cleaning, whereas flat bottoms require large quantities of water to flush out and remove the sludge. During operation, the solids sink to the bottom of the clarifier to form sludge. Depending on the feed input and retention time, this sludge can build up to form relatively thick layers. The microbial activity inside the sludge layers gradually shifts towards anaerobic conditions, stimulating microbial denitrification. This process reduces plant available nitrate and should be avoided, especially if the process water is to be used for hydroponic plant production. Consequently, denitrification can be counterproductive in coupled aquaponics.

The density of the solid waste removed by the clarifier is rather low, compared with other technologies, maintenance is time-consuming, and cleaning the clarifier with freshwater is responsible for the main water loss of the entire system. The required amount of water is affected by its general design, the bottom shape and the accessibility of the PVC baffles to flushing water (Fig. 7.4a, b). Increasing fish stocking densities require higher quantities of water exchange (every day in the week under intensive conditions) to maintain optimal water quality for fish production, which can result in the loss of large amounts of process water, also losing substantial amounts of nutrients required for plant growth. Furthermore, replacement with freshwater introduces calcium and magnesium carbonates which may then precipitate with phosphates. Therefore, the use of such manually operated clarifiers makes predictions on process water composition with respect to optimal plant growth nearly impossible (Palm et al. 2019). It would be more effective to follow

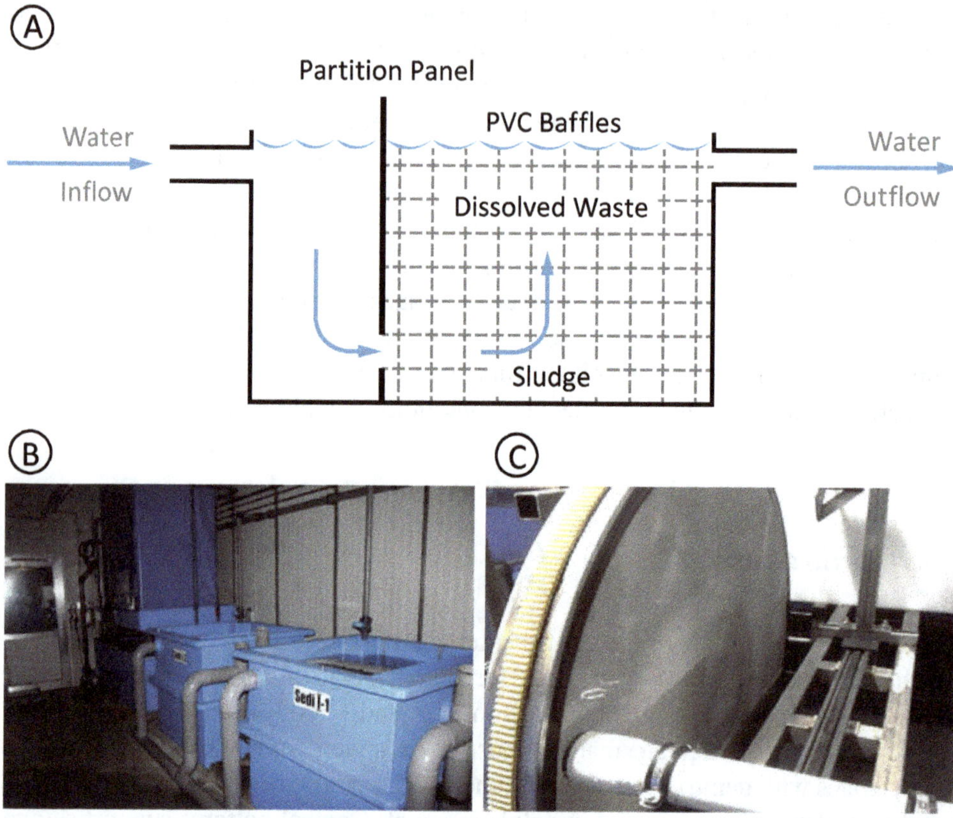

Fig. 7.4 Principle of aquaponic filtration with a sedimenter (**a–b**) and (**c**) disc-filter (PAL-Aquakultur GmbH, Abtshagen, Germany) of commercial African catfish (*Clarias gariepinus*) RAS in the FishGlassHouse (Rostock University, Germany)

Naegle's (1977) example of separating aerobic and anaerobic sludge and gaseous nitrogen discharge with a dual sludge system.

More effective solid waste removal can be achieved by automatic drum- or disk-filters which provide mechanical barriers that hold back solids, which are then removed through rinsing. New developments aim to reduce the use of rinse water through vacuum cleaning technologies, allowing the concentration of total solids in the sludge up to 18% (Dr. Günther Scheibe, PAL-Aquakultur GmbH, Germany, personal communication, Fig. 7.4c). Such effective waste removal has a positive influence on the sludge composition, improving effluent water control in order to better meet the horticultural requirements. Another option is the application of multiple clarifiers (sedimenters) or sludge-removal components in a row.

Biofilters are another essential part of RAS, as they convert ammonia nitrogen via microbial oxidation to nitrate (nitrification). Even though plant roots and the system itself provide surfaces for nitrifying bacteria, the capability to control the water quality is limited. Systems that do not have biofiltration are restricted to mini or hobby installations with low feed inputs. As soon as the biomass of fish and the feed input increases, additional biofilter capacity is required to maintain adequate water quality for fish culture and to provide sufficient nitrate quantities for plant growth.

For domestic and small-scale aquaponics, plant media (gravel or expanded clay for example) can suffice as effective biofilters. However, due to the high potential for clogging and thus the requirement for regular manual cleaning and maintenance, these methods are not suitable for larger-scale commercial aquaponics (Palm et al. 2018). Additionally, Knaus and Palm (2017a) demonstrated that the use of a simple biofilter in a bypass already increased the possible daily feed input in a backyard-coupled aquaponic system by approximately 25%. Modern biofilters that are used in intensive RAS are effective in providing sufficient nitrification capacity for fish and plant production. Because of increased investment costs, such components are more applicable in medium- and larger-scale commercial aquaponic systems.

7.4.1.1 Hydroponics in Coupled Aquaponics

In coupled aquaponics, a wide range of hydroponic subsystems can be used (also see Chap. 4) depending on the scale of operation (Palm et al. 2018). Unless labour has no significant impact on the yield (or profit) and the system is not too large, different hydroponic subsystems can be used at the same time. This is common in domestic and demonstration aquaponics that often use media bed substrate systems (sand, gravel, perlite, etc.) in ebb and flow troughs, DWC channels (deep water culture or raft systems) and even often self-made nutrient film channels (NFT). Most labour-intensive are media substrate beds (sand/gravel) in ebb and flow troughs, which can clog due to the deposition of detritus and often need to be washed (Rakocy et al. 2006). Due to the handling of the substrates, these systems are usually limited in size. On the other hand, DWC hydroponic subsystems require less labour and are less prone to maintenance, allowing them to be adopted for larger planting areas. For this reason, DWC subsystems are mainly found in domestic to small/semi-commercial systems, however, not usually in large-scale aquaponic systems. For larger commercial aquaponic production, the proportion of labour and maintenance in the DWC system is still seen to be too high. Even the use of water resources and energy for pumping are also unfavourable for large-scale systems.

If closed aquaponic systems are designed for profit-oriented production, the use of labour must decrease whilst the production area must increase. This is only possible by streamlining fish production combined with the application of easy-to-use hydroponic subsystems. The nutrient film technique (NFT) can, at present, be considered the most efficient hydroponic system, combining low labour with large plant cultivation areas and a good ratio of water, energy and investment costs. However, not all aquaponic plants grow well in NFT systems and thus it is necessary to find the right plant choice for each hydroponic subsystem, which in turn correlates with the nutrient supply of a specific fish species integrated in a specific hydroponic subsystem design. For coupled aquaponics, the sometimes higher particle load in the water can be problematic by clogging drips, pipes and valves in NFT installations. Hence, large aquaponic systems have to contain professional water management with effective mechanical filtration to avoid recirculation blockages. When the continuous supply of water is ensured through the pipes, the NFT system can be used in all types of coupled aquaponic systems, but is most recommended for production under small/semi-commercial systems and large-scale systems (Palm et al. 2018).

7.5 Scaling Coupled Aquaponic Systems

Typical coupled aquaponic system range from small to medium scale and larger sized systems (Palm et al. 2018). Upscaling remains one of the future challenges because it requires careful testing of the possible fish and plant combinations. Optimal unit sizes can be repeated to form multiunit systems, independent of the scale of production. According to Palm et al. (2018), the range of aquaponic systems were categorized into (1) mini, (2) hobby, (3) domestic and backyard, (4) small/semi-commercial and (5) large(r)-scale systems, as described below:

Mini installations (Fig. 7.5) usually consist of a small fish reservoir such as a fish tank or aquarium on which the plants grow on the surface or within a small hydroponic bed. Conventional aquarium filters, aeration and pumps are usually used. Mini systems are usually 2 m^2 or less in size (Palm et al. 2018). These small aquaponic systems can be used in the home with only few plants for home consumption and planted with plants such as tomatoes, herbs or ornamentals. Such systems add new values to human living space by adding 'nature' back into the family life area which is especially popular in big cities. Some mini systems consist of only a plant vase and one or more fish without filter and pump. However, these systems are only short-term to operate because a regulated filtration is missing.

Hobby aquaponic systems are categorized to reach a maximum size of 10 m^2 (Palm et al. 2018). With a higher fish stocking density, more feed and aeration, a mechanical sedimentation unit (sedimenter/clarifier) is necessary (Fig. 7.6). The sedimenter removes particulate matter –'sludge' such as faeces and uneaten feed

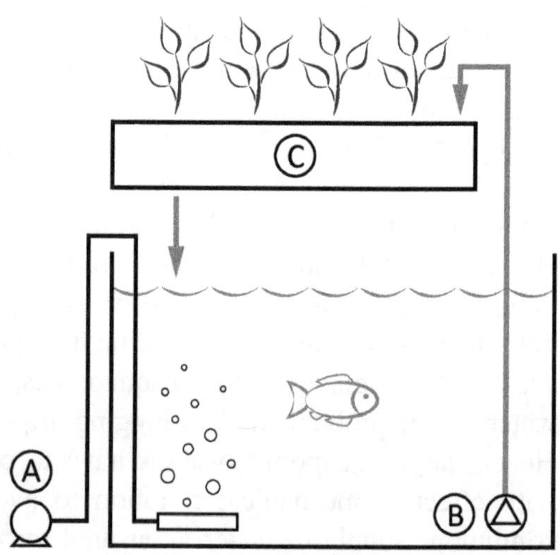

Fig. 7.5 Principle of a domestic coupled mini aquaponic system (< 2 m^2, after Palm et al. 2018) with aeration (**a**) and a pump (**b**), the hydroponics (**c**) act like a biofilter

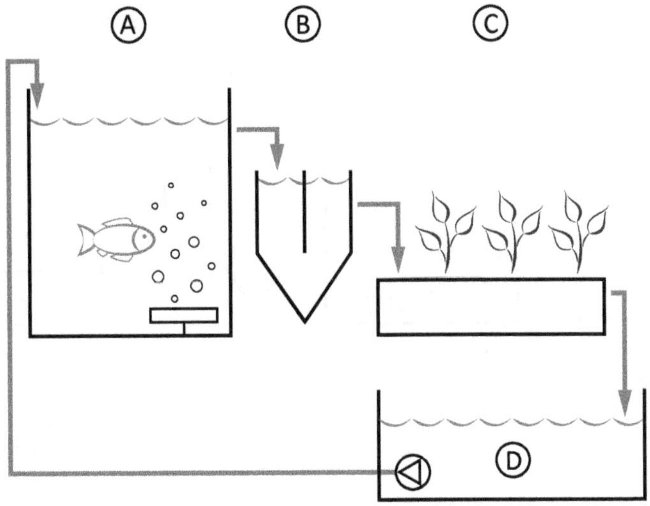

Fig. 7.6 Principle of a coupled domestic hobby aquaponic system (2–10 m^2 after Palm et al. 2018) with (**a**) fish tank and aeration, (**b**) sedimenter or clarifier altered after Nelson and Pade (2007), (**c**) hydroponics bed, e.g. gravel with different crops which acts like a biofilter and (**d**) a sump with the pump

from the system without using energy. The water flows by gravity from the fish tank to the sedimenter and then through the hydroponic tanks and then drops into a sump from where a pump or air lift pumps the water back to the fish tanks. In hobby installations, the plant beds act as a natural microbial filter and often media bed substrates such as sand, fine gravel or perlite are used. Hobby aquaponic systems are more the category of gimmicks that do not target food production. They rather enjoy the functionality of the integrated system. Hobby systems, as the name implies, are usually installed by hobbyists who are interested in growing a variety of aquatic organisms and plants for their own use and for 'fun'.

Domestic/backyard aquaponics has the purpose of external home use production of fish and plants characterized as having a maximum production area of 50 m^2 (Palm et al. 2018). These systems are built by enthusiasts. The construction is technically differentiated with a higher fish production, additional aeration and a higher feed input. The coupled aquaponics principle is applied with the use of one single pump which recirculates the water from a sump (lowest point) to higher standing fish tanks and then by gravity via sedimenter and a biofilter (with aeration and bacteria substrates) to the hydroponic units (Fig. 7.7).

For biofiltration, conventional bed filters can also be used as described in Palm et al. (2014a, b, 2015). In backyard aquaponics, hydroponics could consist alone or together of raft or DWC (deep water culture) troughs, substrate subsystems such as coarse gravel/sand ebb and flow boxes or nutrient film technique (NFT) channels. In the northern hemisphere, in outside installations, production is limited to the spring, summer and early autumn periods because of the weather conditions. With this scale of operation, fish and plants can be produced for private consumption (and production can be extended through small greenhouse production), but direct sales in small quantities are also possible.

Small and semi-commercial scale aquaponic systems are characterized by being up to 100 m^2 (Palm et al. 2018) with production focused on the retail market. More tanks, often with a higher stocking density, additional filters and water treatment

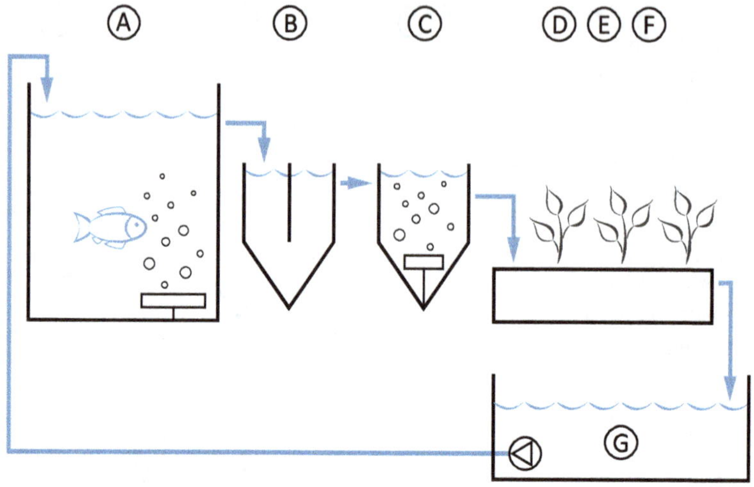

Fig. 7.7 Principle of a coupled domestic backyard aquaponic system, 10–50 m^2 (from Palm et al. 2018) with (**a**) fish tank and aeration, (**b**) sedimenter or clarifier altered after Nelson and Pade (2007), (**c**) biofilter with substrates and aeration, (**d**) hydroponic unit which could consists of combined raft or DWC channels, (**e**) gravel or sand media substrate system, (**f**) nutrient film technique NFT-channels and (**g**) a sump with one pump

systems and a larger hydroponic area with more diverse designs characterize these systems.

Large(r)-scale commercial operations above 100 m^2 (Palm et al. 2018) and reaching many thousands of square metres reach the highest complexity and require careful planning of the water flow and treatment systems (Fig. 7.8). General components are multiple fish tanks, designed as intensive recirculation aquaculture systems (RAS), a water transfer point or a sump allowing water exchange between the fish and plants, and commercial plant production units (aquaponics *s.s./s.l.*). As fish production is meant for intensive stocking densities, components such as additional filtration with the help of drum filters, oxygen supply, UV light treatments for microbial control, automatic controlled feeding and computerization including automatic water quality control classify these systems.

These systems have a multiunit design capable of upscaling under fully closed water recirculation which also allows for staggered production, parallel cultivation of different plants that require different hydroponic subsystems and better control of the different units in the case of disease outbreak and plant pest control.

7.6 Saline/Brackish Water Aquaponics

A relatively new field of research is the evaluation of different salinities of the process water for plant growth. Since freshwater worldwide is in continuously increasing demand and at high prices, some attention has been given to the use of saline/brackish water resources for agriculture, aquaculture and also aquaponics. The use of brackish water is significant as many countries such as Israel have

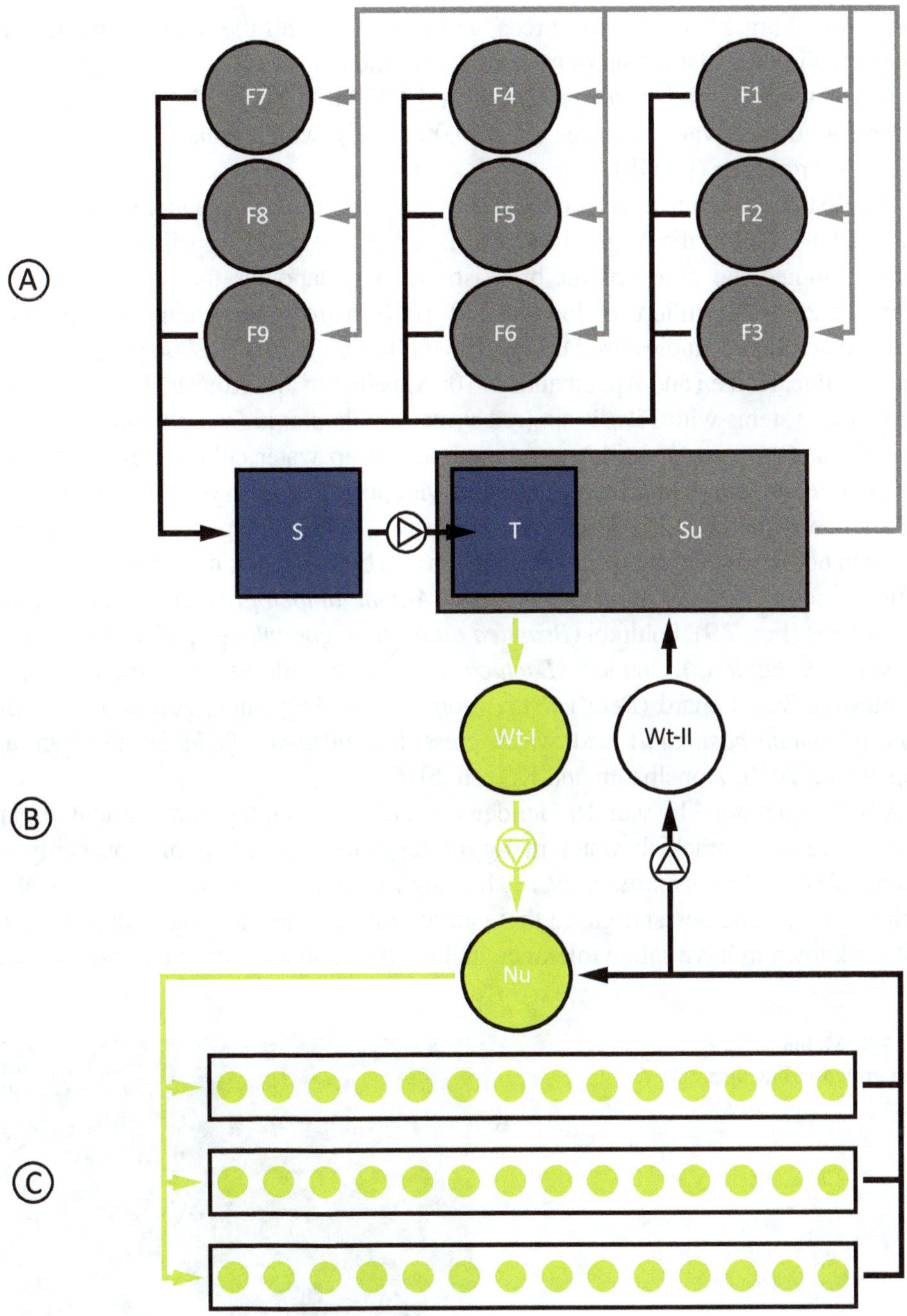

Fig. 7.8 Schema (supervision) of a large-scale aquaponics module adopted after the FishGlassHouse at University of Rostock (Germany) (1000 m^2 total production area, Palm et al. 2018) with (**a**) independent aquaculture unit, (**b**) the water transfer system and (**c**) the independent hydroponic unit; F1-F9 fish tanks, S) sedimenter, P-I pump one (biofilter pump), P-II pump two (aquaculture recirculation pump), T) trickling filter, Su) sump. In the middle, nutrient water transfer system with Wt-I) water transfer tank from the aquaculture unit, P-III) pump three, which pumps the nutrient rich water from aquaculture to C) hydroponics unit on the right with Nu) nutrient tank and an independent hydroponic recirculation system and planting tables (or NFT); P-IV) pump four, which pumps the nutrient low water from the hydroponic unit back to Wt-II) water transfer tank two and to the aquaculture unit for coupled (or decoupled if not used) aquaponic conditions

underground brackish water resources, and more than half the world's underground water is saline. Whilst the amount of underground saline water is only estimated as 0.93% of world's total water resources at 12,870,000 km^3, this is more than the underground freshwater reserves (10,530,000 km^3), which makes up 30.1% of all freshwater reserves (USGS).

The first published research on the use of brackish water in aquaponics was carried out in 2008–2009 in the Negev Desert of Israel (Kotzen and Appelbaum 2010). The authors studied the potential for brackish water aquaponics that could utilize the estimated 200–300 billion m^3 located 550–1000 metres underground in the region. This and additional studies used up to 4708–6800 μS/cm (4000–8000 μS/cm = moderately saline, Kotzen and Appelbaum 2010; Appelbaum and Kotzen 2016) in coupled aquaponic systems with *Tilapia* sp. (red strain of Nile tilapia *Oreochromis niloticus* x blue tilapia *O. aureus* hybrids), combined with deep water culture floating raft and gravel systems. The systems were mirrored with potable water systems as a control. A wide range of herbs and vegetables were grown, with very good and comparative results in both brackish and freshwater systems. In both systems fish health and growth were as good as plant growth of leeks (*Allium ampeloprasum*), celery (*Apium graveolens*) (Fig. 7.9), kohlrabi (*Brassica oleracea v. gongylodes*), cabbage (*Brassica oleracea v. capitata*), lettuce (*Lactuca sativa*), cauliflower (*Brassica oleracea v. botrytis*), Swiss chard (*Beta vulgaris vulgaris*), spring onion (*Allium fistulosum*), basil (*Ocimum basilicum*) and water cress (*Nasturtium officinale*) (Kotzen and Appelbaum 2010; Appelbaum and Kotzen 2016).

A 'mission report' by van der Heijden et al. (2014) on integrating agriculture and aquaculture with brackish water in Egypt suggests that red tilapia (probably red strains of *Oreochromis mossambicus*) has high potential combined with vegetables such as peas, tomatoes and garlic that can tolerate low to moderate salinity. Plants that are known to have saline tolerance include the cabbage family (Brassicas), such

Fig. 7.9 Mature celery plant grown in brackish water

as cabbage (*Brassica oleracea*), broccoli (*Brassica oleracea italica*), kale (*Brassica oleracea var. sabellica*), Beta family, such as *Beta vulgaris* (beetroot), perpetual spinach (*Beta vulgaris subsp. Vulgaris*), and bell peppers (*Capsicum annuum*) and tomatoes (*Solanum lycopersicum*). An obvious plant candidate for brackish water aquaponics is marsh samphire (*Salicornia europaea*) and potentially other 'strand vegetables' such as sea kale (*Crambe maritima*), sea aster (*Tripolium pannonicum*) and sea purslane (*Atriplex portulacoides*). Gunning (2016) noted that in the most arid regions of the word the cultivation of halophytes as an alternative to conventional crops is gaining significant popularity and *Salicornia europea* is becoming increasingly popular on the menus of restaurants and the counters of fishmongers and health-food stores across the country. This is similarly the case across the UK and the EU where most of the produce is exported from Israel and now also Egypt. A distinct advantage of growing marsh samphire is that it is a 'cut and come again' crop which means it can be harvested at intervals of around 1 month. In its natural environment along saline estuaries *Salicornia europaea* grows along a saline gradient from saline through brackish (Davy et al. 2001). In trials by Gunning (2016), plants were grown from seed, whereas Kotzen grew his trial plants from cut stems bought at the supermarket fish counter. Further studies under saline conditions were performed by Nozzi et al. (2016), who studied the effects of dinoflagellate (*Amyloodinum ocellatum*) infection in sea bass (*Dicentrarchus labrax*) at different salinity levels. Pantanella (2012) studied the growth of the halophyte *Salsola soda* (salt cabbage) in combination with the flathead grey mullet (*Mugil cephalus*) under marine conditions of increasing salt contents on an experimental farm at the University of Tuscia (Italy). Marine water resources have also been successful used in coupled aquaponics with the production of European sea bass (*Dicentrarchus labrax*) and salt-tolerant plants (halophytes) such as *Salicornia dolichostachya*, *Plantago coronopus* and *Tripolium pannonicum* in an inner land marine recirculating aquaculture system (Waller et al. 2015).

7.7 Fish and Plant Choices

7.7.1 Fish Production

In larger scale commercial aquaponics fish and plant production need to meet market demands. Fish production allows species variation, according to the respective system design and local markets. Fish choice also depends on their impact onto the system. Problematic coupled aquaponics fish production due to inadequate nutrient concentrations, negatively affecting fish health, can be avoided. If coupled aquaponic systems have balanced fish to plant ratios, toxic nutrients will be absorbed by the plants that are cleaning the water. Since acceptance of toxic substances is species dependent, fish species choice has a decisive influence on the economic success. Therefore, it is important to find the right combination and ratio between the fish and the plants, especially of those fish species with less water polluting activities and plants with high nutrient retention capacity.

The benefits of having a particular fish family in coupled aquaponic systems are not clearly understood with respect to their specific needs in terms of water quality and acceptable nutrient loads. Naegel (1977) found there was no notable negative impact on the fish and fish growth in his use of tilapia (*Tilapia mossambica*) and common carp (*Cyprinus carpio*). The channel catfish (*Ictalurus punctatus*) was also used by Lewis et al. (1978) and Sutton and Lewis (1982) in the USA. It was demonstrated that the quality of the aquaponics water readily met the demands of the different fish species, especially through the use of 'easy-to-produce' fish species such as the blue tilapia (*Oreochromis aureus*, formerly *Sarotherodon aurea*) in Watten and Busch (1984); Nile tilapia (*Oreochromis niloticus*), which was often used in studies with different plant species as a model fish species (Rakocy 1989; Rakocy et al. 2003, 2004; Al-Hafedh et al. 2008; Rakocy 2012; Villarroel et al. 2011; Simeonidou et al. 2012; Palm et al. 2014a, 2014b; Diem et al. 2017); and also tilapia hybrids-red strain (*Oreochromis niloticus x* blue tilapia *O. aureus* hybrids), that were investigated in arid desert environments (Kotzen and Appelbaum 2010; Appelbaum and Kotzen 2016).

There has been an expansion in the types of fish species used in aquaponics, at least in Europe, which is based on the use of indigenous fish species as well as those that have a higher consumer acceptance. This includes African catfish (*Clarias gariepinus*) which was grown successfully under coupled aquaponic conditions by Palm et al. (2014b), Knaus and Palm (2017a) and Baßmann et al. (2017) in northern Germany. The advantage of *C. gariepinus* is a higher acceptance of adverse water parameters such as ammonium and nitrate, as well as there is no need for additional oxygen supply due to their special air breathing physiology. Good growth rates of *C. gariepinus* under coupled aquaponic conditions were further described in Italy by Pantanella (2012) and in Malaysia by Endut et al. (2009). An expansion of African catfish production under coupled aquaponics can be expected, due to unproblematic production and management, high product quality and increasing market demand in many parts of the world.

In Europe, other fish species with high market potential and economic value have recently become the focus in aquaponic production, with particular emphasis on piscivorous species such as the European pikeperch 'zander' (*Sander lucioperca*). Pikeperch production, a fish species that is relatively sensitive to water parameters, was tested in Romania in coupled aquaponics. Blidariu et al. (2013a, b) showed significantly higher P_2O_5 (phosphorous pentoxide) and nitrate levels in lettuce (*Lactuca sativa*) using pikeperch compared to the conventional production, suggesting that the production of pikeperch in coupled aquaponics is possible without negative effects on fish growth by nutrient toxicity. The Cyprinidae (Cypriniformes) such as carp have been commonly used in coupled aquaponics and have generally shown better growth with reduced stocking densities and minimal aquaponic process water flow rates (efficient water use) during experiments in India. The optimal stocking density of koi carp (*Cyprinus carpio* var. koi) was at 1.4 kg/m (Hussain et al. 2014), and the best weight gain and yield of *Beta vulgaris* var. *bengalensis* (spinach) was found with a water flow rate of 1.5 L/min (Hussain et al. 2015). Good fish growth and plant yield of water spinach (*Ipomoea aquatica*) with a maximum percentage of nutrient removal (NO_3-N, PO_4-P, and K) was reported at a minimum water flow rate of 0.8 L/min with polycultured koi carp (*Cyprinus carpio* var. koi) and gold fish (*Carassius auratus*) by Nuwansi et al. (2016). It is interesting to note that plant growth and nutrient removal in koi (*Cyprinus carpio* var. koi) and gold fish (*Carassius auratus*) production (Hussain et al. 2014, 2015) with *Beta vulgaris* var. *bengalensis* (spinach) and water spinach (*Ipomoea aquatica*) increased linearly with a decrease in process water flow between 0.8 L/min and 1.5 L/min. These results suggest that for cyprinid fish culture, lower water flow is recommended as this has no negative impacts on fish growth. In contrast, however, Shete et al. (2016) described a higher flow rate of 500 L h^{-1} (approx. 8 L/min) for common carp and mint (*Mentha arvensis*) production, indicating the need for different water flow rates for different plant species. Another cyprinid, the tench (*Tinca tinca*), was successfully tested by Lobillo et al. (2014) in Spain and showed high fish survival rates (99.32%) at low stocking densities of 0.68 kg m^{-3} without solids removal devices and good lettuce survival rates (98%). Overall, members of the Cyprinidae family highly contribute to the worldwide aquaculture production (FAO 2017); most likely this would also be true under aquaponic conditions and productivity, but the economic situation should be tested for each country separately.

Other aquatic organisms such as shrimp and crayfish have been introduced into coupled aquaponic production. Mariscal-Lagarda et al. (2012) investigated the influence of white shrimp process water (*Litopenaeus vannamei*) on the growth of tomatoes (*Lycopersicon esculentum*) and found good yields in aquaponics with a twofold water sparing effect under integrated production. Another study compared the combined semi-intensive aquaponic production of freshwater prawns (*Macrobrachium rosenbergii* – the Malaysian shrimp) with basil (*Ocimum basilicum*) versus traditional hydroponic plant cultivation with a nutrient solution (Ronzón-Ortega et al. 2012). However, basil production in aquaponics was initially less effective (25% survival), but with increasing biomass of the prawns, the plant biomass also increased so that the authors came to a positive conclusion with the production of basil with *M. rosenbergii.* Sace and Fitzsimmons (2013) reported a better plant growth in lettuce (*Lactuca sativa*), Chinese cabbage (*Brassica rapa pekinensis*) and pakchoi (*Brassica rapa*) with *M. rosenbergii* in polyculture with the Nile tilapia (*O. niloticus*). The cultivation with prawns stabilized the system in terms of the chemical-physical parameters, which in turn improved plant growth, although due to an increased pH, nutrient deficiencies occurred in the Chinese cabbage and lettuce. In general, these studies demonstrate that shrimp production under aquaponic conditions is possible and can even exert a stabilizing effect on the closed loop – or coupled aquaponic principle.

7.7.2 *Plant Production*

The cultivation of many species of plants, herbs, fruiting crops and leafy vegetables have been described in coupled aquaponics. In many cases, the nutrient content of the aquaponics process water was sufficient for good plant growth. A review by

Thorarinsdottir et al. (2015) summarized information on plant production under aquaponic production conditions from various sources. Lettuce (*Lactuca sativa*) was the main cultivated plant in aquaponics and was often used in different variations such as crisphead lettuce (iceberg), butterhead lettuce (bibb in the USA), romaine lettuce and loose leaf lettuce under lower night (3–12 °C) and higher day temperatures (17–28 °C) (Somerville et al. 2014). Many experiments were carried out with lettuce in aquaponics (e.g. Rakocy 1989) or as a comparison of lettuce growth between aquaponics, hydroponics and complemented aquaponics (Delaide et al. 2016). Romaine lettuce (*Lactuca sativa longifolia* cv. Jericho) was also investigated by Seawright et al. (1998) with good growth results similar to standalone hydroponics and an increasing accumulation of K, Mg, Mn, P, Na and Zn with increasing fish biomass of Nile tilapia (*Oreochromis niloticus*). Fe and Cu concentrations were not affected. Lettuce yield was insignificant with different stocking densities of fish (151 g, 377 g, 902 g, 1804 g) and plant biomass between 3040 g (151 g fish) and 3780 g (902 g fish). Lettuce was also cultivated, e.g. by Lennard and Leonard (2006) with Murray Cod (*Maccullochella peelii peelii*), and by Lorena et al. (2008) with the sturgeon 'bester' (hybrid of *Huso huso* female and *Acipenser ruthenus* male) and by Pantanella (2012) with Nile tilapia (*O. niloticus*). As a warm water crop, basil (*Ocimum basilicum*) was reported as a good herb for cultivation under coupled aquaponics and was reported as the most planted crop by 81% of respondents in findings of an international survey (Love et al. 2015). Rakocy et al. (2003) investigated basil with comparable yields under batch and staggered production (2.0; 1.8 kg/m^2) in contrast to field cultivation with a comparatively low yield (0.6 kg/m^2). Somerville et al. (2014) described basil as one of the most popular herbs for aquaponics, especially in large-scale systems due to its relatively fast growth and good economic value. Different cultivars of basil can be grown under higher temperatures between 20 and 25 °C in media beds, NFT (nutrient film technique) and DWC (deep water culture) hydroponic systems. Basil grown in gravel media beds can reach 2.5-fold higher yield combined with tilapia juveniles (*O. niloticus*, 0.30 g) in contrast to *C. gariepinus* (0.12 g) (Knaus and Palm 2017a).

Tomatoes (*Lycopersicon esculentum*) were described by Somerville et al. (2014) as an 'excellent summer fruiting vegetable' in aquaponics and can cope with full sun exposure and temperatures below 40 °C depending on tomato type. However, economic sustainability in coupled aquaponics is disputed due to the reduced competitiveness of aquaponics tomato production compared to high-engineered conventional hydroponic production in greenhouses in, e.g. the Netherlands Improvement Centre of DLV GreenQ in Bleiswijk with tomato yield of 100.6 kg m^{-2} (Hortidaily 2015), or even higher (Heuvelink 2018). Earlier investigations focused on the cultivation of this plant mostly compared to field production. Lewis et al. (1978) reported nearly double the crop of tomatoes under aquaponics compared to field production and the iron deficiency which occurred was fixed by using ethylene diamine tetra-acetic acid. Tomatoes were also produced in different aquaponic systems over the last decades, by Sutton and Lewis (1982) with good plant yields at water temperatures up to 28 °C combined with Channel catfish (*Ictalurus punctatus*), by Watten and Busch (1984) combined with tilapia

(*Sarotherodon aurea*) and a calculated total marketable tomato fruit yield of 9.6 kg/m^{-2}, approximately 20% of recorded yields for decoupled aquaponics (47 kg/m^2/y, Geelen 2016). McMurtry et al. (1993) combined hybrid tilapia (*Oreochromis mossambicus* x *Oreochromis niloticus*) with tomatoes in associated sand biofilters which showed optimal 'plant yield/high total plant yield' of 1:1.5 tank/biofilter ratio (sand filter bed) and McMurtry et al. (1997) with increasing total plant fruit yield with increasing biofilter/tank ratio. It must be stated that the production of tomatoes is possible under coupled aquaponics. Following the principle of soilless plant cultivation in aquaponics sensu stricto after Palm et al. (2018), it is advantageous to partially fertilize certain nutrients such as phosphorous, potassium or magnesium to increase yields (see challenges below).

The cultivation of further plant species is also possible and testing of new crops is continuously being reported. In the UK, Kotzen and Khandaker have tested exotic Asian vegetables, with particular success with bitter gourd, otherwise known as kerala or bitter melon (*Momordica charantia*) (Kotzen pers. comm.). Taro (*Colocasia esculenta*) is another species which is readily grown with reported success both for its large 'elephant ear' like leaves as well as its roots (Kotzen pers. comm.). Somerville et al. (2014) noted that crops such as cauliflower, eggplant, peppers, beans, peas, cabbage, broccoli, Swiss chard and parsley have the potential for cultivation under aquaponics. But there are many more (e.g. celery, broccoli, kohlrabi, chillies, etc.) including plants that prefer to have wet root conditions, including water spinach (*Ipomoea aquatica*) and mint (*Menta* sp.) as well as some halophytic plants, such as marsh samphire (*Salicornia europaea*).

Ornamental plants can also be cultivated, alone or together with other crops (intercropping), e.g. *Hedera helix* (common ivy) grown at the University of Rostock by Palm & Knaus in a coupled aquaponic system. The trials used 50% less nutrients that would be normally supplied to the plants under normal nursery conditions with a 94.3% success rate (Fig. 7.10).

Fig. 7.10 Three quality categories of ivy (*Hedera helix*), grown in a coupled aquaponic system indicating the quality that the nursery trade requires (**a**) very good and directly marketable, (**b**) good and marketable and (**c**) not of high enough quality

Besides the chosen plant and variant, there are two major obstacles that concern aquaponics plant production under the two suggested states of fish production, extensive and intensive. Under extensive conditions, nutrient availability inside the process water is much lower than under commercial plant production, nutrients such as K, P and Fe are deficient, and the conductivity is between 1000 and 1500 μS / cm, which is much less than applied under regular hydroponic production of commercial plants regularly between 3000 and 4000 μS / cm. Plants that are deficient in some nutrients can show signs of leaf necroses and have less chlorophyll compared with optimally fertilized plants. Consequently, selective addition of some nutrients increases plant quality that is required to produce competitive products.

In conclusion, commercial plant production of coupled aquaponics under intensive fish production has the difficulty to compete with regular plant production and commercial hydroponics at a large scale. The non-optimal and according to Palm et al. (2019) unpredictable composition of nutrients caused by the fish production process must compete against optimal nutrient conditions found in hydroponic systems. There is no doubt that solutions need to be developed allowing optimal plant growth whilst at the same time providing the water quality required for the fish.

7.7.3 Fish and Plant Combination Options

Combining fish and plants in closed aquaponics can generate better plant growth (Knaus et al. 2018b) combined with benefits for fish welfare (Baßmann et al. 2017). Inside the process water, large variations in micronutrients and macronutrients may occur with negative effects on plant nutritional needs (Palm et al. 2019). A general analysis of coupled aquaponic systems has shown that there are low nutrient levels within the systems (Bittsanszky et al. 2016) in comparison with hydroponic nutrient solutions (Edaroyati et al. 2017). Plants do not tolerate an under or oversupply of nutrients without effects on growth and quality, and the daily feed input of the aquaponic system needs to be adjusted to the plant's nutrient needs. This can be achieved by regulating the stocking density of the fish as well as altering the fish feed. Somerville et al. (2014) categorized plants in aquaponics according to their nutrient requirements as follows:

1. Plants with low nutrient requirements (e.g. basil, *Ocimum basilicum*)
2. Plants with medium nutritional requirements (e.g. cauliflower, *Brassica oleracea var. Botrytis*)
3. Plants with high nutrient requirements such as fruiting species (e.g. strawberries, *Fragaria* spec.).

Not all plants can be cultured in all hydroponic subsystems with the same yield. The plant choice depends on the hydroponic subsystem if conventional soilless aquaponic systems (e.g. DWC, NFT, ebb and flow; aquaponics sensu stricto' – *s.s.* – in the narrow sense) are used. Under aquaponics farming ('aquaponics sensu lato' – *s.l.* – in a broader sense, Palm et al. 2018), the use of inert soil or with addition of fertilizer applies gardening techniques from horticulture, increasing the possible range of species.

Under hydroponic conditions, the component structures of the subsystems have a decisive influence on plant growth parameters. According to Love et al. (2015), most aquaponic producers used raft and media bed systems and to a smaller amount NFT and vertical towers. Lennard and Leonard (2006) studied the growth of Green oak lettuce (*Lactuca sativa*) and recorded the relationship Gravel bed > Floating raft > NFT in terms of biomass development and yield in combination with the Murray Cod (*Maccullochella peelii peelii*) in Australia. Knaus & Palm (2016–2017, unpublished data) have tested different hydroponic subsystems such as NFT, floating raft and gravel substrate on the growth of different plants in the FishGlassHouse in a decoupled aquaponic experimental design, requiring subsequent testing under coupled conditions. With increasing production density of African catfish (*C. gariepinus*, approx. 20–168 kg/m^3), most of the cultured crops such as cucumbers (*Cucumis sativus*), basil (*Ocimum basilicum*) and pak choi (*Brassica rapa chinensis*) tended to grow better, in contrast to Lennard and Leonard (2006), in gravel and NFT aquaponics (GRAVEL > NFT > RAFT; Wermter 2016; Pribbernow 2016; Lorenzen 2017), and Moroccan mint 'spearmint' (*Mentha spicata*) showed the opposite growth performance (RAFT = NFT > GRAVEL) with highest leaf numbers in NFT (Zimmermann 2017). This demonstrates an advantage of gravel conditions and can be used figuratively also in conventional plant pots with soil substrate under coupled aquaponic conditions. This type of aquaponics was designated as 'horticulture – aquaponics (*s.l.*)' due to the use of substrates from the horticultural sector (soil, coco fibre, peat, etc.) (see Palm et al. 2018). This involves all plant cultivation techniques that allow plants to grow in pots, whereby the substrate in the pot itself may be considered equivalent to a classical gravel substrate for aquaponics. Research by Knaus & Palm (unpublished data) showed variance in the quality of commonly grown vegetables and thus their suitability for growing in this type of aquaponics with soil (Fig. 7.11, Table 7.1). In this type of aquaponics, beans, lambs lettuce and radish did well.

The plant choice (species and strain) and especially the hydroponic subsystem and/or substrate, including peat, peat substitutes, coco fibre, composts, clay, etc. or a mix of them (see Somerville et al. 2014), has a significant impact on the economic success of the venture. The efficiency of some substrates must be tested in media bed hydroponic sub-units (e.g. the use of sand (McMurtry et al. 1990, 1997), gravel (Lennard and Leonard 2004) and perlite (Tyson et al. 2008). The use of other media bed substrates such as volcanic gravels or rock (tuff/tufa), limestone gravel, river bed gravel, pumice stone, recycled plastics, organic substrates such as coconut fibre, sawdust, peat moss and rice trunk have been described by Somerville et al. (2014). Qualitative comparative studies with recommendations, however, are very rare and subject of future research.

7.7.4 Polyponics

The combination of different aquatic organisms in a single aquaponic system can increase total yields. First applied by Naegel (1977), this multispecies production principle was coined from the term polyculture combined with aquaponics in

Fig. 7.11 Experiments with a variety of commonly grown vegetables, under winter conditions in winter 2016/2017 in the FishGlassHouse (University of Rostock, Germany)

Table 7.1 Recommendation for the use of gardening plants in aquaponic farming with the use of 50% of the regular fertilizer in pots with soil

Name	Lat. Name	Possible for aquaponics	Mark	Nutrient regime
Beans	*Phaseolus vulgaris*	Yes	1	Extensive
Peas	*Pisum sativum*	No	2	Intensive
Beet	*Beta vulgaris*	No	2	Both
Tomatoes	*Solanum lycopersicum*	No	2.3	Both
Lamb's lettuce	*Valerianella locusta*	Yes	1	Both
Radish	*Raphanus sativus*	Yes	1	Both
Wheat	*Triticum aestivum*	No	2	Both
Lettuce	*Lactuca sativa*	Yes	1	Intensive

coupled systems as 'polyponic' (polyculture + aquaponics) by Knaus and Palm (2017b). Like IMTA (integrated multitrophic aquaculture), polyponics expands the diversity of the production systems. Using multiple species in one system has both advantages and disadvantages as (a) diversification allows the producer to respond to local market demands but (b) on the other hand, focus is spread across a number of products, which requires greater skill and better management. Published information on polyponics is scarce. However, Sace and Fitzsimmons (2013) reported better plant growth of lettuce, Chinese cabbage and pakchoi in polyculture with freshwater shrimp (*Macrobrachium rosenbergii*) and Nile tilapia (*O. niloticus*) in coupled aquaponics. Alberts-Hubatsch et al. (2017) described the cultivation of noble crayfish (*Astacus astacus*), hybrid striped bass (*Morone saxatilis* x *M. chrysops*), microalgae (*Nannochloropsis limnetica*) and watercress (*Nasturtium officinale*), where crayfish growth was higher than expected, feeding on watercress roots, fish faeces and a pikeperch-designed diet.

Initial investigations at the University of Rostock showed differences in plant growth in two identical 25m^2 backyard-coupled aquaponic units with the production of African catfish (*Clarias gariepinus*) and Nile tilapia (*Oreochromis niloticus*, Palm et al. 2014b). The plant yields of lettuce (*Lactuca sativa*) and cucumber fruits (*Cucumis sativus*) were significantly better in combination with *O. niloticus*. This effect was also seen by Knaus and Palm (2017a) with a 2.5-fold higher yield in basil (*Ocimum basilicum*) and two times more biomass of parsley (*Petroselinum crispum*) combined with *O. niloticus*. Another comparison between *O. niloticus* and common carp (*Cyprinus carpio*) showed a twofold higher gross biomass per plant (g plant^{-1}) of tomatoes (*Solanum lycopersicum*) with tilapia and a slightly increased gross biomass of cucumbers (*Cucumis sativus*) with carp, however, with higher cucumber fruit weight in the *O. niloticus* aquaponic unit (Knaus and Palm 2017b). The yield of mint (*Mentha* x *piperita*) was approximately 1.8 times higher in the tilapia unit, but parsley was 2.4 times higher combined with the carp (Knaus et al. 2018a). The results of these experiments followed the order of plant growth: *O. niloticus* > *C. carpio* > *C. gariepinus*, whilst fish growth showed a reverse order with: *C. gariepinus* > *O. niloticus* > *C. carpio*.

According to these results, the fish choice influences the plant yield and a combination of different fish species and their respective growth performance allows adjustment of a coupled aquaponics to optimal fish and plant yields. During consecutive experiments (*O. niloticus* only, *C. gariepinus* only), a higher basil (*O. basilicum*) biomass yield of 20.44% (Plant Growth Difference – PGD) was observed for *O. niloticus* in contrast to the basil yield with *C. gariepinus* (Knaus et al. 2018b). Thus, *O. niloticus* can be used to increase the plant yield in a general *C. gariepinus* system. This so-called boost effect by tilapia enhances the overall system production output and compensates *i*) poorer plant growth with high fish growth of *C. gariepinus* as well as *ii*) poorer fish growth in *O. niloticus* with a boost to the plant yield. A first commercial polyponic farm has opened in Bali, Indonesia, producing tilapia combined with Asian catfish (*Clarias batrachus*) and conventional farm products.

7.8 System Planning and Management Issues

Coupled aquaponics depends on the nutrients that are provided from the fish units, either a commercial intensive RAS or tanks stocked under extensive conditions in smaller operations. The fish density in the latter is often about 15–20 kg/m^3 (tilapia, carp), but extensive African catfish production can be higher up to 50 kg/m^3. Such different stocking densities have a significant influence on nutrient fluxes and nutrient availability for the plants, the requirement of water quality control and adjustment as well as appropriate management practices.

The process water quality with respect to nutrient concentrations is primarily dependent on the composition of the feed and the respective turnover rates of the fish. The difference between feed input and feed nutrients, assimilating inside the fish or lost through maintenance of the system, equals the maximum potential of

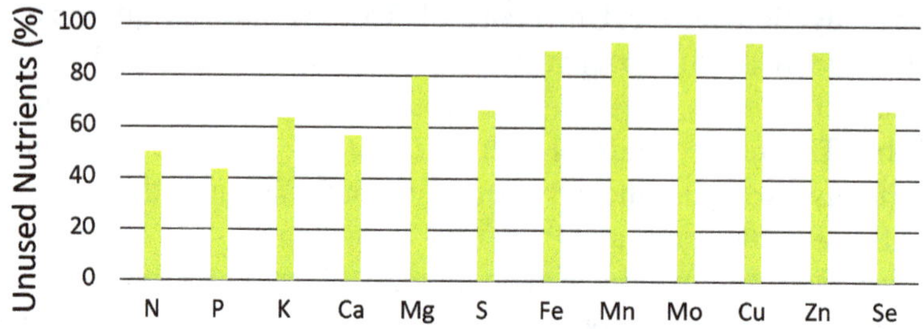

Fig. 7.12 Unused nutrients in African catfish aquaculture that are potentially available for aquaponic plant production (original data)

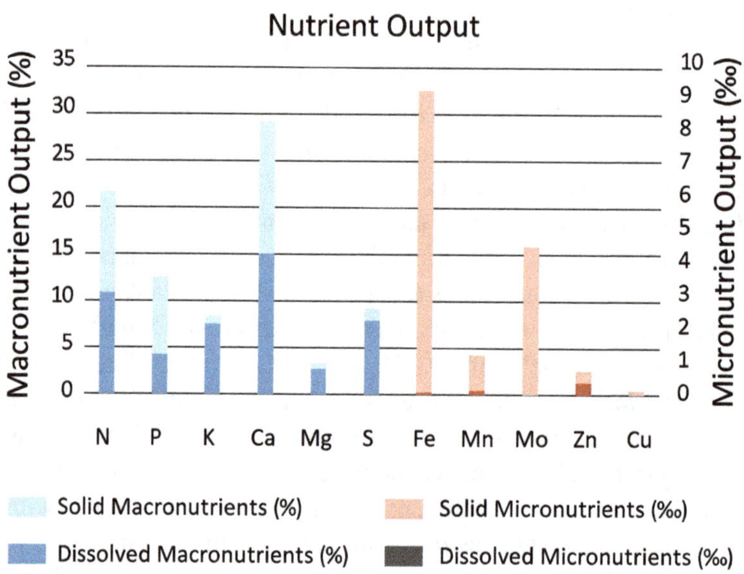

Fig. 7.13 Distribution of macro- and micronutrients inside the process water and the solids. (Data from Strauch et al. (2018))

plant available nutrients from aquaculture. As noted above, the nutrient concentrations should be adjusted to levels, which allow the plants to grow effectively. However, not all fish species are able to withstand such conditions. Consequently, resilient fish species such as the African catfish, tilapia or carp are preferred aquaponic candidates. At the University of Rostock, whole catfish and its standard diet as output and input values were analysed to identify the turnover rates of the macronutrients N, P, K, Ca, Mg and S and the micronutrients Fe, Mn, Mo, Cu, Zn and Se. With the exception of P, more than 50% of the feed nutrients given to the fish are not retained in its body and can be considered potentially available as plant nutrients (Strauch et al. 2018; Fig. 7.12). However, these nutrients are not equally distributed inside the process water and the sediments. Especially macronutrients (N, P, K) accumulate in the process water as well as inside the solid fraction whilst the micronutrients, such as iron, disappear in the solid fraction separated by the clarifier. Figure 7.13 shows the nutrient output per clarifier cleaning after 6 days of

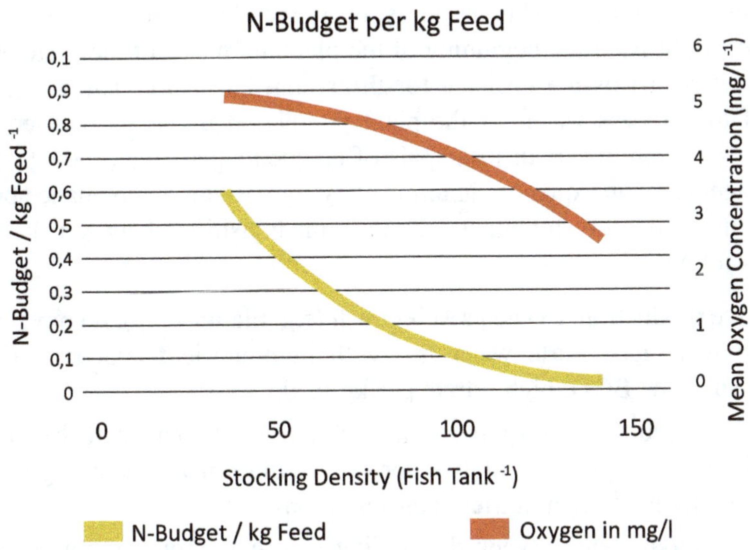

Fig. 7.14 N-budget per kg feed and oxygen level in African catfish aquaculture under three different stocking densities (original data)

sludge collection in an intensive African catfish RAS. The proportions of plant essential nutrients that are bound in the solids relative to the respective amounts that appear dissolved are significant: N = 48%, P = 61%, K = 10%, Ca = 48%, Mg = 16%, S = 11%, Fe = 99%, Mn = 86%, Mo = 100%, Zn = 48% and Cu = 55%.

One key management factor is the availability of oxygen inside the system, which is crucial to keep the concentration of plant available nitrate in the process water high. Conventional clarifiers that are applied in many RAS remove carbon-rich solid wastes from the recirculation but will leave them in contact with the process water until the next cleaning interval of the sedimentation tank. During this time, the carbon-rich organic matter is utilized as a source of energy by denitrifying bacteria, accounting for significant losses of nitrate. It outgasses as nitrogen into the atmosphere and is lost. Under intensive production conditions, large quantities of organic sludge will accumulate inside the sedimentation tanks, with consequences for maintenance, replacement with freshwater and subsequently for the nutrient composition inside the process water. Figure 7.14 illustrates the nutrient concentrations in the holding tanks of African catfish RAS under three different stocking densities (extensive: 35 fish / tank, semi-intensive: 70 fish / tank, intensive: 140 fish/ tank). The higher the stocking density and the lower the resulting oxygen content inside the system, the lower is the plant available nitrate per kg feed inside the system.

In general, with increasing fish intensity, the availability of oxygen inside the system decreases because of the consumption of the oxygen by the fish and aerobic sludge digestion inside the clarifier and the hydroponic subsystems. Oxygen levels can be maintained at higher levels, but this requires additional investment for oxygen monitoring and control. This issue is of tremendous importance for coupled aquaponics, right from the beginning of the planning phase of the systems because the different scenarios are decisive for the planned fish production, the resulting quality of the process water for the plant production units, and consequently for economic sustainability. Four principals of coupled aquaponic production systems with management consequences in terms of system design, maintenance procedures and nutrient availability for plant growth, with transitions between them, can be defined as follows:

- Extensive production, oxygen resilient fish (e.g. tilapia, carp), no oxygen control, O_2 above 6 mg/L, little water use with high nutrient concentrations, small investment, low BOD, high nitrate per kg feed.
- Intensive production, oxygen resilient fish (e.g. African catfish), no oxygen control, O_2 below 6 mg/L, high water use, medium investment, high BOD, low nitrate per kg feed, high nutrient concentrations.
- Extensive production, oxygen demanding fish (e.g. Trout), oxygen control, O_2 above 6–8 mg/L, high water use, medium investment, low BOD, high nitrate per kg feed, low nutrient concentrations.
- Intensive production, oxygen demanding fish (e.g. Trout, pikeperch), oxygen control, O_2 above 6–8 mg/L, high water use, high investment, low BOD, medium nitrate per kg feed.

In addition to the stocking density and the average amount of oxygen inside the system, the plant production regime, i.e. batch or staggered cultivation, has consequences for the plant available nutrients inside the process water (Palm et al. 2019). This is the case especially with fast growing fish, where the feed increase during the production cycle can be so rapid that there needs to be a higher water exchange rate and thus nutrient dilution can increase, with consequences for the nutrient composition and management.

The same oxic or anoxic processes that occur in the RAS as a part of the coupled aquaponic system also occur inside the hydroponic subsystems. Therefore, oxygen availability and possibly aeration of the plant water can be crucial in order to optimize the water quality for good plant growth. The oxygen allows the heterotrophic bacteria to convert organic bound nutrients to the dissolved phase (i.e. protein nitrogen into ammonia) and the nitrifying bacteria to convert the ammonia into nitrate. The availability of oxygen in the water also reduces anoxic microbial metabolism (i.e. nitrate- and/or sulphate-reducing bacteria, Comeau 2008), processes which can have tremendous effects on the reduction of nutrient concentrations. The aeration of the roots also has the advantage that water and nutrients are transported to the root surface, and that particles that settle on the root surface are removed (Somerville et al. 2014).

7.9 Some Advantages and Disadvantages of Coupled Aquaponics

The following discussion reveals a number of key pros and challenges of coupled aquaponics as follows:

Pro: Coupled aquaponic systems have many food production benefits, especially saving resources under different production scales and over a wide range of geographical regions. The main purpose of this production principle is the most efficient and sustainable use of scarce resources such as feed, water, phosphorous as a limited plant nutrient and energy. Whilst, aquaculture and hydroponics (as stand-alone), in comparison to aquaponics are more competitive, coupled aquaponics may have the edge in terms of sustainability and thus a justification of these systems especially when seen in the context of, for example, climate change, diminishing resources, scenarios that might change our vision of sustainable agriculture in future.

Pro: Small-scale and backyard-coupled aquaponics are meant to support local and community-based food production by households and farmers. They are not able to stem high investment costs and require simple and efficient technologies. This applies for tested fish and plant combinations in coupled aquaponics.

Pro: The plants in contemporary coupled aquaponics have the similar role in treating waste as constructed wetlands do in the removal of waste from water (Fig. 7.15). The plants in the hydroponic unit in coupled aquaponics therefore fulfil the task of purifying the water and can be considered a 'biological advanced unit of water purification' in order to reduce the environmental impact of aquaculture.

Challenge: It has been widely accepted that using only fish feed as the input for plant nutrition is often qualitatively and quantitatively insufficient in comparison to conventional agriculture production systems (e.g. N-P-K hydroponics manure) (Goddek et al. 2016), limiting the growth of certain crops in coupled aquaponics.

Pro: Coupled aquaponic systems have a positive influence on fish welfare. Most recent studies demonstrate that in combination with cucumber and basil, the agonistic behaviour of African catfish (*C. gariepinus*) was reduced (Baßmann et al. 2017, 2018). More importantly, comparing injuries and behavioural patterns with the control, aquaponics with high basil density influenced African catfish even more positively. Plants release substances into the process water like phosphatases (Tarafdar and Claassen 1988; Tarafdar et al. 2001) that are able to hydrolyse biochemical phosphate compounds around the root area and exude organic acids (Bais et al. 2004). Additionally, microorganisms on the root surfaces play an important role through the excretion of organic substances increasing the solubilization of minerals making them available for plant nutrition. It is evident that the environment of the rhizosphere, the 'root exudate', consists of many organic compounds such as organic acid anions, phytosiderophores, sugars, vitamins, amino acids, purines, nucleosides, inorganic ions, gaseous molecules, enzymes and root border cells (Dakora and Phillips

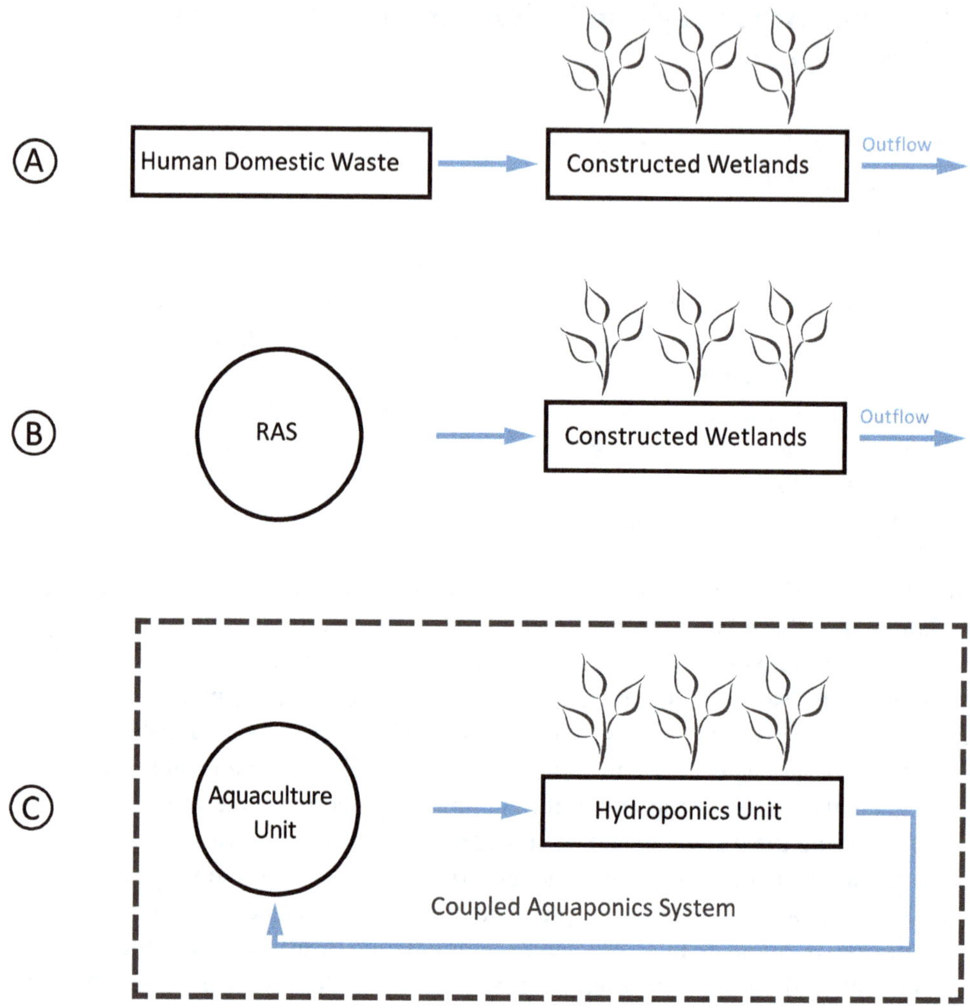

Fig. 7.15 Development of coupled aquaponic systems from (**a**) domestic waste constructed wetlands (CW) and (**b**) CW in combination with recirculating aquaculture systems (RAS) to (**c**) hydroponic units in coupled aquaponic systems

2002), which may influence the health of aquatic organisms in coupled aquaponic systems. This symbiotic relationship is not available in either pure aquaculture or decoupled aquaponics. However, considerable research still needs to be undertaken to understand the responsible factors for better fish welfare.

Pro: Aquaponics can be considered as an optimized form of the conventional agricultural production especially in those areas where production factors caused by the environmental conditions are particularly challenging, e.g. in deserts or highly populated urban areas (cities). Coupled aquaponic systems can be easily adjusted to the local conditions, in terms of system design and scale of operation.

Challenge: Coupled aquaponic also show disadvantages, due to often unsuitable component ratio conditions of the fish and plant production. In order to avoid consequences for fish welfare, coupled aquaponic systems must balance the feed input, stocking density as well as size of the water treatment units and hydroponics. So far knowledge of component ratios in coupled aquaponics is still limited, and modelling to overcome this problem is at the beginning. Rakocy (2012) suggested 57 g of feed/day per square meter of lettuce growing area and a composite ratio of 1 m^3 of fish-rearing tank to 2 m^3 of pea gravel that allows a production of 60 kg / m^3 tilapia. Based on the UVI-system, the size ratios themselves were perceived as a disadvantage since a relatively large ratio of plant growing area to fish surface area of at least 7:3 must be achieved for adequate plant production. On the other hand, system designs of coupled systems are highly variable, often not comparable, and the experiences made cannot be easily transferred to another system or location. Consequently, far more research data is needed in order to identify the best possible production ratios finally also enabling upscaling of coupled aquaponic systems through multiplying optimal designed basic modules (also see Chap. 11).

Challenge: Adverse water quality parameters have been stated to negatively affect fish health. As Yavuzcan Yildiz et al. (2017) pointed out, nutrient retention of plants should be maximized to avoid negative effects of water quality on fish welfare. It is important to select adequate fish species that can accept higher nutrient loads, such as the African catfish (*C. gariepinus*) or the Nile tilapia (*O. niloticus*,). More sensible species such as the Zander or pikeperch (*Sander lucioperca*) might be also applied in aquaponics because they prefer nutrient enriched or eutrophic water bodies with higher turbidity (Jeppesen et al. 2000; Keskinen and Marjomäki 2003; [see Sect. 7.7.1. Fish production]). So far, there is scant data allowing precise statements on fish welfare impairments. With plants generally needing high potassium concentrations between 230 and 400 mg/L inside the process water, 200–400 mg/L potassium showed no negative influence on African catfish welfare (Presas Basalo 2017). Similarly, 40 and 80 mg/L ortho-P in the rearing water had no negative impact on growth performance, feed efficiency and welfare traits of juvenile African catfish (Strauch et al. 2019).

Challenge: Another issue is the potential transmission of diseases in terms of food safety, to people through the consumption of plants that have been in contact with fish waste. In general, the occurrence of zoonoses is minor because closed aquaponics are fully controlled systems. However, germs can accumulate in the process water of the system components or in the fish gut. *Escherichia coli* and *Salmonella* spp. (zoonotic enteric bacteria) were identified as indicators of faecal contamination and microbial water quality, however, they were detected in aquaponics only in very small quantities (Munguia-Fragozo et al. 2015). Another comparison of smooth-textured leafy greens between aquaponics, hydroponics and soil-based production showed no significant differences in aerobic plate counts (APC, aerobic bacteria), Enterobacteriaceae, non-pathogenic *E. coli* and *Listeria*, suggesting a comparable contamination level with pathogens (Barnhart et al. 2015). *Listeria* spp. was most frequent (40%) in hydroponics with de-rooted plants (aquaponic plants with roots 0%, aquaponic plants without roots <10%), but not necessarily the harmful *L. monocytogenes* species. It was suggested that the source of the bacteria may be due to the lack of hygiene management, with little relevance to aquaponics as such. Another infectious bacterium, *Fusobacteria* (*Cetobacterium*), was detected by Schmautz et al. (2017) in the fish faeces with a high prevalence of up to 75%. Representatives of *Fusobacteria* are responsible for human diseases (hospital germ, abscesses, infections), reproducing in biofilms or as part of the fish intestines. Human infections with *Fusobacteria* from aquaponics have not yet been recorded but may be possible by neglecting the required hygiene protocols.

In general, there is rather little information about diseases caused by the consumption of fish and plants originating from coupled aquaponic systems. In Wilson (2005), Dr. J.E. Rakocy stated that there was no recorded human disease outbreak in 25 years of coupled aquaponic production. However, a washing procedure of the plant products should be used to reduce the number of bacteria as a precaution. A chlorine bath (100 ppm) followed by a potable water rinse was recommended by Chalmers (2004). If this methodology is used and the contact of the plants or plant products with the recirculating process water is avoided, the likelihood of contamination with human pathogenic bacteria can be strongly reduced. This is a necessary precaution not only for coupled but also for all other forms of aquaponics.

References

Alberts-Hubatsch H, Ende S, Schuhn A, von der Marwitz C, Wirtz A, Fuchs V, Henjes J, Slater M (2017) Integration of hybrid striped bass *Morone saxatilis* x *M. chrysops*, noble crayfish *Astacus Astacus*, watercress *Nasturtium officinale* and microalge *Nannochloropsis limnetica* in an experimental aquaponic system. Dubrovnik meeting abstract, EAS 2017 Oostende, Belgium

Al-Hafedh YS, Alam A, Beltagi MS (2008) Food production and water conservation in a recirculating aquaponic system in Saudi Arabia at different ratios of fish feed to plants. J World Aquacult Soc 39(4):510–520. https://doi.org/10.1111/j.1749-7345.2008.00181.x

Appelbaum S, Kotzen B (2016) Further investigations of aquaponics using brackish water resources of the Negev desert. Ecocycles Sci J Eur Ecocycles Soc 2(2):26–35. https://doi.org/10.19040/ecocycles.v2i2s.53

Bais HP, Park SW, Weir TL, Callaway RM, Vivanco JM (2004) How plants communicate using the underground information superhighway. Trends Plant Sci 9(1):26–32

Barnhart C, Hayes L, Ringle D (2015) Food safety hazards associated with smooth textured leafy greens produced in aquaponic, hydroponic, and soil-based systems with and without roots in retail. University of Minnesota Aquaponics. The University of Minnesota Aquaponics. 17 p

Baßmann B, Brenner M, Palm HW (2017) Stress and welfare of African catfish (*Clarias gariepinus* Burchell, 1822) in a coupled aquaponic system. Water 9(7):504. https://doi.org/10.3390/w9070504

Baßmann B, Harbach H, Weißbach S, Palm HW (2018) Effect of plant density in coupled aquaponics on the welfare status of African catfish (*Clarias gariepinus*). J World Aquacult Soc. (in press)

Bittsanszky A, Uzinger N, Gyulai G, Mathis A, Villarroel M, Kotzen B, Komives T (2016) Nutrient supply of plants in aquaponic systems. Ecocycles Sci J Eur Ecocycles Soc 2(2):17–20. https://doi.org/10.19040/ecocycles.v2i2.57

Blidariu F, Drasovean A, Grozea A (2013a) Evaluation of phosphorus level in green lettuce conventional grown under natural conditions and aquaponic system. Bull Univ Agric Sci Vet Med Cluj-Napoca Anim Sci Biotechnol:128–135

Blidariu F, Alexandru D, Adrian G, Isidora R, Dacian L (2013b) Evolution of nitrate level in green lettuce conventional grown under natural conditions and aquaponic system. Sci Pap Anim Sci Biotechnol 46(1):244–250

Chalmers GA (2004) Aquaponics and food safety. Alberta April, Lethbridge, 77 p

Comeau Y (2008) Microbial metabolism. In: Biological wastewater treatment: principles, modeling and design, Cap, 2. IWA/Cambridge University Press, London, pp 9–32

Dakora FD, Phillips DA (2002) Root exudates as mediators of mineral acquisition in low-nutrient environments. In: Food security in nutrient-stressed environments: exploiting plants' genetic capabilities. Springer, Dordrecht, pp 201–213

Danaher JJ, Shultz RC, Rakocy JE, Bailey DS (2013) Alternative solids removal for warm water recirculating raft aquaponic systems. J World Aquacult Soc 44(3):374–383. https://doi.org/10.1111/jwas.12040

Davy AJ, Bishop GF, Costa CSB (2001) Salicornia L. (*Salicornia pusilla* J. Woods, *S. ramosissima* J. Woods, *S. europaea* L., *S. obscura* PW ball & Tutin, *S. nitens* PW ball & Tutin, *S. fragilis* PW ball & Tutin and *S. dolichostachya* moss). J Ecol 89(4):681–707

Delaide B, Goddek S, Gott J, Soyeurt H, Jijakli MH (2016) Lettuce (*Lactuca sativa* L. var. Sucrine) growth performance in complemented aquaponic solution outperforms hydroponics. Water 8 (10):467

Delaide B, Delhaye G, Dermience M, Gott J, Soyeurt H, Jijakli MH (2017) Plant and fish production performance, nutrient mass balances, energy and water use of the PAFF box, a small-scale aquaponic system. Aquac Eng 78:130–139

Diem TNT, Konnerup D, Brix H (2017) Effects of recirculation rates on water quality and *Oreochromis niloticus* growth in aquaponic systems. Aquac Eng 78:95–104. https://doi.org/10.1016/j.aquaeng.2017.05.002

Edaroyati MP, Aishah HS, Al-Tawaha AM (2017) Requirements for inserting intercropping in aquaponics system for sustainability in agricultural production system. Agron Res 15 (5):2048–2067. https://doi.org/10.15159/AR.17.070

Endut A, Jusoh A, Ali N, Wan Nik WNS, Hassan A (2009) Effect of flow rate on water quality parameters and plant growth of water spinach (*Ipomoea aquatica*) in an aquaponic recirculating system. Desalin Water Treat 5:19–28. https://doi.org/10.5004/dwt.2009.559

FAO (2017) In: Subasinghe R (ed) World aquaculture 2015: a brief overview, FAO fisheries and aquaculture circular no. 1140. FAO, Rome

Geelen C (2016) Dynamic model of an INAPRO demonstration aquaponic system. Thesis biobased chemistry and technology. Wageningen University Agrotechnology and Food Sciences. 50 p

Goddek S, Vermeulen T (2018) Comparison of Lactuca sativa growth performance in conventional and RAS-based hydroponic systems. Aquac Int:1–10. https://doi.org/10.1007/s10499-018-0293-8

Goddek S, Espinal C, Delaide B, Jijakli M, Schmautz Z, Wuertz S, Keesman K (2016) Navigating towards decoupled aquaponic systems: a system dynamics design approach. Water 8:303. https://doi.org/10.3390/w8070303

Graber A, Junge R (2009) Aquaponic systems: nutrient recycling from fish wastewater by vegetable production. Desalination 246(1–3):147–156

Graber A, Antenen N, Junge R (2014) The multifunctional aquaponic system at ZHAW used as research and training lab. In: Conference VIVUS: transmission of innovations, knowledge and practical experience into everyday practice. Strahinj: Biotehniški center, Naklo. http://www.bc-naklo.si/uploads/media/29-Graber-Antenen-Junge-Z.pdf

Gunning D (2016) Cultivating *Salicornia europaea* (Marsh Samphire), Irish Sea Fisheries Board, http://www.bim.ie/media/bim/content/news,and,events/BIM,Cultivating,,Salicornia,europaea,-,Marsh,Samphire.pdf

Heuvelink E (2018) Tomatoes. In: Heuvelink E (ed) Crop production science in horticulture. Wageningen University/CABI, Wallingford, 388 p

Hortidaily (2015) Dutch start third trial on tomatoes with 100% LED-lighting: 100,6 Kg/m2 at the Improvement Centre in Bleiswijk. What's next? https://www.hortidaily.com/article/6022598/kg-m-at-the-improvement-centre-in-bleiswijk-what-s-next/

Hussain T, Verma AK, Tiwari VK, Prakash C, Rathore G, Shete AP, Nuwansi KKT (2014) Optimizing koi carp, *Cyprinus carpio* var. koi (Linnaeus, 1758), stocking density and nutrient recycling with spinach in an aquaponic system. J World Aquacult Soc 45(6):652–661. https://doi.org/10.1111/jwas.12159

Hussain T, Verma AK, Tiwari VK, Prakash C, Rathore G, Shete AP, Saharan N (2015) Effect of water flow rates on growth of *Cyprinus carpio* var. koi (*Cyprinus carpio* L., 1758) and spinach plant in aquaponic system. Aquac Int 23(1):369–384. https://doi.org/10.1007/s10499-014-9821-3

Jeppesen E, Jensen JP, SØndergaard M, Lauridsen T, Landkildehus F (2000) Trophic structure, species richness and biodiversity in Danish lakes: changes along a phosphorus gradient. Freshw Biol 45(2):201–218

Keskinen T, Marjomäki TJ (2003) Growth of pikeperch in relation to lake characteristics: total phosphorus, water colour, lake area and depth. J Fish Biol 63(5):1274–1282

Khandaker M, Kotzen B (2018) The potential for combining living wall and vertical farming systems with aquaponics with special emphasis on substrates. Aquac Res 49(4):1454–1468

Knaus U, Palm HW (2017a) Effects of fish biology on ebb and flow aquaponical cultured herbs in northern Germany (Mecklenburg Western Pomerania). Aquaculture 466:51–63. https://doi.org/10.1016/j.aquaculture.2016.09.025

Knaus U, Palm HW (2017b) Effects of the fish species choice on vegetables in aquaponics under spring-summer conditions in northern Germany (Mecklenburg Western Pomerania). Aquaculture 473:62–73. https://doi.org/10.1016/j.aquaculture.2017.01.020

Knaus U, Segade Á, Robaina L (2015) Training School 1 – Aquaponic trials: improving water quality and plant production through fish management and diet. 25–29 May 2015, Universidad de Las Palmas de Gran Canaria, Spain. Cost Action FA1305, The EU Aquaponics Hub – Realising Sustainable Integrated Fish and Vegetable Production for the EU

Knaus U, Appelbaum S, Palm HW (2018a) Significant factors affecting the economic sustainability of closed backyard aquaponics systems. Part IV: autumn herbs and polyponics. AACL Bioflux 11(6):1760–1775

Knaus U, Appelbaum S, Castro C, Sireeni J, Palm HW (2018b) Growth performance of basil in a small-scale aquaponic system with the production of tilapia (*Oreochromis niloticus*) and African catfish (*Clarias gariepinus*). (In preparation)

König B, Junge R, Bittsanszky A, Villarroel M, Komives T (2016) On the sustainability of aquaponics. Ecocycles Sci J Eur Ecocycles Soc 2(1):26–32. https://doi.org/10.19040/ecocycles.v2i1.50

Kotzen B, Appelbaum S (2010) An investigation of aquaponics using brackish water resources in the Negev Desert. J Appl Aquac 22(4):297–320. https://doi.org/10.1080/10454438.2010.527571

Lehmonen R, Sireeni J (2017) Comparison of plant growth and quality in hydroponic and aquaponic systems. Bachelor's thesis, University of Jyväskylä, Jyväskylä, Finland. 27 p

Lennard WA, Leonard BV (2004) A comparison of reciprocating flow versus constant flow in an integrated, gravel bed, aquaponic test system. Aquac Int 12(6):539–553. https://doi.org/10.1007/s10499-004-8528-2

Lennard WA, Leonard BV (2006) A comparison of three different hydroponic sub-systems (gravel bed, floating and nutrient film technique) in an aquaponic test system. Aquac Int 14(6):539–550. https://doi.org/10.1007/s10499-006-9053-2

Lewis WM, Yopp JH, Schramm HL Jr, Brandenburg AM (1978) Use of hydroponics to maintain quality of recirculated water in a fish culture system. Trans Am Fish Soc 107(1):92–99. https://doi.org/10.1577/1548-8659(1978)107<92:UOHTMQ>2.0.CO;2

Lobillo JR, Fernández-Cabanás VM, Carmona E, Candón FJL (2014) Manejo básico y resultados preliminares de crecimiento y supervivencia de tencas (*Tinca tinca* L.) y lechugas (*Lactuca sativa* L.) en un prototipo acuapónico. ITEA 110(2):142–159. https://doi.org/10.12706/itea.2014.009

Lorena S, Cristea V, Oprea L (2008) Nutrients dynamic in an aquaponic recirculating system for sturgeon and lettuce (*Lactuca sativa*) production. Lucrări Ştiinţifice-Zootehnie şi Biotehnologii, Universitatea de Ştiinţe Agricole şi Medicină Veterinară a Banatului Timişoara 41(2):137–143

Lorenzen L (2017) Vergleich des Wachstums von chinesischem Blätterkohl (*Brassica rapa chinensis*) in drei verschiedenen Hydroponiksubsystemen unter aquaponischen Bedingungen. Agrar- und Umweltwissenschaftliche Fakultät, Lehrstuhl Aquakultur und Sea-Ranching, Universität Rostock. Masterarbeit. 63 p. [in German]

Love DC, Fry JP, Li X, Hill ES, Genello L, Semmens K, Thompson RE (2015) Commercial aquaponics production and profitability: findings from an international survey. Aquaculture 435:67–74. https://doi.org/10.1016/j.aquaculture.2014.09.023

Loyacano HA, Grosvenor RB (1973) Effects of Chinese waterchestnut in floating rafts on production of channel catfish in plastic pools. Proc Annu Conf Southeast Assoc Game Fish Comm 27:471–473

Mariscal-Lagarda MM, Páez-Osuna F, Esquer-Méndez JL, Guerrero-Monroy I, del Vivar AR, Félix-Gastelum R (2012) Integrated culture of white shrimp (*Litopenaeus vannamei*) and tomato (*Lycopersicon esculentum* mill) with low salinity groundwater: management and production. Aquaculture 366:76–84. https://doi.org/10.1016/j.aquaculture.2012.09.003

McMurtry MR, Nelson PV, Sanders DC, Hodges L (1990) Sand culture of vegetables using recirculated aquacultural effluents. Appl Agric Res 5(4):280–284

McMurtry MR, Sanders DC, Patterson RP, Nash A (1993) Yield of tomato irrigated with recirculating aquacultural water. J Prod Agric 6(3):428–432. https://doi.org/10.2134/jpa1993.0428

McMurtry MR, Sanders DC, Cure JD, Hodson RG, Haning BC, Amand PCS (1997) Efficiency of water use of an integrated fish/vegetable co-culture system. J World Aquacult Soc 28 (4):420–428. https://doi.org/10.1111/j.1749-7345.1997.tb00290.x

Munguia-Fragozo P, Alatorre-Jacome O, Rico-Garcia E, Torres-Pacheco I, Cruz-Hernandez A, Ocampo-Velazquez RV, Garcia-Trejo JF, Guevara-Gonzalez RG (2015) Perspective for aquaponic systems: "omic" technologies for microbial community analysis. BioMed Res Int 2015: 480386. Hindawi Publishing Corporation. BioMed Research International. 2015, 10 pages. https://doi.org/10.1155/2015/480386

Naegel LC (1977) Combined production of fish and plants in recirculating water. Aquaculture 10 (1):17–24. https://doi.org/10.1016/0044-8486(77)90029-1

Nelson RL, Pade JS (2007) Aquaponic equipment the clarifier. Aquaponics J 4(47):30–31

Nozzi V, Strofaldi S, Piquer IF, Di Crescenzo D, Olivotto I, Carnevali O (2016) *Amyloodinum ocellatum* in *Dicentrarchus labrax*: study of infection in salt water and freshwater aquaponics. Fish Shellfish Immunol 57:179–185. https://doi.org/10.1016/j.fsi.2016.07.036

Nuwansi KKT, Verma AK, Prakash C, Tiwari VK, Chandrakant MH, Shete AP, Prabhath GPWA (2016) Effect of water flow rate on polyculture of koi carp (*Cyprinus carpio* var. koi) and goldfish (*Carassius auratus*) with water spinach (*Ipomoea aquatica*) in recirculating aquaponic system. Aquac Int 24(1):385–393. https://doi.org/10.1007/s10499-015-9932-5

Palm HW, Seidemann R, Wehofsky S, Knaus U (2014a) Significant factors influencing the economic sustainability of closed aquaponic systems. part I: system design, chemo-physical parameters and general aspects. AACL Bioflux 7(1):20–32

Palm HW, Bissa K, Knaus U (2014b) Significant factors affecting the economic sustainability of closed aquaponic systems. part II: fish and plant growth. AACL Bioflux. 7(3):162–175

Palm HW, Nievel M, Knaus U (2015) Significant factors affecting the economic sustainability of closed aquaponic systems. part III: plant units. AACL Bioflux 8(1):89–106

Palm HW, Strauch S, Knaus U, Wasenitz B (2016) Das FischGlasHaus – eine Innovationsinitiative zur energie- und nährstoffeffizienten Produktion unterschiedlicher Fisch- und Pflanzenarten in Mecklenburg-Vorpommern ("Aquaponik in MV"). Fischerei & Fischmarkt in Mecklenburg-Vorpommern 1/2016–16:38–47 [in German]

Palm HW, Wasenitz B, Knaus U, Bischoff A, Strauch SM (2017) Two years of aquaponics research in the FishGlassHouse – lessons learned. Dubrovnik meeting abstract, Aquaculture Europe 2017, EAS Oostende Belgium

Palm HW, Knaus U, Appelbaum S, Goddek S, Strauch SM, Vermeulen T, Jijakli MH, Kotzen B (2018) Towards commercial aquaponics: a review of systems, designs, scales and nomenclature. Aquac Int 26(3):813–842

Palm HW, Knaus U, Wasenitz B, Bischoff-Lang AA, Strauch SM (2019) Proportional up scaling of African catfish (*Clarias gariepinus* Burchell, 1822) commercial recirculating aquaculture systems disproportionally affects nutrient dynamics. Aquac Int 26(3):813–842. https://doi.org/10.1007/s10499-018-0249-z

Pantanella E (2012) Nutrition and quality of aquaponic systems. Ph.D. thesis. Università degli Studi della Tuscia. Viterbo, Italy. 124 p

Presas Basalo F (2017) Does water potassium concentration in aquaponics affect the performance of African catfish *Clarias gariepinus* (Burchell, 1822)? Master's thesis, Department of Animal Sciences Aquaculture and Fisheries Group, Wageningen University and University of Rostock Department of Aquaculture and Sea-Ranching. 75 p

Pribbernow M (2016) Vergleich des Wachstums von Basilikum (*Ocimum basilicum* L.) in drei verschiedenen Hydroponik-Subsystemen unter aquaponischer Produktion. Agrar- und Umweltwissenschaftliche Fakultät, Lehrstuhl Aquakultur und Sea-Ranching, Universität Rostock. Masterarbeit. 98 p. [in German]

Rakocy JE (1989) Hydroponic lettuce production in a recirculating fish culture system. In: Island perspectives. Vol. 3. agricultural experiment station, University of the Virgin Islands. pp 5–10

Rakocy JE (2012) Chapter 14: aquaponics – integrating fish and plant culture. In: Tidwell JH (ed) Aquaculture production systems. Wiley-Blackwell, Oxford, pp 344–386

Rakocy JE, Shultz RC, Bailey DS (2000) Commercial aquaponics for the Caribbean: proceedings of the Gulf and Caribbean Fisheries Institute [Proc. Gulf Caribb. Fish. Inst.]. no. 51, pp 353–364

Rakocy J, Shultz RC, Bailey DS, Thoman ES (2003) Aquaponic production of tilapia and basil: comparing a batch and staggered cropping system. In: South Pacific soilless culture conference-SPSCC 648. pp 63–69

Rakocy JE, Bailey DS, Shultz RC, Thoman ES (2004) Update on tilapia and vegetable production in the UVI aquaponic system. In: New dimensions on farmed Tilapia: proceedings of the sixth international symposium on Tilapia in Aquaculture, held September. pp 12–16

Rakocy JE, Masser MP, Losordo TM (2006) Recirculating aquaculture tank production systems: aquaponics-integrating fish and plant culture. SRAC Publication 454:1–16

Rakocy JE, Bailey DS, Shultz RC, Danaher JJ (2011) A commercial-scale aquaponic system developed at the University of the Virgin Islands. In: Proceedings of the 9th international symposium on Tilapia in Aquaculture. pp 336–343

Ronzón-Ortega M, Hernández-Vergara MP, Pérez-Rostro CI (2012) Hydroponic and aquaponic production of sweet basil (*Ocimum basilicum*) and giant river prawn (*Macrobrachium rosenbergii*). Trop Subtrop Agroecosyst 15(2):63–71

Sace CF, Fitzsimmons KM (2013) Vegetable production in a recirculating aquaponic system using Nile tilapia (*Oreochromis niloticus*) with and without freshwater prawn (*Macrobrachium rosenbergii*). Acad J Agric Res 1(12):236–250

Saha S, Monroe A, Day MR (2016) Growth, yield, plant quality and nutrition of basil (*Ocimum basilicum* L.) under soilless agricultural systems. Ann Agric Sci 61(2):181–186

Savidov N (2005) Evaluation of aquaponics technology in Alberta, Canada. Aquaponics Journal 37:20–25

Schmautz Z, Graber A, Jaenicke S, Goesmann A, Junge R, Smits TH (2017) Microbial diversity in different compartments of an aquaponics system. Arch Microbiol 199(4):613–620

Seawright DE, Stickney RR, Walker RB (1998) Nutrient dynamics in integrated aquaculture–hydroponics systems. Aquaculture 160(3–4):215–237. https://doi.org/10.1016/S0044-8486(97)00168-3

Shete AP, Verma AK, Chadha NK, Prakash C, Peter RM, Ahmad I, Nuwansi KKT (2016) Optimization of hydraulic loading rate in aquaponic system with common carp (*Cyprinus carpio*) and mint (*Mentha arvensis*). Aquac Eng 72:53–57. https://doi.org/10.1016/j.aquaeng.2016.04.004

Simeonidou M, Paschos I, Gouva E, Kolygas M, Perdikaris C (2012) Performance of a small-scale modular aquaponic system. AACL Bioflux. 5(4):182–188

Skar SLG, Liltved H, Drengstig A, Homme JM, Kledal PR, Paulsen H, Björnsdottir R, Oddson S, Savidov N (2015) Aquaponics NOMA (Nordic Marin) – new innovations for sustainable aquaculture in the Nordic countries. Nordic Innovation Publication 2015:06, 108 p

Somerville C, Cohen M, Pantanella E, Stankus A, Lovatelli A (2014) Small-scale aquaponic food production. Integrated fish and plant farming, FAO fisheries and aquaculture technical paper no. 589. FAO, Rome. (262 pp)

Strauch SM, Wenzel LC, Bischoff A, Dellwig O, Klein J, Schüch A, Wasenitz B, Palm HW (2018) Commercial African catfish (*Clarias gariepinus*) recirculating aquaculture systems: assessment of element and energy pathways with special focus on the phosphorus cycle. Sustainability 2018 (10):1805. https://doi.org/10.3390/su10061805

Strauch SM, Bischoff AA, Bahr J, Baßmann B, Oster M, Wasenitz B, Palm HW (2019, submitted) Effects of ortho-phosphate on growth performance, welfare and product quality of juvenile African catfish (*Clarias gariepinus*). Fishes 4:3. https://doi.org/10.3390/fishes4010003

Sutton RJ, Lewis WM (1982) Further observations on a fish production system that incorporates hydroponically grown plants. Progress Fish Cult 44(1):55–59. https://doi.org/10.1577/1548-8659(1982)44[55,FOOAFP]2.0.CO;2

Tarafdar JC, Claassen N (1988) Organic phosphorus compounds as a phosphorus source for higher plants through the activity of phosphatases produced by plant roots and microorganisms. Biol Fertil Soils 5(4):308–312

Tarafdar JC, Yadav RS, Meena SC (2001) Comparative efficiency of acid phosphatase originated from plant and fungal sources. J Plant Nutr Soil Sci 164(3):279–282

Thorarinsdottir RI (2014) Implementing commercial aquaponics in Europe – first results from the Ecoinnovation project EcoPonics. Aquaculture Europe 14, October 14–17 2014, San Sebastian, Spain

Thorarinsdottir RI, Kledal PR, Skar SLG, Sustaeta F, Ragnarsdottir KV, Mankasingh U, Pantanella E, van de Ven R, Shultz RC (2015) Aquaponics guidelines. 64 p

Tran H (2015) Aquaponics-coupled and decoupled systems and the water quality needs of each. World Aquaculture 2015 – Meeting Abstract Jeju, Korea, May 27, 2015. World Aquaculture Society PO Box 397 Sorrento, LA 70778–0397 (USA)

Tyson RV, Simonne EH, Treadwell DD, White JM, Simonne A (2008) Reconciling pH for ammonia biofiltration and cucumber yield in a recirculating aquaponic system with perlite biofilters. Hortscience 43(3):719–724

USGS. The USGS Water Science School. https://water.usgs.gov/edu/gallery/global-water-volume.html

Van der Heijden PGM, Roest CWJ, Farrag F, ElWageih H, Sadek S, Hartgers EM, Nysingh SL (2014) Integrated agri-aquaculture with brackish waters in Egypt: mission report (March 9–March 17, 2014) (No. 2526). Alterra Wageningen UR

Villarroel M, Rodriguez Alvariño JM, Duran Altisent JM (2011) Aquaponics: integrating fish feeding rates and ion waste production for strawberry hydroponics. Span J Agric Res 9 (2):537–545

Waller U, Buhmann AK, Ernst A, Hanke V, Kulakowski A, Wecker B, Orellana J, Papenbrock J (2015) Integrated multi-trophic aquaculture in a zero-exchange recirculation aquaculture system for marine fish and hydroponic halophyte production. Aquac Int 23(6):1473–1489. https://doi.org/10.1007/s10499-015-9898-3

Watten BJ, Busch RL (1984) Tropical production of tilapia (*Sarotherodon aurea*) and tomatoes (*Lycopersicon esculentum*) in a small-scale recirculating water system. Aquaculture 41 (3):271–283. https://doi.org/10.1016/0044-8486(84)90290-4

Wermter L (2016) Comparison of three different hydroponic sub-systems of *Cucumis sativus* L. grown in an aquaponic system. Department of Agronomy. Faculty of Agricultural and Environmental Sciences (AUF), Professorship of Crop production and Aquaculture and Sea-Ranching. University of Rostock, Germany. Masterthesis. 40 p

Wilson G (2005) Greenhouse aquaponics proves superior to inorganic hydroponics. Aquaponics J 39(4):14–17

Yavuzcan Yildiz H, Robaina L, Pirhonen J, Mente E, Domínguez D, Parisi G (2017) Fish welfare in aquaponic systems: its relation to water quality with an emphasis on feed and faeces-a review. Water 9(1):13

Zimmermann J (2017) Vergleich des Wachstums von Marokkanischer Minze (*Mentha spicata*) in drei verschiedenen Hydroponik Subsystemen unter aquaponischer Produktion. Agrar- und Umweltwissenschaftliche Fakultät, Lehrstuhl Aquakultur und Sea-Ranching, Universität Rostock. Masterarbeit. 71 p. [in German]

8

Decoupling of Units in Aquaponics Systems

Simon Goddek, Alyssa Joyce, Sven Wuertz, Oliver Körner, Ingo Bläser, Michael Reuter, and Karel J. Keesman

Abstract Traditional aquaponics systems were arranged in a single process loop that directs nutrient-rich water from fish to the plants and back. Given the differing specific nutrient and environmental requirements of plants and fish, such systems presented a compromise to the ideal conditions for rearing of both, thus reducing the efficiency and productivity of such coupled systems. More recently, designs that allow for decoupling of units provide for a more finely tuned regulation of the process water in each of the respective units while also allowing for better recycling of nutrients from sludge. Suspended solids from the fish (e.g. faeces and uneaten feed) need to be removed from the process water before water can be directed to plants in order to prevent clogging of hydroponic systems, a step that represents a significant loss of total nutrients, most importantly phosphorus. The reuse of sludge and mobilization of nutrients contained within that sludge present a number of engineering challenges that, if addressed creatively, can dramatically increase the efficiency and sustainability of aquaponics systems. One solution is to separate, or

S. Goddek (✉) · K. J. Keesman
Mathematical and Statistical Methods (Biometris), Wageningen University, Wageningen, The Netherlands
e-mail: simon.goddek@wur.nl; simon@goddek.nl

A. Joyce
Department of Marine Science, University of Gothenburg, Gothenburg, Sweden
e-mail: alyssa.joyce@gu.se

S. Wuertz
Department Ecophysiology and Aquaculture, Leibniz-Institute of Freshwater Biology and Inland Fisheries, Berlin, Germany
e-mail: wuertz@igb-berlin.de

O. Körner
Leibniz-Institute of Vegetable and Ornamental Crops (IGZ), Grossbeeren, Germany
e-mail: koerner@igzev.de

I. Bläser · M. Reuter
aquaponik manufaktur GmbH, Issum, Germany
e-mail: ingo.blaeser@aquaponik-manufaktur.de

when there are pathogens or production problems, to isolate components of the system, thus maximizing overall control and efficiency of each component, while reducing compromises between the conditions and species-specific requirements of each subsystem. Another potential innovation that is made possible by the decoupling of units involves introducing additional loops wherein bioreactors can be used to treat sludge. An additional distillation loop can ensure increased nutrient concentrations to the hydroponics unit while, at the same time, reducing adverse effects on fish health from high nutrient levels in the RAS unit. Several studies have documented the aerobic and anaerobic digestion performance of bioreactors for treating sludge, but the benefits of the digestate on plant growth are not well-researched. Both remineralization and distillation components consequently have a high unexplored potential to improve decoupled aquaponics systems.

Keywords Decoupled aquaponics · Multi-loop aquaponics · System dynamics · System design · Anaerobic digestion · Desalination

8.1 Introduction

As discussed in Chaps. 5 and 7, single-loop aquaponics systems are well-researched, but such systems have a suboptimal overall efficiency (Goddek et al. 2016; Goddek and Keesman 2018). As aquaponics scales up to industrial-level production, there has been emphasis on increasing the economic viability of such systems. One of the best opportunities to optimize production in terms of harvest yield can be accomplished by uncoupling the components within an aquaponics system to ensure optimal growth conditions for both fish and plants. Decoupled systems differ from coupled systems insomuch as they separate the water and nutrient loops of both the aquaculture and hydroponics unit from each another and thus provide a control of the water chemistry in both systems. Figure 8.1 provides a schematic overview of a traditional coupled system (A), a decoupled two-loop system (B), and a decoupled multi-loop system (C). However, there is considerable debate whether decoupled aquaponics systems are economically advantageous over more traditional systems, given that they require more infrastructure. In order to answer that question, it is necessary to consider different system designs in order to identify their strengths and weaknesses.

The concept of a coupled one-loop aquaponics system as shown in Fig. 8.1a can be regarded as the traditional basis of all aquaponics systems in which water recirculates freely between the aquaculture and hydroponics units, while nutrient-rich sludge is discharged. One of the key drawbacks of such systems is that it is necessary to make trade-offs in the rearing conditions of both subsystems in terms of pH, temperature, and nutrient concentrations (Table 8.1).

In contrast, decoupled or two-loop aquaponics systems separate the aquaculture and aquaponics units from each other (Fig. 8.1b). Here, the sizing of the hydroponic unit is a critical aspect, because ideally it needs to assimilate the nutrients provided by the fish unit directly or via sludge mineralization (e.g. extracting nutrients from

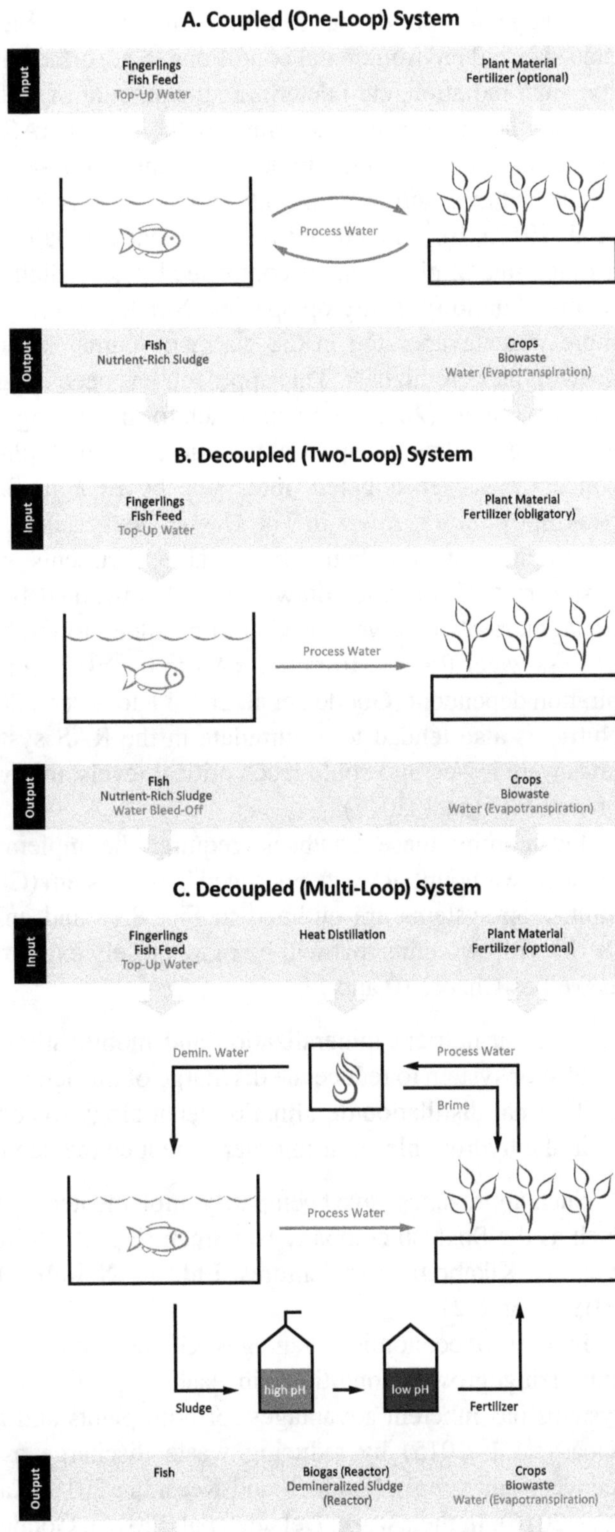

Fig. 8.1 The evolution of aquaponics systems. (**a**) shows a traditional one-loop aquaponics system, (**b**) a simple decoupled aquaponics system, and (**c**) a decoupled multi-loop aquaponics system. The blue font stands for water input, output, and flows and the red for waste products

the sludge and providing it to the plants in a soluble form). Indeed, both the plant area size and environmental conditions (e.g. surface, leaf area index, relative humidity, solar radiation, etc.) determine the amount of water that can be evapotranspired and are the main factors determining the rate of RAS water replacement. The water sent from the RAS to the hydroponic unit is consequently replaced by clean water which reduces nutrient concentrations and thus improves water quality (Monsees et al. 2017a, b). The amount of water that can be replaced depends on evapotranspiration rate of plants that is controlled by net radiation, temperature, wind velocity, relative humidity, and crop species. Notably, there is a seasonal dependency, with more water evaporated in the warmer, sunnier seasons which is also when plant growth rates are highest. This approach has been suggested by Goddek et al. (2015) and Kloas et al. (2015) as an approach for improving the design of one-loop systems and better utilizing capacity to assure optimal plant growth performance. The concept has been adopted, inter alia, by *ECF* in Berlin, Germany, and the now bankrupt *UrbanFarmers* in The Hague, Netherlands.

Despite potential benefits, initial experiments with a decoupled single-loop design met with serious drawbacks. This resulted from the high amounts of additional nutrients that were needed to be added to the hydroponic loop given that the process water flowing from the RAS to the hydroponics loop is purely evapotranspiration dependent (Goddek et al. 2016; Kloas et al. 2015; Reyes Lastiri et al. 2016). Nutrients also tended to accumulate in the RAS systems when evapotranspiration rates were lower, and could reach critical levels, thus requiring periodic bleeding off of water (Goddek 2017).

Overcoming these drawbacks required the implementation of additional loops to reduce the amount of waste produced in the system (Goddek and Körner 2019). Such multi-loop systems are outlined in Fig. 8.1c and enhance the two-loop approach (8.1b) with two units that will be more closely explored in the next two subchapters as well as Chaps. 10 and 11:

1. Efficient nutrient mineralization and mobilization, using a two-stage anaerobic reactor system to reduce the discharge of nutrients from the system via fish sludge
2. Thermal distillation/desalination technology to concentrate the nutrient solution in the hydroponics unit in order to reduce the need for additional fertilizers

Such approaches have been partly implemented by various aquaponics producers such as the Spanish company NerBreen (Fig. 8.1) (Goddek and Keesman 2018) as well as Kikaboni AgriVentures Ltd. in Nairobi, Kenya, (van Gorcum et al. 2019) (Fig. 8.2).

In terms of economic advantages (Goddek and Körner 2019; Delaide et al. 2016), optimizing growth conditions in each respective loop of decoupled aquaponics systems has inherent advantages for both plants and fish (Karimanzira et al. 2016; Kloas et al. 2015) by reducing waste discharge as well as improving nutrient recovery and supply (Goddek and Keesman 2018; Karimanzira et al. 2017; Yogev et al. 2016). In their work, Delaide et al. (2016), Goddek and Vermeulen (2018), and Woodcock (pers. Comm.) show that decoupled aquaponics systems achieve better growth performance than their respective one-loop aquaponics and hydroponics control groups. Despite this, there are various problems that still need to be resolved,

Fig. 8.2 Pictures of existing multi-loop system in (1) Spain (NerBreen) and (2) Kenya (Kikaboni AgriVentures Ltd.). Whereas the NerBreen System is located in a controlled environment, the Kikaboni System is using a semi-open foil-tunnel system

including technical issues such as system scaling, parameter optimization, and engineering choices for greenhouse technologies for different regional scenarios. In the rest of this chapter, we will focus on some of the current developments to provide an overview of ongoing challenges, as well as promising developments in the field.

8.2 Mineralization Loop

In RAS, solid and nutrient-rich sludge must be removed from the system to maintain water quality. By adding an additional sludge recycling loop, accumulating RAS wastes can be converted into dissolved nutrients for reuse by plants rather than discarded (Emerenciano et al. 2017). Within bioreactors, microorganisms can break down this sludge into bioavailable nutrients, which can subsequently be delivered to plants (Delaide et al. 2018; Goddek et al. 2018; Monsees et al. 2017a, b). Many one-loop aquaponics systems already include aerobic (Rakocy et al. 2004) and anaerobic (Yogev et al. 2016) digesters to transform nutrients that are trapped in the fish sludge and make them bioavailable for plants. However, integrating such a system into a coupled one-loop aquaponics system has several disadvantages:

1. The dilution factor for nutrient-rich effluents is much higher when discharging them to a single-loop system in relation to discharging them to the hydroponics unit only. Effectively, nutrients diluted by entering in contact with large volumes of fish rearing water.
2. Fish are unnecessarily exposed to the mineralization reactor's effluents; e.g. the effluents of anaerobic reactors can include volatile fatty acids (VFAs) and ammonia that might potentially harm the fish; such reactors also represent an additional source for potential introduction of pathogens.
3. Around 90% of the nutrients trapped in the sludge can be recovered when RAS sludge is maintained at a pH of 4 (Jung and Lovitt 2011). Such a low pH is not possible when operating bioreactors at a pH around 7 (Goddek et al. 2018), which is the usual trade-off pH value within one-loop aquaponics systems.

With respect to pH, Fig. 8.3 shows the approximate pH values of the respective process water flows in a multi-loop aquaponics system (e.g. as presented in Fig. 8.1c). Figure 8.3 also shows the impact of mineralization reactors on the performance of the system as a whole, based on the anaerobic reactors proposed by Goddek et al. (2018). Such a system represents only one possible solution for treating sludge, with alternative approaches discussed in Chap. 10. The decrease in pH of the process water flowing from the RAS subsystem into the hydroponics loop as shown in Fig. 8.3 demonstrates acidification in the nutrient concentration loop (i.e. demineralized water has a pH of 7). Thus, the effluent has a lower pH than the RAS outlet, which reduces the need to adjust the pH for optimal plant growth conditions.

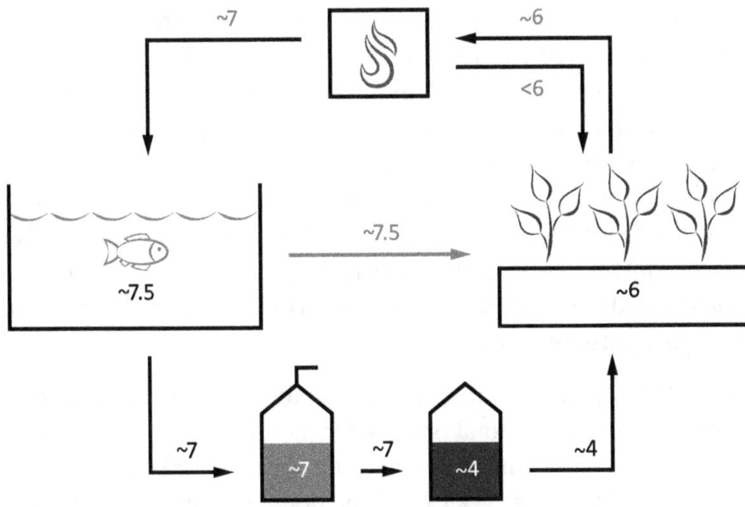

Fig. 8.3 Approximate pH of the water within the different system components as well as the process water. The '~' indicated an approximation

Table 8.1 Overview of optimal growth conditions for fish and plants and preferred operational conditions for sludge nutrient recycling treatment

Subsystem	Species/ function	pH	Temperature (°C)	Nitrate (NO_3) (mg/L)
Recirculating aquaculture system (RAS)	*Oreochromis niloticus* (Nile tilapia)	7–9 (Ross 2000)	27–30 (El-Sayed 2006)	<100–200 (Dalsgaard et al. 2013)
	Oncorhynchus mykiss (rainbow trout)	6.5–8.5 (FAO 2005)	15 (Coghlan and Ringler 2005)	<40 (Davidson et al. 2011; Schrader et al. 2013)
Hydroponics	*Lactuca sativa* (lettuce)	5.5–6.5 (Resh 2012)	21–25 (Resh 2012)	730 (Resh 2012)
	Lycopersicon esculentum (tomato)	6.3–6.5 (Resh 2002)	18–24 (Resh 2002)	666 (Sonneveld and Voogt 2009)
Anaerobic reactor	Methanogenesis	6.8–7-4 (de Lemos Chernicharo 2007)	30–35 (Alvarez and Lidén 2008; de Lemos Chernicharo 2007)	–
	Sludge mobilization	4.0 (Jung and Lovitt 2011)	n/a	–

The two-stage reactor system works as follows:

- *In the first stage* (pH around 7 to provide optimal conditions for methanogenesis; Table 8.1), the organic matter is broken down to sustain a high degree of methane production (i.e. carbon removal). Mirzoyan and Gross (2013) reported a total suspended solids reduction of around 90%, using upflow anaerobic sludge blanket reactor technology. This has the benefit that (1) biogas is harvested as a renewable energy source and (2) fewer VFAs are produced in the second stage. The sludge retention time in the first stage should be several months, before removing the accumulated nutrients in the sludge (e.g. calcium phosphate aggregation) within the second stage.

- *In the second stage*, nutrients in suspended solids are effectively mobilized and become available for plant uptake. This mobilization is the most effective in a low-pH environment (Goddek et al. 2018; Jung and Lovitt 2011). Once the pH of acidic reactors is decreased, it usually remains stable; thus less pH regulation is required in the hydroponic unit.

The effluents that are rich in nutrients may require some post-treatment depending on the amount of measured total suspended solids and VFAs. However, it is important to keep in mind that ammonia can stimulate plant growth, e.g. leafy greens, when it accounts for 5–25% of the total nitrogen concentration (Jones 2005). However, fruit vegetables such as tomatoes or sweet peppers are particularly sensitive to ammonia in the nutrient solution. An aerobic post-effluent treatment or a well-aerated hydroponics sump would be required in systems growing those types of crops.

8.2.1 Determining Water and Nutrient Flows

For system sizing (Sect. 8.4), the amount of water flowing from the RAS system via the reactor(s) to the hydroponics unit (Q_{MIN}) needs to be known (Eq. 8.1):

$$Q_{\mathrm{MIN}}\ (\mathrm{kg/day}) = \frac{n_{\mathrm{feed}} \times k_{\mathrm{sludge}}}{\pi_{\mathrm{sludge}}} \tag{8.1}$$

where n_{feed} is the amount of fish feed in kg, k_{sludge} is the proportion coefficient of fish feed ending up as sludge, and π_{sludge} is the proportion of total solids (i.e. sludge) in the sludge water flow entering the mineralization loop.

The sludge concentration can be increased by adding a gravity separation device prior to the bioreactors, directing the 'clear' supernatant back to the RAS system. This formula can also be used to get an input for sizing the reactor based on the hydraulic retention time (Chap. 10). Between 20 and 40% of the fish feed ends up as total suspended solids in the RAS-derived sludge (Timmons and Ebeling 2013). As an example, it has been found that tilapia sludge contains around 55% of nutrients that were added to the system via feed (Neto and Ostrensky 2013; Yavuzcan Yildiz et al. 2017) which represents a valuable resource for crop growth.

The main nutrients that can be recovered via a mineralization process are N and P. As P (one of the major components of sludge) is the most valuable macronutrient in terms of cost and availability for crop production, it should be the first element to be optimized in the aquaponic system.

The mineralization rate of the mineralization loop is calculated as follows:

$$\text{Mineralization (g/day)} = (n_{\text{feed}} \times 1000) \times \pi_{\text{feed}} \times \pi_{\text{sludge}} \times \eta_{\text{min}} \tag{8.2}$$

where n_{feed} is the feed input to the system (in kg); $\pi_{\text{feed}}$is the proportion of the nutrient in the feed formulation;$\pi_{\text{sludge}}$is the proportion of a specific feed-derived element ending up in the sludge; and $\eta_{\text{min}}$is the mineralization and mobilization efficiency of the reactor system.

The last step would be to determine the concentration of the respective element in the effluent of the mineralization loop:

$$\text{Nutrient concentration (mg/L)} = \frac{\text{Mineralization} \times 1000}{Q_{\text{MIN}}} \tag{8.3}$$

Example 8.1
Our RAS system is fed with 10 kg of fish feed per day. We assume that 25% of the fed feed ends up as sludge. In our system, we use a Radial Flow Settler (RFS) to concentrate the sludge to 1% dry matter. Consequently, the flow from the RAS to HP via the mineralization loop is calculated as follows:

$$Q_{\text{MIN}}\ \text{(kg/day)} = \frac{10\text{kg} \times 0.25}{0.01} = 250 \approx 250\text{kg/day}$$

We decide to size our system on P. The P content of our feed (in most cases provided by the feed manufacturer) is 1.5% and 55% of it ends up in the sludge (Neto and Ostrensky 2013). We assume that our reactors achieve a mineralization efficiency of 90% for this element. Therefore, the grams of P transferred to the hydroponics unit each day can be determined:

$$\text{Mineralization (g/day)} = (10\text{kg} \times 1000) \times 0.55 \times 0.015 \times 0.9 = 74.25$$

The concentration of the effluent is consequently:

$$\text{Nutrient concentration (mg/L)} = \frac{74{,}25\text{g} \times 1000}{250\text{L}} = 297\ \text{mg/L}$$

This concentration of P in the effluent in the example box above is approximately six times higher than in most hydroponics nutrient solutions. The research of Goddek et al. (2018) underpins this theoretical number, and they report that their RAS sludge contained 150 and 200 mg/L of P for two independent systems, respectively (1% TSS sludge), with a fish feed P content of 0.83% in dry matter feed for the latter (200 mg/L).

8.3 Distillation/Desalination Loop

In decoupled aquaponics systems, there is a one-way flow from the RAS to the hydroponics unit. In practice, plants take up water supplied by RAS, which in turn is topped up with fresh (i.e. tap or rain) water. The necessary outflow from the RAS unit is equal to the difference between the water leaving the HP system via plants (and via the distillation unit) and the water entering the hydroponics unit from the mineralization reactor, if the system includes a reactor (Fig. 8.4). A simplified summary is that the long-term water flux requirement from RAS to HP is equal to the crop water consumption by evapotranspiration and plant water storage in the plant biomass.

However, in terms of mass balances, the amount of nutrients leaving the hydroponics system via the plants needs to be replaced to assure a constant equilibrium. This poses a dilemma, as the maximum tolerable nutrient concentration in RAS is much lower than what is necessary in HP. The high nutrient flows ($\rho_{RAS} \times Q_{RAS}$) for HP can thus not be accomplished by the low RAS nutrient concentrations. Instead, without a distillation/desalination loop, the nutrient concentration would increase in the RAS while decreasing in the hydroponics system. A possible remedy is to discharge RAS water (and thus also nutrients) to decrease the nutrient concentration there and add fertilizer to the hydroponics nutrient solution. In terms of environmental and economic impact, this solution is less satisfying and does not serve the aim of a closed loop combined production.

The implementation of a distillation unit as shown in Fig. 8.3 represents a potential solution for this dilemma. Such distillation technologies (e.g. thermal

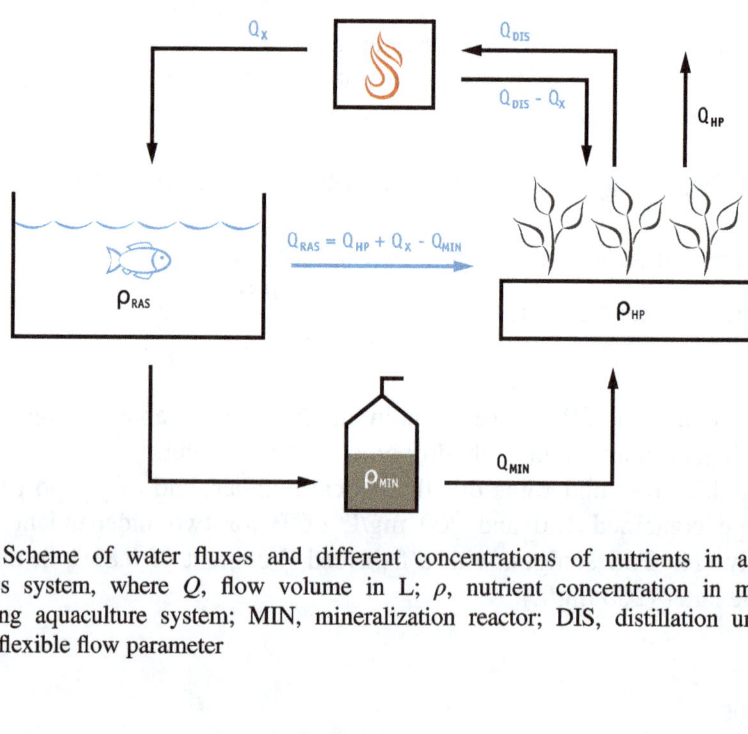

Fig. 8.4 Scheme of water fluxes and different concentrations of nutrients in a decoupled aquaponics system, where Q, flow volume in L; ρ, nutrient concentration in mg/L; RAS, recirculating aquaculture system; MIN, mineralization reactor; DIS, distillation unit; and X, unknown/flexible flow parameter

membrane distillation) have the potential to separate dissolved salts and nutrients from water (Shahzad et al. 2017; Subramani and Jacangelo 2015). In the context of multi-loop aquaponics systems, and as an alternative to additional fertilization and water bleed-off with corresponding extra costs, this technology could not only provide fresh water to the system but also achieve desired nutrient concentrations for the respective subsystems (Goddek and Keesman 2018).

For the implementation (i.e. sizing) of such a distillation unit, simple mass balance equations can be used. The remaining system, however, must be sized beforehand (either via rules of thumb or via mass balance equations; see Sect. 8.5), because the nutrients that enter the system should be in equilibrium with the bioavailable nutrients taken up by the crop (***Note:*** the sweet spot of decoupled systems is its flexibility. Consequently, one can also oversize the hydroponics part of the system although that will necessitate the use of more fertilizer). The easiest way to estimate nutrient uptake is to use the assumption that nutrients are taken up/absorbed much the same as dissolved ions in irrigation water (i.e. no element-specific chemical, biological or physical resistances). Consequently, to maintain equilibrium, all nutrients taken up by the crop as contained in the nutrient solution need to be added back to the hydroponics system (Eq. 8.4).

$$\phi_{\mathrm{RAS}} + \phi_{\mathrm{MIN}} - \phi_{\mathrm{HP}} = 0 \tag{8.4}$$

where ϕ_{RAS} is the nutrient flow from the RAS system to the hydroponics system, ϕ_{MIN} is the nutrient flow from the mineralization unit to the hydroponics system and ϕ_{HP} is the nutrient plant uptake. For this equation, it is assumed that the distillation system has an efficiency of close to 100%. Thus, Q_{DIS} goes back to the hydroponics subsystem.

Consequently:

$$(\rho_{\mathrm{HP}} \times Q_{\mathrm{HP}}) = (\rho_{\mathrm{RAS}} \times Q_{\mathrm{RAS}}) + (\rho_{\mathrm{MIN}} \times Q_{\mathrm{MIN}}) \tag{8.5}$$

where Q is the flow volume in L, and $\rho$is the nutrient concentration in mg/L.

As stated above, the flow from RAS to the hydroponics unit is the difference of the sum of the water flows leaving the hydroponics system (i.e. $Q_{\mathrm{HP}} + Q_{\mathrm{X}}$) and the inflow from the bioreactor (Q_{MIN}), i.e. $Q_{\mathrm{RAS}} = Q_{\mathrm{HP}} + Q_{\mathrm{X}} - Q_{\mathrm{MIN}}$, which leads us to the following equation:

$$\begin{aligned}(\rho_{\mathrm{HP}} \times Q_{\mathrm{HP}}) = {} & (\rho_{\mathrm{RAS}} \times Q_{\mathrm{HP}}) + (\rho_{\mathrm{RAS}} \times Q_{\mathrm{X}}) - (\rho_{\mathrm{RAS}} \times Q_{\mathrm{MIN}}) \\ & + (\rho_{\mathrm{MIN}} \times Q_{\mathrm{MIN}})\end{aligned} \tag{8.6}$$

The targeted variable is the distillation flow (Q_{X}) that is required to maintain the nutrient concentration equilibrium in the hydroponics system. For this, Eq. 8.6 is solved for Q_{X}in the following steps:

$$(\rho_{RAS} \times Q_X) = (\rho_{HP} \times Q_{HP}) - (\rho_{MIN} \times Q_{MIN}) - (\rho_{RAS} \times Q_{HP}) + (\rho_{RAS} \times Q_{MIN}) \tag{8.7}$$

$$Q_X = \frac{\rho_{HP} \times Q_{HP}}{\rho_{RAS}} - \frac{\rho_{MIN} \times Q_{MIN}}{\rho_{RAS}} - Q_{HP} + Q_{MIN} \tag{8.8}$$

Note that the distillation flow $Q_X$is highly dynamic and depends on the evapotranspiration rate of the plants, which is climate-dependent. The dynamic outcome, however, can be used for sizing the distillation unit. To calculate the required inflow into the distillation unit, the following formula can be used:

$$Q_{DIS} = Q_X \times \frac{100}{\eta_{DIS}} \tag{8.9}$$

where Q is the flow volume in L and η the demineralization efficiency of the used device (in %).

Distillation technology can hence drastically reduce the water and environmental (i.e. fertilizer usage) footprint of multi-loop aquaponics systems. However, aquaponics systems become even more complex when considering their implementation. Even though this additional loop might not make any sense for small-scale systems, it has the potential to take larger commercial systems to a new level. Yet, one has to consider that thermal distillation technology requires high amounts of thermal energy and might not be economically reasonable everywhere. Regions with high global solar radiation levels or geothermal energy sources might be the most suitable for this technology. The economical sustainability of such systems is consequently also location dependent.

Another point to bear in mind is the high temperature of distilled water and brine from the distillation unit. Depending on the environmental conditions and the fish species used, the hot distillation water could be used to heat up the RAS water; the brine, however, needs to cool down before re-entering the HP subsystem.

8.4 Sizing Multi-loop Systems

Sizing an aquaponics system requires balancing the nutrient input and -output. Here, we basically apply the same principle as sizing a one-loop system. Yet, this approach is a bit more complicated, but will be fully illustrated with the aid of an example.

Figure 8.5 illustrates the mass balance diagram for our system approach. In the optimal situation, the system has only one input and output. However, in practice, one will have to add additional nutrients to the hydroponics part to optimize plant growth. This model can be used to size the system, e.g. based on phosphorus, which is a non-renewable resource (Chap. 2). The input to the system (m_{feed}) is the fraction

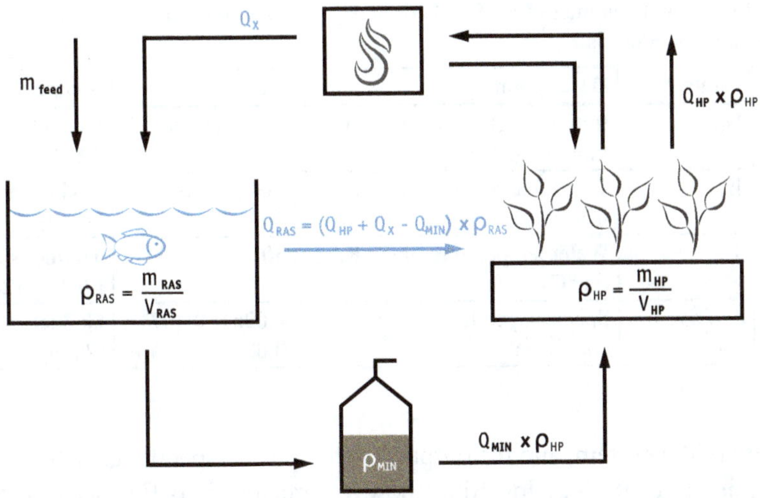

Fig. 8.5 Scheme that shows the mass balance within a four-loop aquaponics system; where m_{feed} are the dissolved nutrients added to the system via feed. Add labels: Q_{DIS} - Q_X to distillate returned to HP; 'sludge' for nutrients entering reactor

of a nutrient that the fish excrete in a dissolved form. The remainder accumulates in the fish as biomass or ends up as sludge (see previous section). The output is the plant nutrient uptake. Determining nutrient uptake of plants depends on many factors and is very complex; the easiest way to give a rough estimate is to consider plant respiration as the main driver of nutrient uptake (Goddek and Körner 2019).

Evapotranspiration rate is highly climate dependent and is either directly or indirectly influenced by absorbed shortwave radiation, relative humidity, temperature, and CO_2 concentration. Due to the high complexity of a multi-loop system, we assume that the plants are located in a climate-controlled greenhouse, and therefore we only need to consider global radiation as the dynamic variable determining how much shortwave radiation is absorbed. In other words, we first need to determine how much of the added nutrients become available for the plants, and then determine how much the plants actually take up.

8.4.1 Feed Input

The fish feed rate depends on the total biomass in the system and the feed conversion ratio (FCR). Timmons and Ebeling (2013) provide a simple approach for determining fish growth rates for different fish species. However, we recommend taking industrial data to determine the biomass more precisely. Lupatsch and Kissil (1998) (Eq. 8.10) provide a general growth formula, for which Goddek and Körner (2019) determined the growth coefficients by curve fitting using the mathematical software environment MATLAB (internal function 'fitnlm') with empirical data for Nile tilapia (*Oreochromis niloticus*). Additional initial and final weights, water

Table 8.2 Fish growth parameters for Eq. 8.10 for a given water temperature (T). W_0 and W_f can be adjusted to one's own needs

Function	Parameters	Description	Value	Source
Fish growth	W_0	Initial weight of the tilapia fingerlings (in g)	For example, 55	Goddek and Körner (2019)
	W_f	Target harvest weight of the fish (in g)	For example, 600	Goddek and Körner (2019)
	T	Water temperature of the RAS (in °C)	30	Timmons and Ebeling (2013)
	αw; βw; γw	Species-specific growth coefficients	0.0261; 0.4071; 0.0827	Goddek and Körner (2019)

temperature of the system, and the output for the species-specific growth coefficients can be found in Table 8.2. Inserting these parameters into Eq. 8.10 gives us the weight at a specific day for this fish species.

$$W_t = \left[W_0^{1-\beta_W} + (1-\beta_W)\alpha_W \exp\{\gamma_W T\}t\right]^{\frac{1}{1-\beta_W}} \tag{8.10}$$

where W_t (g) is the fish weight at a specific time (days), W_0 (g) is the initial fish weight, T is the water temperature (in °C), αw βw and γw are species-specific growth coefficients (no units), and t is the time in days.

Based on the output of the equation above, we were able to determine how much feed the fish will require per growth stage. Most of the times, the feed rate (X% of the body weight) or FCR is mentioned by the species-specific feed manufacturer. However, Timmons and Ebeling (2013) provide a rough guideline for FCR for tilapia: 0.7–0.9 for tilapia that weigh *less* than 100 g and 1.2–1.3 for tilapia that weigh *more* than 100 g. This is done via the following equation.

$$\text{Feed rate (g)} = \text{FCR} \times (WG_t \times m_{\text{fish}}) \tag{8.11}$$

where FCR is the feed conversion ratio, WG_t is the weight gain (per day), and m_{fish} is the amount of fish in the fish tank.

The weight gain (WG) per day can be determined with Eq. 8.10 by subtracting the weight of, e.g. day 10 from the weight of day 11. This can be done for each tank. Figure 8.6 shows the fish feed input to the system for tilapia using the equations above. The average feed input per day after the system is totally cycled is 165 kg.

8.4.2 Nutrient Availability

Neto and Ostrensky (2013) report a soluble N excretion of 33% and a soluble P excretion of 17% of feed input when rearing Nile tilapia (*Oreochromis niloticus,* L.).

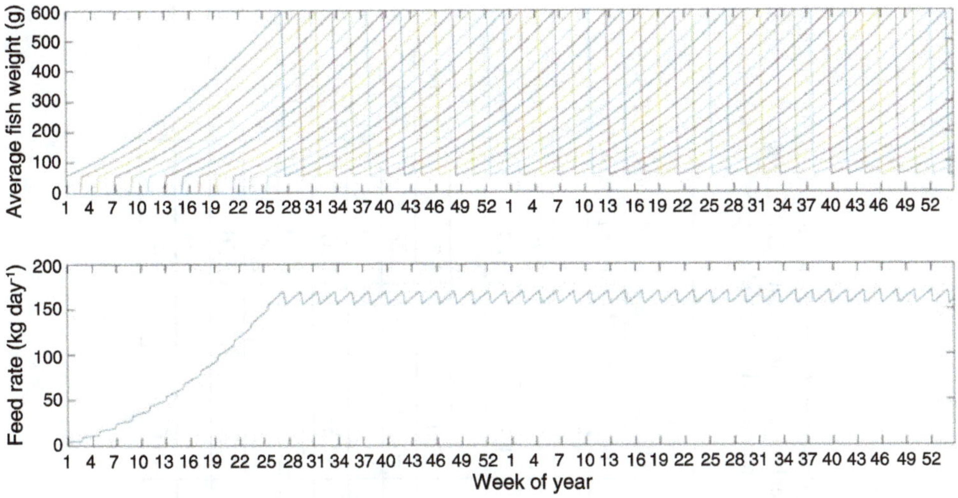

Fig. 8.6 Example of biomass balance for tilapia reared in 13 tanks in cohorts with a total volume (including biofilter and sump) of 482.000 L at a max. Total biomass of 80 t for a period of 2 years including start-up phase with average fish weight (**a**) (each line represents one tank/cohort) and the daily total feed rate (**b**) (data taken from Goddek and Körner 2019)

These are the nutrients that finally accumulate in the RAS system and can be taken up by the plants.

8.4.3 Plant Uptake

Table 8.3 gives an overview of the crop-specific evapotranspiration (ET_c) rates that are linked to global radiation. One mm of ET per square meter equals 1 L. For simple sizing, one should take the annual daily average (see next section).

8.4.4 Balancing the Subsystems

Balancing the loops is necessary for sizing the system. The input should be equal to the output (Fig. 8.5). In a decoupled aquaponics system incorporating a bioreactor unit, we have two nutrient inflow streams: (1) the fraction of the feed that is excreted to the RAS system in a soluble form and (2) the fraction of the nutrients in the fish sludge that the bioreactor(s) manage to mineralize and mobilize. The major outflow stream (apart from the periodic removal of demineralized sludge) of nutrients is the nutrient uptake of the plants. The differential Eq. 8.12. expresses this balance:

Table 8.3 Overview of outside global radiation levels subarctic, temperate maritime, and arid conditions (based on Goddek and Körner 2019) and their respective crop evapotranspiration (*ET_c, mm day-1*) rates for *lettuce* and *tomato* grown in a controlled greenhouse environment of 20°C and 80% relative humidity. Lettuce was cultivated with continuous planting year round; tomato was planted in January and removed in December (Faroe Islands and the Netherlands) or July and June (Namibia)

	Faroe Islands			The Netherlands			Namibia		
	Global radiation	ET_c/lettuce	ET_c/tomato	Global radiation	ET_c/lettuce	ET_c/tomato	Global radiation	ET_c/lettuce	ET_c/tomato
Month	mol m^{-2} day^{-1}	kg m^{-2} day^{-1}		mol m^{-2} day^{-1}	kg m^{-2} day^{-1}		mol m^{-2} day^{-1}	kg m^{-2} day^{-1}	
January	1.4	0.78	0.52	4.5	0.78	0.53	54.2	2.74	4.55
February	5.2	0.85	1.38	9.1	0.93	1.40	53.7	2.70	4.47
March	13.7	1.20	2.12	17.0	1.28	2.14	51.2	2.42	3.96
April	30.6	1.90	3.05	27.9	1.82	2.90	40.2	3.05	5.38
May	39.2	2.29	3.57	32.2	2.40	3.74	30.0	2.70	4.59
June	39.6	2.33	3.60	36.6	2.52	3.91	30.5	2.28	3.80
July	34.5	2.17	3.37	36.4	2.54	3.94	32.1	2.40	1.76
August	21.3	1.67	2.73	31.7	2.28	3.51	37.5	2.61	3.92
September	13.2	1.20	2.04	23.1	1.75	2.77	43.2	2.00	3.02
October	6.0	0.91	1.77	13.3	1.17	1.94	51.6	2.07	3.09
November	2.1	0.78	1.60	6.2	0.87	1.62	57.9	2.30	3.58
December	0.4	0.79	1.66	3.5	0.77	1.52	59.8	2.44	3.95
Average	17.3	1.41	2.28	20.6	1.59	2.50	45.6	2.47	3.83

$$\text{Mineralization (Eq.8.2)} + m_{\text{feed}} = \frac{Q_{\text{HP}} \times \rho_{\text{HP}}}{1000} \tag{8.12}$$

$$\left(n_{\text{feed}} \times 1000 \times \pi_{\text{feed}} \times \pi_{\text{sludge}} \times \eta_{\min}\right) + m_{\text{feed}} = \frac{Q_{\text{HP}} \times \rho_{\text{HP}}}{1000} \tag{8.13}$$

where n_{feed} is the average feed (in kg) entering the RAS system, $\pi_{\text{feed}}$is the proportion of the nutrient in the feed formulation, π_{sludge} is the proportion of a specific feed-derived element ending up in the sludge, and $\eta_{\min}$ is the mineralization and mobilization efficiency of the reactor system, m_{feed} is the average amount of a nutrient that the fish defecate in a dissolved form, Q_{HP} is the average total evapotranspiration, and $\rho_{\text{HP}}$is the target (i.e. optimal) nutrient concentration for a specific nutrient in the hydroponic subsystem.

However, to be able to determine the required area, there are two variables that need to be redefined in order to solve this equation. Equation 8.14 shows how to calculate the soluble nutrient excretion. In Eq. 8.15, we show that the average total evapotranspiration is a product of the area and the plant-specific evapotranspiration rate (here shown as an average) per m^2.

$$m_{\text{feed}} = n_{\text{feed}} \times \pi_{\text{feed}} \times \eta_{\text{excr}} \tag{8.14}$$

where η_{excr} represents the fraction of the nutrient excreted by the fish in a soluble form.

$$Q_{\text{HP}} = A \times ET_{\text{c}} \tag{8.15}$$

where Q_{HP} represents the average total evapotranspiration (in L), A the area, and ET_{c} the average crop-specific evapotranspiration in mm/m^2 (i.e. L/m^2).

Solving Eq. 8.13 by incorporating Eqs. 8.14 and 8.15 to find A, we are able to calculate the required plant area with respect to the average feed input (Eq. 8.15).

$$A = \frac{\left(n_{\text{feed}} \times 1000 \times \pi_{\text{feed}} \times \eta_{\text{excr}} \times 1000\right) + \left(n_{\text{feed}} \times 1000 \times \pi_{\text{feed}} \times \pi_{\text{sludge}} \times \eta_{\min} \times 1000\right)}{ET_{\text{c}} \times \rho_{\text{HP}}} \tag{8.16}$$

Example 8.2

For this example, we want to size (i.e. balance) the system with respect to P. We assume that the RAS component of our system requires an average daily feed input of 150 kg. The manufacturer reports the P content of the fish feed to be 1%. We estimate the P ending up in the sludge to be 55% and the P that fish excrete in a soluble form to be 17%. The bioreactors perform quite well and mineralize around 85% of the P.

On the output side, we calculated the average crop-specific evapotranspiration rate for lettuce (by, e.g. using the FAO Penman-Monteith equation). At our location, it is around 1.3 mm/day (i.e. 1.3 L/day). The optimal P composition of the nutrient solution is reported to be 50 mg/L (Resh 2013). Finding the area of plant cultivation needed to uptake the P produced by the system is then solved by:

$$A = \frac{\left(\underline{n_{\text{feed}}} \times 1000 \times \pi_{\text{feed}} \times \eta_{\text{excr}} \times 1000\right) + \left(\underline{n_{\text{feed}}} \times 1000 \times \pi_{\text{feed}} \times \pi_{\text{sludge}} \times \eta_{\min} \times 1000\right)}{\underline{ET_{\text{c}}} \times \rho_{\text{HP}}}$$

$$A = \frac{(150000 \times 0.01 \times 0.17 \times 1000) + (150000 \times 0.01 \times 0.55 \times 0.85 \times 1000)}{1.3 \times 50}$$

$$= \frac{255000 + 701250}{65}$$

$$= 14711\text{m}^2 = 1.47\text{ha}$$

The example above shows that the majority of the P in the hydroponics unit originates from the bioreactors. Thus, implementation of a bioreactor within a decoupled system has a very high impact on P sustainability. By contrast, in order to size simple one-loop aquaponics systems, a rule of thumb is usually applied. For leafy plants approx. 40–50 g and for fruity plants approx. 50–80 g of feed is required per m^2 cultivation area (FAO 2014). When looking at the feed input in the given example above = 150 kg, and dividing it by 45 (the average of the leafy plant approximation), the proposed cultivation area is around 3750 m^2. Leaving out the sludge mineralization, our example would suggest a cultivation area of 3333 m^2 when sizing the system on P.

8.4.5 *Role of the Distillation Unit*

The role of the distillation unit is to keep the nutrient concentration of the RAS system and the hydroponics system at their respective desired levels. Since nutrient accumulation and the corresponding specific nutrient density are dynamic in RAS systems (i.e. depending on the ET_c rates) that depend on the Q_{HP} and Q_X flow (Fig. 8.5), the size of the distillation unit cannot be determined using a differential

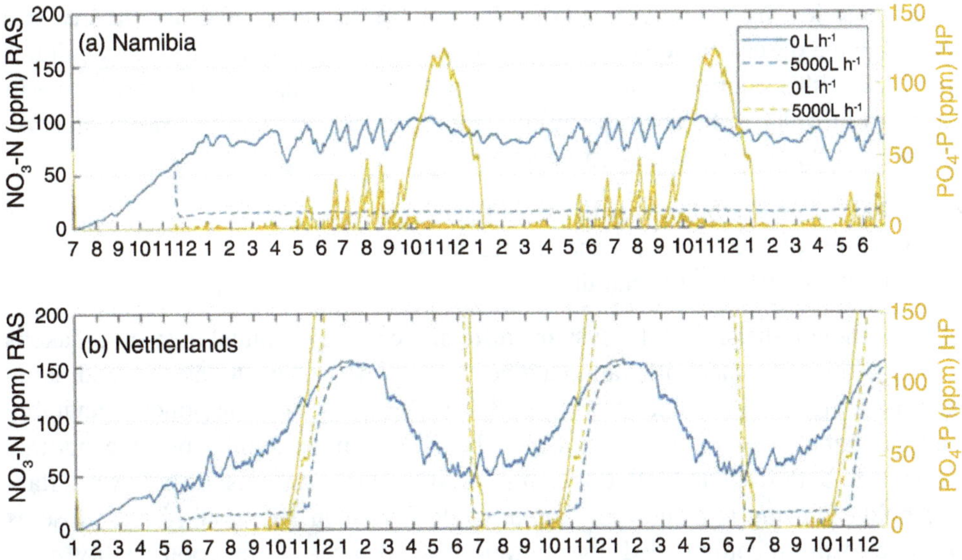

Fig. 8.7 Simulations comparing NO3-N concentration in the RAS water system on the impact of distillation flows (no, solid line; 5000 L h-1, dashed line) on hydroponics (yellow, ---) and RAS (blue, ---) nutrient solution concentrations in (**a**) Namibia and (**b**) the Netherlands, i.e. in low and high latitudes (Namibia 22.6°S and the Netherlands, 52.1°N, respectively) within a 36-month period (including the system run-up phase) using local climate data and climate-adjusted greenhouses as model input

equation. Thus, a time series model is required to determine the nutrient concentration in the RAS over time. The nutrient concentration at a specific time is necessary to be able to execute mass balance equations within the system (Sect. 8.3).

For the system to be balanced (i.e. input = output), we can give a general guideline on the required capacity of such a distillation unit. The objective is to avoid nutrient accumulation in the RAS system. Figure 8.7a, b shows the impact of distillation flows on the hydroponics and RAS nutrient solution *without a mineralization loop in two different latitudes*. Both systems have the same feed input (in average 158.6 kg day^{-1}; see Fig. 8.6). However, by taking the environmental conditions and climate-adjusted greenhouses into account, the necessary and optimal hydroponic area differs between geographical locations (see Chap. 11). Hydroponic systems with low potential evaporation rates, as are common in locations at high latitudes (i.e. far from the equator) would need larger cultivation areas than places closer to the equator. At the same time, a higher annual variation in irradiation and thus transpiration is common in these regions, thus a higher demand on seasonal variability on water and nutrients is present (see Fig. 8.7). In greenhouse cultivation, however, supplementary lighting may be necessary, and in countries such as Norway, vegetable cultivation without supplementary lighting hardly takes place. In addition, the total crop leaf surface makes a difference; crops with a high leaf area per unit ground area (i.e. leaf area index) transpire more than crops with smaller leaf areas, and a distinct difference can be seen between tomato and lettuce crops. All of these factors need to be considered when planning and sizing the aquaponic system.

In the following we provide an overview of the optimized hydroponic area size for the above described aquaponics systems: The cultivation area for monocultures simulated with scenarios in steps of 250 m^2 to find the fitting area of either lettuce or tomatoes in order to balance the system appropriately was *without supplementary lighting* (for lettuce or tomato, respectively):

- 17.000 m^2 or 11.750 m^2 for Faroe Islands
- 15.500 m^2 or 11.000 m^2 for the Netherlands
- 8750 m^2 or 6500 m^2 for Namibia

Even though the size of the systems differs, the average annual nutrient uptake is similar. However, when integrating a digester system, we have to take the additional nutrient source into consideration (Fig. 8.1c). Changing one component inevitably leads to imbalances of the system, yet the system must aim to provide optimal nutrients to both RAS and HP. For example, NO3-N in RAS must be below a certain threshold <200 mg L^{-1} for, e.g. tilapia, while PO4-P in HP should be as close as possible to the recommended concentration of 50 mg L^{-1} for good-quality plant cultivation. Thus, simulation studies help determine sizing of components in a decoupled closed multi-loop aquaponics system in order to achieve optimal nutrient supplies for both fish and plants. For that purpose, Goddek and Körner (2019) created a numeric aquaponics simulator.

However, planning an aquaponic system involves some basic system understanding in order to reach a balance that minimizes the unwanted peaks in nutrient demand and supply. Since the driving force for nutrient dynamics is the evapotranspiration of the crop (ET_c in the HP system), that is largely driven by microclimate and absorbed light. In a perfectly balanced system, this would be a fully automated and controlled (see Sect. 8.5 Monitoring and Control) environment with 24-h lighting. Plants need a certain dark period of about 4–6 h, so the best-balanced system is realistically to carry out aquaponics in closed plant factories solely with artificial light sources. This, however, demands high electrical input and investment costs and is only feasible with very high product prices. Therefore, we recommend greenhouse production with supplementary lighting (if necessary and if it pays off) as a practical and economically feasible way of building an aquaponic unit. Placing both plants and fish in the same physical construction results in additional synergies including reduced heating and increased plant growth through elevated CO_2 (Körner et al. 2017).

In addition to these technical issues, plant cultivation procedures (the practical horticultural part of the system) have to be adjusted to the needs of aquaponics such that there is a constant crop nutrient demand (assuming same climate and light) as shown in Table 8.3. Cultivation of lettuce and other leafy greens are carried out continuously (Körner et al. 2018), while larger crops, like fruit vegetables such as tomato, cucumber, or sweet pepper, are usually sown in winter, and the first harvest is often in late winter/early spring followed by removal of plants and another crop sown for harvest in winter again. Without interplanting, i.e. either various crop types in the same system or batches of fruit vegetables planted throughout the year in order to sustain nutrient demand, periods of low nutrient demand and high nutrient levels

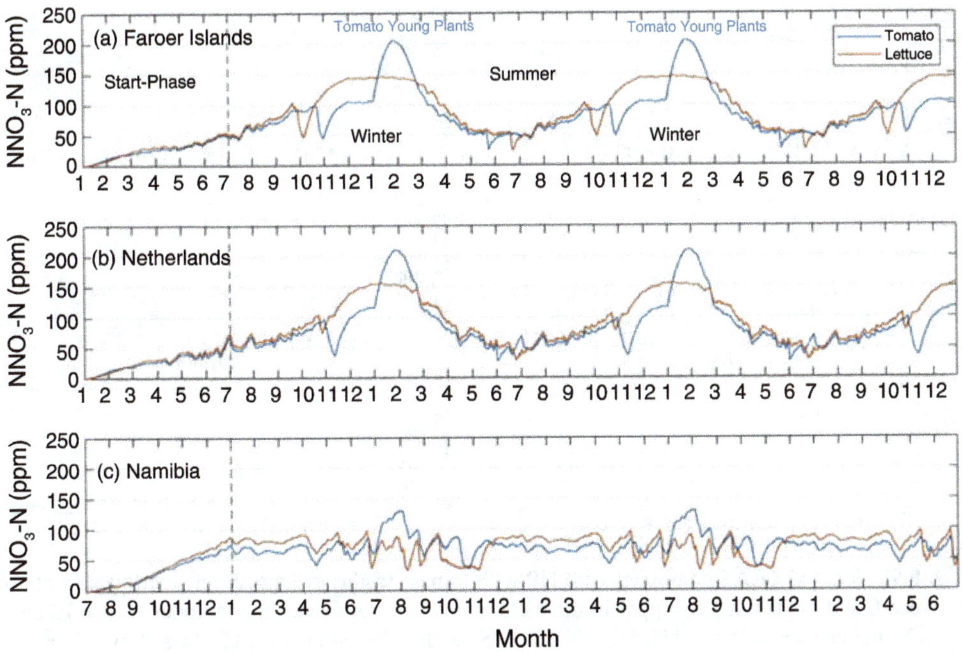

Fig. 8.8 NO3-N in RAS combined with HP growing tomato or lettuce in three climate zones and decreasing latitudes (Faroe Islands 62.0°N, the Netherlands 52.1°N, Namibia 22.6°S) with optimized area for hydroponics (see above) in a 36-month simulation using local climate data and climate-adjusted greenhouses as model input

will occur. Based on Goddek and Körner (2019), we show the variation of NO_3-N in RAS for tomato (often not adjusted in aquaponics) and lettuce when no supplementary lighting is used for three climate zones (Faroe Islands, the Netherlands, and Namibia) (Fig. 8.8). System balance is achievable by increasing the daily light integral (i.e. sum of mol light received during a 24-h period) with dynamic supplementary lighting control (Körner et al. 2006).

Applying distillation/desalination technologies can contribute to significant reductions in nutrient levels in the RAS while adjusting levels in the HP system closer to optima, i.e. the unit concentrates nutrients to levels required by plants. Figure 8.9 illustrates the effect of a desalination unit on RAS NO_3-N concentration when applying between 0 and 5000 L h^{-1} and $system^{-1}$. It is obvious that with increasing desalination flux, the NO_3-N concentration in the RAS system is decreasing. The unit, however, is controlled by the demand of PO_4 in the HP system. Peaks need to be avoided and, as stated above, this can be achieved by creating a stable climatic environment with dynamic light controls. It is obvious that in climate regions with fewer annual differences in solar radiation, there is less variation in ET_c and the complete system is more stable. Installing lamps and keeping a daily light integral of at least 10 mol m^{-2} can compensate for seasonal variations. Interplanting and mixed crop production help level the peak resulting from the traditional tomato cultivation protocol with young plants in winter when both

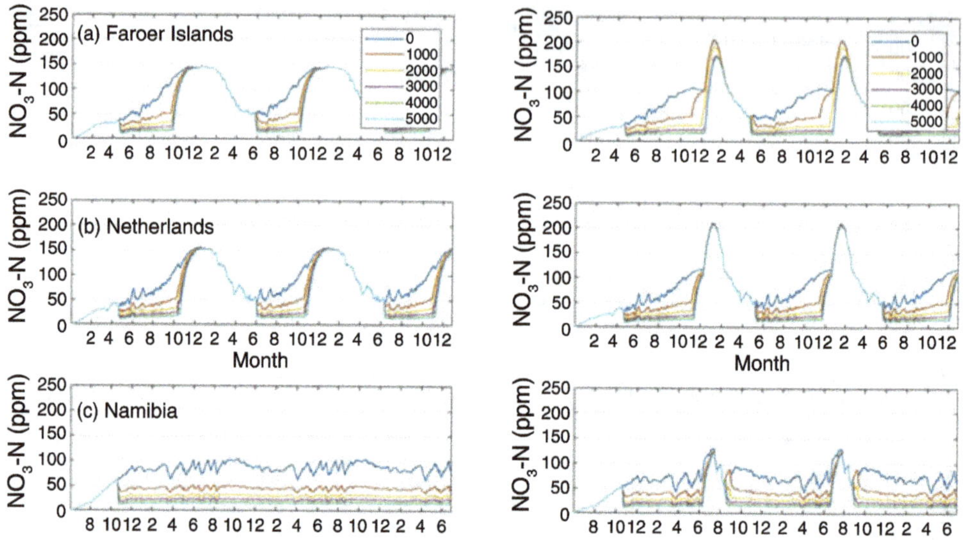

Fig. 8.9 NO3-N in RAS combined with HP with tomato (right) or lettuce (left) with desalination between 0 and 5000 L h^{-1} supply in three climate zones and decreasing latitudes (Faroe Islands 62.0°N, the Netherlands 52.1°N, Namibia 22.6°S) with adjusted area for HP (see above) in a 36-month simulation using local climate data and climate-adjusted greenhouses as model input

climate (low radiation) and cultivation (small plants, low potential ET_c) contribute to nutrient accumulation.

8.5 Monitoring and Control

In classical feedback control, like PI or PID (Proportional-Integral-Derivative) control, the controlled variables (CV) are directly measured, compared with a setpoint, and subsequently fed back to the process via a feedback control law.

In Fig. 8.10, the signals, without the time argument, are denoted by a small letter, where y is the controlled variable (CV) which is compared with the reference (setpoint) signal r. The tracking error $\boldsymbol{\varepsilon}$ (i.e. $r - y$) is fed into the controller, either in hardware or software, from which the control input u, also known as the manipulated variable (MV), is generated. The input u directly affects the process (P) from which an output (y) results. The sampled output is subsequently compared with r, which closes the loop. In practice, this loop continues until the controller is switch off. There exists extensive literature on feedback control (Doyle et al. 1992; Morris 2001; Ogata 2010), and this has been a subject of research for many years, starting with the works of Bode (1930) and Nyquist (1932).

In RAS, typical CVs are temperature, pH, and dissolved oxygen (DO) concentration, for which reliable sensors exist. Consequently, feedback control of these water quality parameters can be easily realized. However, in practice, most often, the input and output signals are disturbed by noise processes, such as

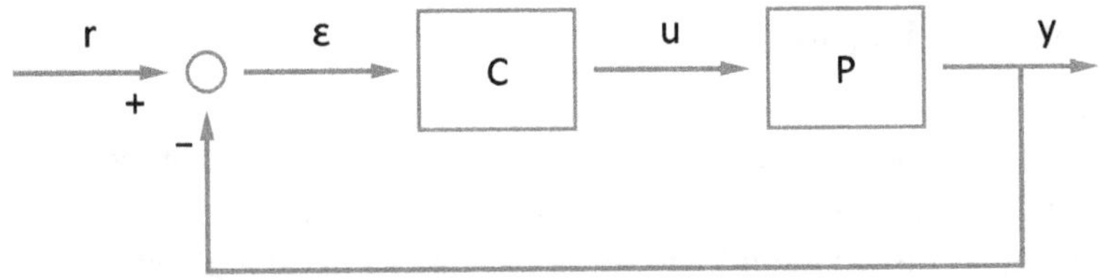

Fig. 8.10 Feedback control with controller (C) and process (*P*). r reference signal, eps tracking error, u input signal, y output sigma

unknown random inputs and measurement noise. Moreover, the overall process (*P*) may change over time as a result of growth, maturation, senescence, etc. Fish feed is another input into the RAS and its effect on fish growth cannot be directly seen or measured. For these parameters, model-based controllers (e.g. feedforward, model predictive, and optimal control) are typically introduced to predict the response of a change in the control input. However, fish feed is commonly added on the basis of values found in tables or recipes, but this rule-based control may need some adjustment in real practice to act as a feedback controller. Fish behaviours in RAS are a classical feedback control measure as fish react physiologically to environmental changes with variations in movement, location, receptiveness to feed, etc.

Hydroponic production usually takes place in protected environments such as greenhouses or plant factories where both the root and aerial environment need to be controlled. On-off controllers that predictively model optimal aerial environments have been proven superior in experimental research, but commercialization has been slow, whereas feedback controllers are standard in most climate-controlled greenhouses. However, the actuator varies with the type of controller with heating valves and vents typically feedback-controlled but lighting usually having an ON-OFF mechanism and only a few being dimmable. Controllers that rely on sensor or data input can respond to fast growth in a protected environment and result in high-quality produce with high market prices that improves its cost benefits. Many commercial greenhouses still have the classical centrally located sensor hanging 1–2 m above the crop and covering several hundred square meters is still in use, but multiple wireless sensors covering smaller areas are being introduced although much of the detailed data cannot be used because rather large climatic zones are controlled by the same actuators. Advances in sensor technology (e.g. microclimate temperature sensors, image processors, real-time gas-exchange or chlorophyll fluorescence measurements) connected to modelling software could use decision-support systems and become automated control systems.

In typical bioreactor systems, temperature, pH, dissolved oxygen in aerated systems, and gas fluxes in anaerobic systems are continuously measured and adjusted with available temperature, pH, and dissolved oxygen controllers. In addition to this, both hydraulic (HRT) and sludge retention (SRT) times are also frequently set by controlling (waste)water flows and biomass waste flows, respectively.

8.6 Economic Impact

Technologies that generate less profit, but are better for the environment usually only get implemented when the operators either receive an incentive in the form of subsidies or policies force them to do so. In the case of one-loop aquaponics systems, the appeal lies in the novel technology and the system's approach to sustainable resource use rather than its economic potential. However, recent publications provide evidence for production gains: leafy greens grow better in decoupled environments than in sterile hydroponic systems (Delaide et al. 2016; Goddek and Vermeulen 2018) and lettuce in decoupled aquaponics systems had a growth advantage of approximately 40% compared to state-of-the-art hydroponic approaches.

Even though higher growth rates can be expected, multi-loop aquaponics systems are still far more complex than hydroponics systems and significant initial investments are required for implementation. Most geographic locations require a high-tech greenhouse to control environmental conditions (i.e. a relative humidity of 80%, constant temperatures of around 20°C). Renewable energy sources can be used for cooling and heating, but currently such systems are only profitable when setting up on a large scale (i.e. > 1 ha) where good market conditions prevail.

8.7 Environmental Impact

Based on Example 8.2, there is evidence that treating sludge in digesters can have a beneficial impact on nutrient reutilization, especially phosphorus. Bioreactor systems, such as a sequential two-stage UASB reactor system, can increase the phosphorus recycling efficiency up to 300% (Chap. 10). Previously, in Chap. 2, we discussed the phosphorus paradox in relation to both phosphate scarcity and problems with eutrophication. Bioreactors have significant advantages for increased nutrient recovery from sludge, thus helping to close the nutrient cycling loop within aquaponics systems. However, further research is needed to refine such systems to optimize the bioavailability of specific nutrients. Figures 8.11, 8.12, and 8.13 show the input, output, and waste streams of stand-alone aquaculture and hydroponics systems compared with a decoupled aquaponics system. It can be seen that the decoupled approach constitutes a promising agricultural concept for a waste reduction and recycling system.

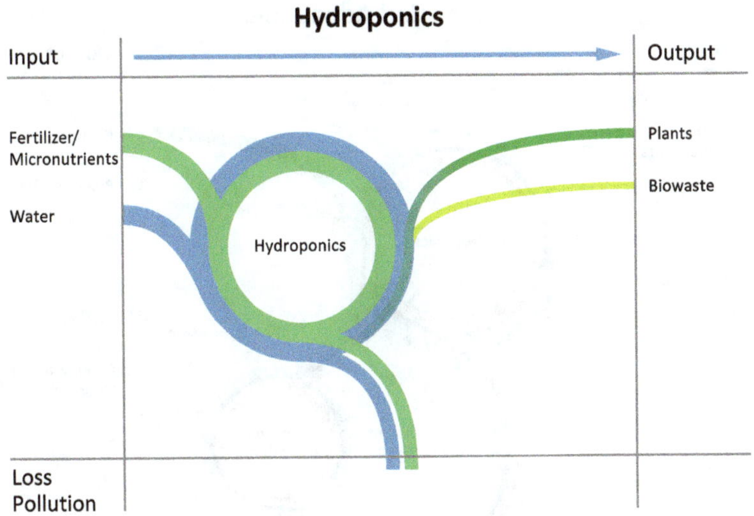

Fig. 8.11 Input, output, and loss streams in a stand-alone hydroponic system

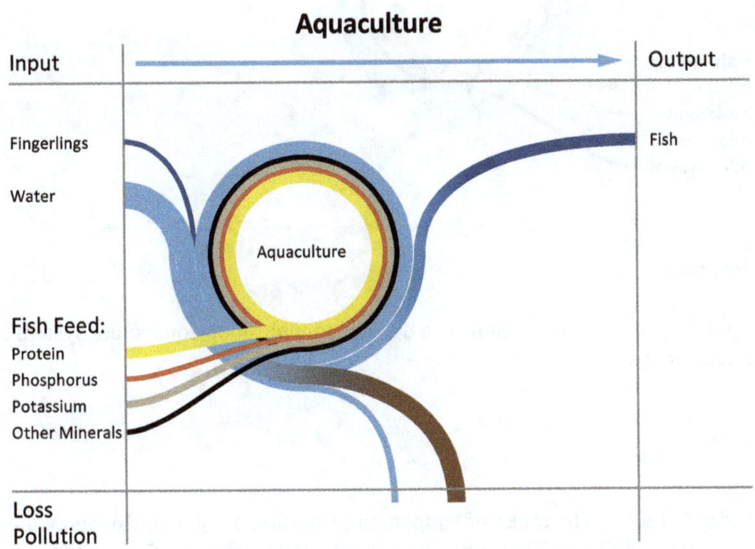

Fig. 8.12 Input, output, and loss streams in a stand-alone aquaculture system

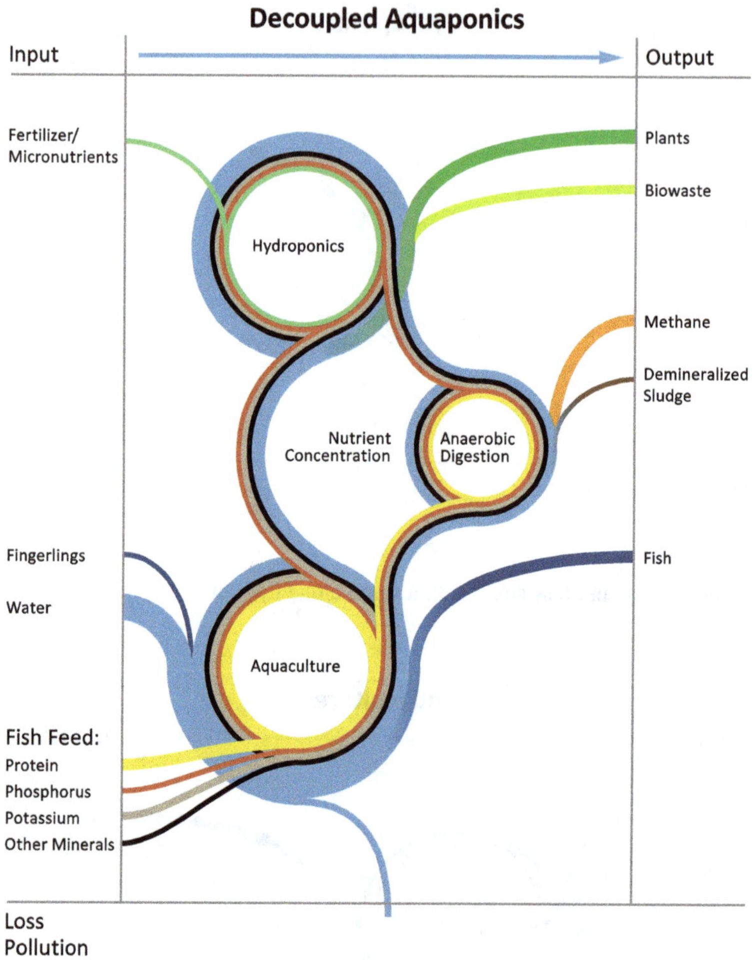

Fig. 8.13 Input, output, and loss streams in a decoupled multi-loop aquaponics system comprising an anaerobic reactor system

References

Alvarez R, Lidén G (2008) The effect of temperature variation on biomethanation at high altitude. Bioresour Technol 99:7278–7284. https://doi.org/10.1016/j.biortech.2007.12.055

Bode HW (1930) A method of impedance correction. Bell Syst Tech J 9:794–835. https://doi.org/10.1002/j.1538-7305.1930.tb02326.x

Coghlan SM, Ringler NH (2005) Temperature-dependent effects of rainbow trout on growth of Atlantic Salmon Parr. J Great Lakes Res 31:386–396. https://doi.org/10.1016/S0380-1330(05)70270-7

Dalsgaard J, Lund I, Thorarinsdottir R, Drengstig A, Arvonen K, Pedersen PB (2013) Farming different species in RAS in Nordic countries: current status and future perspectives. Aquac Eng 53:2–13

Davidson J, Good C, Welsh C, Summerfelt ST (2011) Abnormal swimming behavior and increased deformities in rainbow trout Oncorhynchus mykiss cultured in low exchange water recirculating aquaculture systems. Aquac Eng 45:109–117

de Lemos Chernicharo CA (2007) Anaerobic reactors, 4th edn. IWA Publishing, New Delhi

Delaide B, Goddek S, Gott J, Soyeurt H, Jijakli M (2016) Lettuce (Lactuca sativa L. var. Sucrine) growth performance in complemented aquaponic solution outperforms hydroponics. Water 8:467. https://doi.org/10.3390/w8100467

Delaide B, Goddek S, Keesman KJ, Jijakli MH (2018) Biotechnologie, agronomie, société et environnement = Biotechnology, agronomy, society and environment: BASE. https://popups.uliege.be:443/1780-4507 22, 12

Doyle JC, Francis BA, Tannenbaum A (1992) Feedback control theory. Macmillan Pub. Co

El-Sayed A-FM (2006) Tilapia culture. CABI Publishing, Oxfordshire

Emerenciano M, Carneiro P, Lapa M, Lapa K, Delaide B, Goddek S (2017) Mineralizacão de sólidos. Aquac Bras: 21–26

FAO (2005) Cultured aquatic species information programme. Oncorhynchus mykiss. [WWW Document]. FAO Fish. Aquac. Dep. [online]. URL http://www.fao.org/fishery/culturedspecies/Oncorhynchus_mykiss/en. Accessed 21 Sep 2015

FAO (2014) Small-scale aquaponic food production: integrated fish and plant farming. FAO, Rome

Goddek S (2017) Opportunities and challenges of multi-loop aquaponic systems. Wageningen University, Wageningen. https://doi.org/10.18174/412236

Goddek S, Keesman KJ (2018) The necessity of desalination technology for designing and sizing multi-loop aquaponics systems. Desalination 428:76–85. https://doi.org/10.1016/j.desal.2017.11.024

Goddek S, Körner O (2019) A fully integrated simulation model of multi-loop aquaponics: a case study for system sizing in different environments. Agric Syst 171:143–154

Goddek S, Vermeulen T (2018) Comparison of Lactuca sativa growth performance in conventional and RAS-based hydroponic systems. Aquac Int 26:1–10. https://doi.org/10.1007/s10499-018-0293-8

Goddek S, Delaide B, Mankasingh U, Ragnarsdottir K, Jijakli H, Thorarinsdottir R (2015) Challenges of sustainable and commercial aquaponics. Sustainability 7:4199–4224. https://doi.org/10.3390/su7044199

Goddek S, Espinal C, Delaide B, Schmautz Z, Jijakli H (2016) Navigating towards decoupled Aquaponic systems (DAPS): a system dynamics design approach. Water 8:7

Goddek S, Delaide BPL, Joyce A, Wuertz S, Jijakli MH, Gross A, Eding EH, Bläser I, Reuter M, Keizer LCP, Morgenstern R, Körner O, Verreth J, Keesman KJ (2018) Nutrient mineralization and organic matter reduction performance of RAS-based sludge in sequential UASB-EGSB reactors. Aquac Eng 83:10–19. https://doi.org/10.1016/J.AQUAENG.2018.07.003

Jones BJ (2005) Hydroponics – a practical guide for the soilless grower, 2nd edn. CRC Press, Boca Raton

Jung IS, Lovitt RW (2011) Leaching techniques to remove metals and potentially hazardous nutrients from trout farm sludge. Water Res 45:5977–5986. https://doi.org/10.1016/j.watres.2011.08.062

Karimanzira D, Keesman KJ, Kloas W, Baganz D, Rauschenbach T (2016) Dynamic modeling of the INAPRO aquaponic system. Aquac Eng 75:29–45. https://doi.org/10.1016/j.aquaeng.2016.10.004

Karimanzira D, Keesman K, Kloas W, Baganz D, Rauschenbach T (2017) Efficient and economical way of operating a recirculation aquaculture system in an aquaponics farm. Aquac Econ Manag 21:470–486. https://doi.org/10.1080/13657305.2016.1259368

Kloas W, Groß R, Baganz D, Graupner J, Monsees H, Schmidt U, Staaks G, Suhl J, Tschirner M, Wittstock B, Wuertz S, Zikova A, Rennert B (2015) A new concept for aquaponic systems to improve sustainability, increase productivity, and reduce environmental impacts. Aquac Environ Interact 7:179–192. https://doi.org/10.3354/aei00146

Körner O, Andreassen AU, Aaslyng JM (2006) Simulating dynamic control of supplementary lighting. Acta Hortic 711:151–156. https://doi.org/10.17660/ActaHortic.2006.711.17

Körner O, Gutzmann E, Kledal PR (2017) A dynamic model simulating the symbiotic effects in aquaponic systems. Acta Hortic 1170:309–316. https://doi.org/10.17660/ActaHortic.2017.1170.37

Körner O, Pedersen JS, Jaegerholm J (2018) Simulating lettuce production in multi layer moving gutter systems. Acta Hortic 1227:283–290

Lupatsch I, Kissil GW (1998) Predicting aquaculture waste from gilthead seabream (*Sparus aurata*) culture using a nutritional approach. Aquat Living Resour 11:265–268. https://doi.org/10.1016/S0990-7440(98)80010-7

Mirzoyan N, Gross A (2013) Use of UASB reactors for brackish aquaculture sludge digestion under different conditions. Water Res 47:2843–2850. https://doi.org/10.1016/j.watres.2013.02.050

Monsees H, Keitel J, Paul M, Kloas W, Wuertz S (2017a) Potential of aquacultural sludge treatment for aquaponics: evaluation of nutrient mobilization under aerobic and anaerobic conditions. Aquac Environ Interact 9:9–18. https://doi.org/10.3354/aei00205

Monsees H, Kloas W, Wuertz S, Cao X, Zhang L, Liu X (2017b) Decoupled systems on trial: eliminating bottlenecks to improve aquaponic processes. PLoS One 12:e0183056. https://doi.org/10.1371/journal.pone.0183056

Morris KA (2001) Introduction to feedback control. Harcourt/Academic Press, San Diego

Neto RM, Ostrensky A (2013) Nutrient load estimation in the waste of Nile tilapia Oreochromis niloticus (L.) reared in cages in tropical climate conditions. Aquac Res 46:1309–1322. https://doi.org/10.1111/are.12280

Nyquist H (1932) Regeneration theory. Bell Syst Tech J 11:126–147. https://doi.org/10.1002/j.1538-7305.1932.tb02344.x

Ogata K (2010) Modern control engineering, 5th edn. Pearson, Delhi

Rakocy JE, Shultz RC, Bailey DS, Thoman ES (2004) Aquaponic production of tilapia and basil: comparing a batch and staggered cropping system. Acta Hortic 648:63–69

Resh HM (2002) Hydroponic tomatoes. CRC Press, Boca Raton

Resh HM (2012) Hydroponic food production: a definitive guidebook for the advanced home gardener and the commercial hydroponic grower. CRC Press, Boca Raton

Resh HM (2013) Hydroponic food production: a definite guidebook for the advanced home gardener and the commercial hydroponic grower, 7th edn. CRC Press, Hoboken

Reyes Lastiri D, Slinkert T, Cappon HJ, Baganz D, Staaks G, Keesman KJ (2016) Model of an aquaponic system for minimised water, energy and nitrogen requirements. Water Sci Technol. wst2016127 74:30–37. https://doi.org/10.2166/wst.2016.127

Ross LG (2000) Environmental physiology and energetics. In: McAndrew BJ (ed) Tilapias: biology and exploitation. Springer, Dordrecht, pp 89–128

Schrader KK, Davidson JW, Summerfelt ST (2013) Evaluation of the impact of nitrate-nitrogen levels in recirculating aquaculture systems on concentrations of the off-flavor compounds geosmin and 2-methylisoborneol in water and rainbow trout (Oncorhynchus mykiss). Aquac Eng 57:126–130. https://doi.org/10.1016/j.aquaeng.2013.07.002

Shahzad MW, Burhan M, Ang L, Ng KC (2017) Energy-water-environment nexus underpinning future desalination sustainability. Desalination 413:52–64. https://doi.org/10.1016/j.desal.2017.03.009

Sonneveld C, Voogt W (2009) Nutrient management in substrate systems. In: Plant nutrition of greenhouse crops. Springer, Dordrecht, pp 277–312. https://doi.org/10.1007/978-90-481-2532-6_13

Subramani A, Jacangelo JG (2015) Emerging desalination technologies for water treatment: a critical review. Water Res 75:164–187. https://doi.org/10.1016/j.watres.2015.02.032

Timmons MB, Ebeling JM (2013) Recirculating aquaculture, 3rd edn. Ithaca Publishing Company LLC, Ithaca

van Gorcum B, Goddek S, Keesman KJ (2019) Gaining market insights for aquaponically produced vegetables in Kenya. Aquac Int:1–7. https://link.springer.com/article/10.1007/s10499-019-00379-1

Yavuzcan Yildiz H, Robaina L, Pirhonen J, Mente E, Domínguez D, Parisi G (2017) Fish welfare in aquaponic systems: its relation to water quality with an emphasis on feed and faeces—a review. Water 9:13. https://doi.org/10.3390/w9010013

Yogev U, Barnes A, Gross A (2016) Nutrients and energy balance analysis for a conceptual model of a three loops off grid. Aquaponics Water 8:589. https://doi.org/10.3390/W8120589

9

Nutrients in Aquaponics and their Cycling

Mathilde Eck, Oliver Körner, and M. Haïssam Jijakli

Abstract In aquaponics, nutrients originate mainly from the fish feed and water inputs in the system. A substantial part of the feed is ingested by the fish and either used for growth and metabolism or excreted as soluble and solid faeces, while the rest of any uneaten feed decays in the tanks. While the soluble excretions are readily available for the plants, the solid faeces need to be mineralised by microorganisms in order for its nutrient content to be available for plant uptake. It is thus more challenging to control the available nutrient concentrations in aquaponics than in hydroponics. Furthermore, many factors, amongst others pH, temperature and light intensity, influence the nutrient availability and plant uptake. Until today, most studies have focused on the nitrogen and phosphorus cycles. However, to ensure good crop yields, it is necessary to provide the plants with sufficient levels of all key nutrients. It is therefore essential to better understand and control nutrient cycles in aquaponics.

Keywords Aquaponics · Nutrient cycling · Solubilisation · Microbiological processes

9.1 Introduction

Aquaponic systems offer various advantages when it comes to producing food in an innovative and sustainable way. Besides the synergistic effects of increased aerial CO_2 concentration for greenhouse crops and decreased total heat energy consumption when cultivating fish and crops in the same space (Körner et al. 2017),

M. Eck (✉) · M. H. Jijakli
Integrated and Urban Plant Pathology Laboratory, Université de Liège, Agro-Bio Tech, Gembloux,, Belgium
e-mail: mathilde.eck@uliege.be; mh.jijakli@uliege.be

O. Körner
Leibniz-Institute of Vegetable and Ornamental Crops (IGZ), Grossbeeren, Germany
e-mail: koerner@igzev.de

aquaponics has two main advantages for nutrient cycling. First, the combination of a recirculating aquaculture system with hydroponic production avoids the discharge of aquaculture effluents enriched in dissolved nitrogen and phosphorus into already polluted groundwater (Buzby and Lin 2014; Guangzhi 2001; van Rijn 2013), and second, it allows for the fertilisation of the soilless crops with what can be considered an organic solution (Goddek et al. 2015; Schneider et al. 2004; Yogev et al. 2016) instead of using fertilisers of mineral origin made from depleting natural resources (Schmautz et al. 2016; Chap. 2). Furthermore, aquaponics yields comparable plant growth as compared with conventional hydroponics despite the lower concentrations of most nutrients in the aquaculture water (Graber and Junge 2009; Bittsanszky et al. 2016; Delaide et al. 2016), and production can be even better than in soil (Rakocy et al. 2004). Increased CO_2 concentrations in the aerial environment and changes in the biomes of the root zone are thought to be main reasons for this. In addition, the mineral content and the nutritional quality of tomatoes grown aquaponically have been reported to be equivalent or superior to the mineral content of conventionally grown ones (Schmautz et al. 2016).

Despite having two attractive assets (i.e. the recycling of aquaculture effluents and the use of organic fertilisers), the use of aquaculture effluents increases the challenge of monitoring the nutrients within the solution. Indeed, it is harder to control the composition of a solution where the nutrients originate from a biological degradation of organic matter than to follow the evolution of the nutrients' concentration in a precisely dosed hydroponic solution based on mineral compounds (Bittsanszky et al. 2016; Timmons and Ebeling 2013). Moreover, a plant's nutritional needs vary during the growth period in accordance with physiological stages, and it is necessary to meet these needs to maximise yields (Bugbee 2004; Zekki et al. 1996; Chap. 4).

In order to recycle aquaculture effluents to produce plant biomass, it is necessary to optimise the recycling rates of phosphorus and nitrogen (Goddek et al. 2016; Graber and Junge 2009; Chap. 1). Several factors can influence this, such as the fish species, fish density, water temperature, the type of plants and the microbial community (ibid.). Therefore, it is of prime importance to understand the functioning of the nutrient cycles in aquaponics (Seawright et al. 1998). This chapter aims at explaining the origins of the nutrients in an aquaponic system, describing the nutrient cycles and analysing the causes of nutrient losses.

9.2 Origin of Nutrients

The major sources of nutrients in an aquaponic system are the fish feed and the water added (containing Mg, Ca, S) (see Sect. 9.3.2.) into the system (Delaide et al. 2017; Schmautz et al. 2016) as further elaborated in Chap. 13. With respect to fish feed, there are two main types: fishmeal-based and plant-based feed. Fishmeal is the classic type of feed used in aquaculture where lipids and proteins rely on fish meal and fish oil (Geay et al. 2011). However, for some time now, concerns regarding the sustainability of such feed have been raised and attention drawn towards plant-based diets (Boyd 2015; Davidson et al. 2013; Hua and Bureau 2012; Tacon and Metian 2008). A meta-analysis conducted by Hua and Bureau (2012) revealed that the use of plant proteins in fish feed can influence the growth of fish if incorporated in high proportions. Indeed, plant proteins can have an impact on the digestibility and levels of anti-nutritional factors of the feed. In particular, phosphorus originating from plants and thus in the form of phytates does not benefit, for example, salmon, trout and several other fish species (Timmons and Ebeling 2013). It is not surprising that this observation is highly dependent on the fish species and on the quality of the ingredients (Hua and Bureau 2012). However, little is known of the impact of varying fish feed composition on crop yields (Yildiz et al. 2017).

Classical fish feed is composed of 6–8 macro ingredients and contains 6–8% organic nitrogen, 1.2% organic phosphorus and 40–45% organic carbon (Timmons and Ebeling 2013) with around 25% protein for herbivorous or omnivorous fish and around 55% protein for carnivorous fish (Boyd 2015). Lipids can be fish or plant based as well (Boyd 2015).

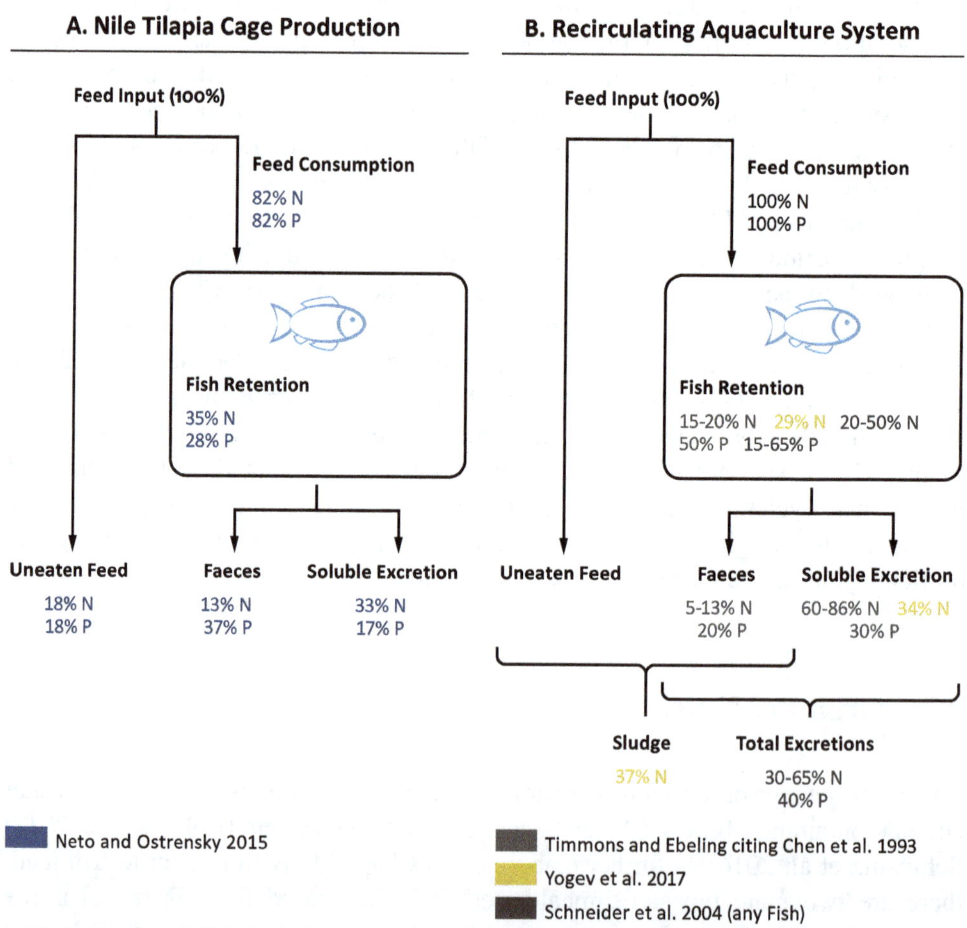

Fig. 9.1 Environmental flow of nitrogen and phosphorus in % for (**a**) Nile tilapia cage production (after Neto and Ostrensky 2015) and (**b**) RAS production (from a variety of sources)

Once fish feed is added into the system, a substantial part of it is eaten by the fish and either used for growth and metabolism or excreted as soluble and solid faeces, while the rest of the given feed decays in the tanks (Goddek et al. 2015; Schneider et al. 2004) (Fig. 9.1). In this case, the feed leftovers and metabolic products are partly dissolved in the aquaponic water, thus enabling the plants to uptake nutrients directly from the aquaponic solution (Schmautz et al. 2016).

In most cultivation systems (Chaps. 7 and 8), nutrients can be added to complement the aquaponic solution and ensure a better matching with the plants' needs (Goddek et al. 2015). Indeed, even when the system is coupled, it is possible to add iron or potassium (which are often lacking) without harming the fish (Schmautz et al. 2016).

9.2.1 *Fish Feed Leftovers and Fish Faeces*

Ideally, all the given feed should be consumed by the fish (Fig. 9.1). However, a small part (less than 5% (Yogev et al. 2016)) is often left to decompose in the system and contributes to the nutrient load of the water (Losordo et al. 1998; Roosta and Hamidpour 2013; Schmautz et al. 2016), thus consuming dissolved oxygen and releasing carbon dioxide and ammonia (Losordo et al. 1998), amongst other things. The composition of fish feed leftovers depends on the composition of the feed.

Logically enough, the composition of fish faeces depends on the fish's diet which also has an impact on the water quality (Buzby and Lin 2014; Goddek et al. 2015). However, the nutrient retention in fish biomass is highly dependent on fish species, feeding levels, feed composition, fish size and system temperature (Schneider et al. 2004). At higher temperatures, for example, fish metabolism is accelerated and thus results in more nutrients contained in the solid fraction of the faeces (Turcios and Papenbrock 2014). The proportion of excreted nutrients also depends on the quality and digestibility of the diet (Buzby and Lin 2014). The digestibility of the fish feed, the size of the faeces and the settling ratio should be carefully considered to ensure a good balance in the system and to maximise crop yields (Yildiz et al. 2017). Indeed, while it is a priority that fish feed should carefully be chosen to suit fish needs, the feed components could also be selected to suit plant's requirements when it makes no difference to fish (Goddek et al. 2015; Licamele 2009; Seawright et al. 1998).

9.3 Microbiological Processes

9.3.1 *Solubilisation*

Solubilisation consists of the breaking down of the complex organic molecules composing fish waste and feed leftovers into nutrients in the form of ionic minerals which plants can absorb (Goddek et al. 2015; Somerville et al. 2014). In both aquaculture (Sugita et al. 2005; Turcios and Papenbrock 2014) and aquaponics, solubilisation is conducted mainly by heterotrophic bacteria (van Rijn 2013; Chap. 6) which have not yet been fully identified (Goddek et al. 2015). Some studies have started deciphering the complexity of these bacteria communities (Schmautz et al. 2017). In current aquaculture, the most commonly observed bacteria are *Rhizobium* sp., *Flavobacterium* sp., *Sphingobacterium* sp., *Comamonas* sp., *Acinetobacter* sp., *Aeromonas* sp. and *Pseudomonas* sp. (Munguia-Fragozo et al. 2015; Sugita et al. 2005). An example of the major role of bacteria in aquaponics could be the transformation of insoluble phytates into phosphorus (P) made available for plant uptake through the production of phytases which are particularly present in *γ-proteobacteria* (Jorquera et al. 2008). (More research needs to be done in this area). Other nutrients than P can also be trapped as solids and evacuated from the system with the sludge. Efforts are thus being made to remineralise this sludge with UASB-EGSB reactors in order to reinject nutrients into the aquaponic system (Delaide 2017; Goddek et al. 2016; Chap. 10). Furthermore, different minerals are not released at the same rate, depending on the composition of the feed (Letelier-Gordo et al. 2015), thus leading to more complicated monitoring of their concentration in the aquaponic solution (Seawright et al. 1998).

9.3.2 Nitrification

The main nitrogen source in an aquaponic system is the fish feed and the proteins it contains (Goddek et al. 2015; Ru et al. 2017; Wongkiew et al. 2017; Yildiz et al. 2017). Ideally, 100% of this feed should be eaten by the fish. However, is has been observed that fish only use about 30% of the nitrogen contained in the given feed (Rafiee and Saad 2005). The ingested feed is partly used for assimilation and metabolism (Wongkiew et al. 2017), while the rest is excreted either through the gills or as urine and faeces (Ru et al. 2017). The nitrogen which is excreted through the gills is mainly in the form of ammonia, NH_3 (Wongkiew et al. 2017; Yildiz et al. 2017), while urine and faeces are composed of organic nitrogen (Wongkiew et al. 2017) which is transformed into ammonia by proteases and deaminases (Sugita et al. 2005). In general, the fish excrete nitrogen under the form of TAN, i.e. NH_3 and NH_4^+. The balance between NH_3 and NH_4^+ depends mostly on the pH and temperature. Ammonia is the major waste produced by fish catabolism of the feed proteins (Yildiz et al. 2017).

Nitrification is a two-step process during which the ammonia NH_3 or ammonium NH_4^+ excreted by the fish is transformed first into nitrite NO_2^- and then into nitrate NO_3^- by specific aerobic chemosynthetic autotrophic bacteria. A high availability of dissolved oxygen is required as nitrification consumes oxygen (Carsiotis and Khanna 1989; Madigan and Martinko 2007; Shoda 2014). The first step of this transformation is carried out by ammonia-oxidising bacteria (AOB) such as *Nitrosomonas*, *Nitrosococcus*, *Nitrosospira*, *Nitrosolobus* and *Nitrosovibrio*. The second step is conducted by nitrite-oxidising bacteria (NOB) such as *Nitrobacter*, *Nitrococcus*, *Nitrospira* and *Nitrospina* (Rurangwa and Verdegem 2013; Timmons and Ebeling 2013; Wongkiew et al. 2017). *Nitrospira* is currently deduced to be a complete nitrifier, i.e. to be involved in the production of both nitrite and nitrate (Daims et al. 2015). The same bacteria can be found both in aquaculture and aquaponic systems (Wongkiew et al. 2017). These bacteria are mainly found in biofilms fixed to the media composing the biofilter but can also be observed in the other compartments of the system (Timmons and Ebeling 2013).

Nitrification is of prime importance in aquaponics as ammonia and nitrite are quite toxic for fish: 0.02–0.07 mg/L of ammonia–nitrogen are sufficient to observe damage in warm water fish, and nitrite–nitrogen should be kept under 1 mg/L (Losordo et al. 1998; Timmons and Ebeling 2013). Ammonia affects the central nervous system of the fish (Randall and Tsui 2002; Timmons and Ebeling 2013), while nitrite induces problems with oxygen fixation (Losordo et al. 1998). Nitrate–nitrogen is, however, tolerated by the fish up to 150–300 mg/L (Goddek et al. 2015; Graber and Junge 2009; Yildiz et al. 2017).

Nitrification mostly takes place in biofilters (Losordo et al. 1998; Timmons and Ebeling 2013). Therefore, when starting a system, it is recommended to run the system without fish in order to allow the slowly growing population of nitrifying bacteria to establish (Timmons and Ebeling 2013; Wongkiew et al. 2017). It is also necessary to avoid, as far as possible, the presence of organic matter in the biofilters in order to prevent the growth of highly competitive heterotrophic bacteria (Timmons and Ebeling 2013). Alternatively, commercial mixes of nitrifying bacteria can be added to the system, prior to stocking, to hasten the colonisation process (Kuhn et al. 2010). Nevertheless, small aquaponic systems without biofilter also exist. In these systems, nitrifying bacteria form biofilms of the available surfaces (e.g. hydroponic compartment walls, inert media when using the media bed technique) (Somerville et al. 2014).

9.4 Mass Balance: What Happens to Nutrients once They Enter into the Aquaponic System?

9.4.1 Context

The functioning of aquaponic systems is based on a dynamic equilibrium of the nutrient cycles (Somerville et al. 2014). It is therefore necessary to understand these cycles in order to optimise the management of the systems. Plants growing hydroponically have specific requirements, which should be met during their various growing stages (Resh 2013). Therefore, nutrient concentrations in the different compartments of the system must be closely monitored, and nutrients should be supplemented to prevent deficiencies (Resh 2013; Seawright et al. 1998) either in the system water or via foliar application (Roosta and Hamidpour 2011).

According to Delaide et al. (2016), in some cases, supplementing an aquaponic solution with mineral nutrients in order to reach the same nutrient concentrations as in hydroponics could lead to higher yields than those achieved in hydroponics. The first step to take towards a balanced system is the correct design and relative sizing of the compartments (Buzby and Lin 2014). If the hydroponic compartment is too small compared to the fish tanks, then the nutrients will accumulate in the water and could reach toxic levels. The feed rate ratio (i.e. the amount of fish feed in the system based on the plant-growing surface and the plant type) is often used for the first sizing of the system (Rakocy et al. 2006; Somerville et al. 2014). However, according to Seawright et al. (1998), it is not possible to reach a plant/fish ratio which will enable an optimal match of plants' needs if only fish feed is used as an input. To make sure that the system is well balanced and functions properly, monitoring methods are usually based on the nitrogen cycle (Cerozi and Fitzsimmons 2017; Somerville et al. 2014), but to ensure optimal functioning of the system, it is necessary to monitor more closely the balance of the other macronutrients (N, P, K, Ca, Mg, S) and micronutrients (Fe, Zn, B, Mn, Mo, Cu) (Resh 2013; Somerville et al. 2014; Sonneveld and Voogt 2009) as well. Recent studies (Delaide et al. (2017), Schmautz et al. (2015, 2016)) have started tackling this topic. Schmautz et al. (2015, 2016) compared the impact of three different hydroponic layouts (i.e. nutrient film technique (NFT), floating raft and drip irrigation) on the nutrient uptake of aquaponic tomatoes. Drip irrigation was the system which produced slightly better yields with tomatoes. The mineral content of the fruits (P, K, Ca, Mg) was equivalent to the conventional values even though the iron and zinc contents were higher. The leaves however had lower levels of P, K, S, Ca, Mg, Fe, Cu and Zn than in conventional agriculture. Delaide et al. (2016) followed the cycles of macro- and micronutrients in a coupled aquaponic system. They observed that K, P, Fe, Cu, Zn, Mn and Mo were lacking in their aquaponic solution, while N, Ca, B and Na were quickly accumulated. Graber and Junge (2009) noted that their aquaponic solution contained three times less nitrogen and ten times less phosphorus than a hydroponic solution. As for potassium (K), it was 45 times lower compared to hydroponics. Nevertheless, they obtained yields as similar yields even though the quality was poorer due to a lack of potassium (K).

Factors Influencing the Nutrient Cycles

Light intensity, root zone temperature, air temperature, nutrient availability, growth stage and growth rate all influence a plant's nutrient uptake (Buzby and Lin 2014). Experiments conducted by Schmautz et al. (2016) and Lennard and Leonard (2006) showed that the hydroponic method could also play a role in a plant's nutrient uptake capacity, and it is therefore necessary to match the growing system to the type of vegetables being grown. NFT and DWC (deep water culture – raft) are thus suitable for leafy greens, whereas drip irrigation on rockwool slabs is more suitable for fruity vegetables (Resh 2013).

9.4.2 *Macronutrient Cycles*

Carbon (C)
Carbon is provided to the fish via the feed (Timmons and Ebeling 2013) and to the plants via CO_2 fixation. Fish can use 22% of the carbon contained in the fish feed for biomass increase and metabolism. The rest of the ingested carbon is either expirated under the form of CO_2 (52%) or excreted in a dissolved (0.7–3%) and solid (25%) form (Timmons and Ebeling 2013). The expirated CO_2 can be used by plants for their own carbon source as well (Körner et al. 2017). The uneaten part of the feed carbon is left to decompose in the system. The type of carbohydrates found in fish feed (e.g. starch or non-starch polysaccharides) can also influence the digestibility of the feed and the biodegradability of the waste in an aquaculture or aquaponic system (Meriac et al. 2014).

Nitrogen (N)
Nitrogen is absorbed by the plants either in the nitrate or ammonium form (Sonneveld and Voogt 2009; Xu et al. 2012) depending on the concentration and plant's physiology (Fink and Feller 1998 cited by Wongkiew et al. 2017). Associations between plants and microorganisms should not be overlooked as plants affect the presence of the microorganisms in aquaponics, and microorganisms can play a significant role in the nitrogen uptake capacity of plants (Wongkiew et al. 2017). The uptake of nitrogen by plants is affected by the ambient carbon dioxide concentration (Zhang et al. 2008 cited by Wongkiew et al. 2017).

Phosphorus (P)
Phosphorus is one of the essential elements for plant growth and can be absorbed under its ionic orthophosphate form ($H_2PO_4^-$, HPO_4^{2-}, PO_4^{3-}) (Prabhu et al. 2007; Resh 2013). Little is known about the dynamics of phosphorus in aquaponics. The main input of phosphorus in the system is the fish feed (Cerozi and Fitzsimmons 2017; Delaide et al. 2017; Schmautz et al. 2015), and in un-supplemented systems (Chap. 7), phosphorus tends to be limiting and thus can impede plant growth (Graber and Junge 2009; Seawright et al. 1998). According to Rafiee and Saad (2005), fish can use up to 15% of the phosphorus contained in the feed. In a system growing lettuce, Cerozi and Fitzsimmons (2017) noticed that the amount of phosphorus provided by the fish feed can be sufficient or insufficient depending on the growth stage. Up to 100% of phosphorus present in the fish water can be recycled in the plant biomass, depending on the design of the system. Graber and Junge (2009) observed a 50% recycling, while Schmautz et al. (2015) reported that 32% of the phosphorus could be found in the fruit and 28% in the leaves. The solubility of phosphorus depends on the pH, and a higher pH will foster the precipitation of phosphorus, thus rendering it unavailable for the plants (Yildiz et al. 2017). Phosphorus can precipitate as struvite (magnesium ammonium phosphate) (Le Corre et al. 2005) and/or hydroxyapatite (Cerozi and

Fitzsimmons 2017; Goddek et al. 2015). These insoluble complexes are removed via solid fish sludge from the system. Schneider et al. (2004) reported that 30–65% of the phosphorus contained in the fish feed remains unavailable to plants as it is fixed in the solid excretions which are then removed through mechanical filtration. Yogev et al. (2016) estimated that this loss can be up to 85%. One option to prevent this massive loss of P via solid sludge is to add a digestion compartment to the aquaponic system. During aerobic or anaerobic digestion, the P is released into the digestate and could be re-introduced into the circulating water (Goddek et al. 2016).

Potassium (K)
Delaide et al. (2017) found that the major source of K in their system was the fish feed. Fish can use up to 7% of the K contained in the fish feed (Rafiee and Saad 2005). However, potassium is not necessary for fish which leads to a low potassium composition of the fish feed and to even lower levels of potassium available for the plants (Graber and Junge 2009; Seawright et al. 1998; Suhl et al. 2016). To supply potassium, a KOH pH buffer is often used as the pH often decreases in aquaponics due to nitrification (Graber and Junge 2009). In an aquaponic system planted with tomatoes, potassium accumulated mainly in the fruits (Schmautz et al. 2016).

Magnesium (Mg), Calcium (Ca) and Sulphur (S)
The main source for Mg, Ca and S is tap water which facilitates the absorption by the plants as the nutrients are already available (Delaide et al. 2017). Calcium is however present in insufficient levels in aquaponics (Schmautz et al. 2015; Seawright et al. 1998) and is added under the form of calcium hydroxide $Ca(OH)_2$ (Timmons and Ebeling 2013). According to Rafiee and Saad (2005), fish can use on average 26.8% of the calcium and 20.3% of the magnesium present in the feed. Sulphur is often at low levels in aquaponic systems (Graber and Junge 2009; Seawright et al. 1998).

9.4.3 Micronutrient Cycles

Iron (Fe), manganese (Mn) and zinc (Zn) derive mainly from the fish feed, while boron (B) and copper (Cu) derive from the tap water (Delaide et al. 2017). In aquaponics, key micronutrients are often present but at too low levels (Delaide et al. 2017), and supplementation from external sources of nutrients is then necessary (Chap. 8). Iron deficiencies occur very often in aquaponics (Schmautz et al. 2015; Seawright et al. 1998; Fitzsimmons and Posadas; 1997 cited by Licamele 2009), mostly because of the non-availability of the ferric ion form. This deficiency can be solved by the use of bacterial siderophore (i.e. organic iron-chelating compounds) produced by genera such as *Bacillus* or *Pseudomonas* (Bartelme et al. 2018) or by iron supplementation with chemical chelated iron to avoid iron precipitation.

9.4.4 Nutrient Losses

Reducing nutrient loss is a constant challenge facing aquaponics practitioners. Nutrient loss occurs in several ways, e.g. the settlement of the sludge (37% of faeces and 18% of uneaten feed) (Neto and Ostrensky 2015), water losses, denitrification, ammonia volatilisation, etc. (Wongkiew et al. 2017). As an example, Rafiee and Saad (2005) note that 24% of the iron, 86% of the manganese, 47% of the zinc, 22% of the copper, 16% of the calcium, 89% of the magnesium, 6% of the nitrogen, 6% of the potassium and 18% of the phosphorus contained in the fish feed were contained in the sludge. The sludge can hold up to 40% of the nutrients present in the feed input (Yogev et al. 2016).

Denitrification can lead to a loss of 25–60% of the nitrogen (Hu et al. 2015; Zou et al. 2016). Denitrification is also linked to anoxic conditions (Madigan and Martinko 2007; van Lier et al. 2008) and low carbon levels and is responsible for the transformation of nitrate into nitrite, nitric oxide (NO), nitrous oxide (N_2O) and eventually nitrogen gas (N_2) with flows into the atmosphere (ibid.). Denitrification is conducted by several facultative heterotrophic bacteria such as *Achromobacter*, *Aerobacter*, *Acinetobacter*, *Bacillus*, *Brevibacterium*, *Flavobacterium*, *Pseudomonas*, *Proteus* and *Micrococcus* sp. (Gentile et al. 2007; Michaud et al. 2006; Wongkiew et al. 2017). Some bacteria can perform both nitrification and denitrification if dissolved oxygen levels are below 0.3 mg/L (Fitzgerald et al. 2015; Wongkiew et al. 2017). The loss of nitrogen can also occur via anaerobic ammonium oxidation (ANAMMOX), i.e. the oxidation of ammonium into dinitrogen gas in the presence of nitrite (Hu et al. 2011).

Another important loss of nitrogen which should be available for the plants is the consumption of the nitrogen by the heterotrophic aerobic bacteria present in the aquaponic systems. Indeed, the nitrogen used by these bacteria is lost to nitrifying bacteria, and nitrification is thus impeded (Blancheton et al. 2013). These bacteria are particularly present when the C/N ratio increases as they are more competitive and more able to colonise the media than the autotrophic nitrifying bacteria (Blancheton et al. 2013; Wongkiew et al. 2017).

9.4.5 Nutrient Balance Systems Dynamics

The nutrient concentration of the two major subsystems in an aquaponic system, i.e. the fish tanks (aquaculture) and the hydroponic solution, needs to be balanced for each of their needs. In closed aquaponic systems, nutrients are transported from fish to plants more or less directly through filters (usually a drum filter or a settler and then a biofilter) for nitrification. However, the nutrient needs of crops and the supplied nutrients from the aquaponic subsystem are not in balance. In multi-loop and decoupled systems (Chap. 8), it is easier to provide optimal conditions for both, fish and plant sections. Through modelling of the systems (Chap. 11), the optimal size of the hydroponic area to the fish tanks, biofilters and other equipment can be calculated (Goddek and Körner 2019). This is particularly important with decoupled multi-loop systems which comprise various types of equipment. For example, UASB (upflow anaerobic sludge blanket) (Goddek et al. 2018) or desalination units (Goddek and Keesman 2018) need to be sized carefully as discussed in Chap. 8. The basic mismatch of nutrients supplied by the fish environment and the needs of the crops needs to be rectified and balanced. For the purpose of up-concentrating the nutrients, Goddek and Keesman (2018) have described an appropriate desalination unit (Chap. 8). This approach, however, only solves part of the problem, as the perfectly balanced system is driven by a non-dynamic evaporation rate achievable only in closed chambers and perfectly working plants. The reality, however, is that the evapotranspiration of the crop (ET_c) in greenhouse-based aquaponics systems is highly dependent on multiple factors such as physical climate and biological variables. ET_c is calculated per area of the ground surface covered by the crop and is calculated for different levels in the canopy (z) by integrating irradiative net fluxes, boundary layer resistance, stomata resistance and the vapour pressure deficit, in the canopy (Körner et al. 2007) using the Penman–Monteith equation. This equation, nevertheless, only calculates the water flux through the crop. Nutrient uptake can either be calculated simply by assuming that all diluted nutrients in the water are taken up by the crop. In reality though, uptake of nutrients is a highly complicated matter. Different nutrients have different states, changing with parameters such as pH. Meanwhile nutrient availability to the plants strongly depends on pH and the relationships of nutrients to each other (e.g. K/Ca availability). In addition, the microbiome in the root zone plays an important role (Orozco-Mosque et al. 2018) which is not yet implemented in models although some models differentiate between phloem and xylem pathways. Thus, the vast amount of nutrients is not modelled in detail for aquaponic nutrient balancing and sizing of systems. The easiest way to estimate nutrient uptake is the assumption that nutrients are taken up/absorbed as dissolved in irrigation water and apply to the above-explained ET_c calculation approach and assuming that no element-specific chemical, biological or physical resistances exist. Consequentially, to maintain equilibrium, all nutrients taken up by the crop as contained in the nutrient solution need to be added back to the hydroponic system.

9.5 Conclusions

9.5.1 Current Drawbacks of Nutrient Cycling in Aquaponics

In hydroponics, the nutrient solution is accurately determined and the nutrient input into the system is well understood and controlled. This makes it relatively easy to adapt the nutrient solution for each plant species and for each growth stage. In aquaponics, according to the definition (Palm et al. 2018), the nutrients have to originate at least at 50% from uneaten fish feed, fish solid faeces and fish soluble excretions, thus making the monitoring of the nutrient concentrations available for plant uptake more difficult. A second drawback is the loss of nutrients through several pathways such as sludge removal, water renewal or denitrification. Sludge removal induces a loss of nutrients as several key nutrients such as phosphorus often precipitate and are then trapped in the evacuated solid sludge. Water renewal, which has to take place even if in small proportions, also adds to the loss of nutrients from the aquaponic circuit. Finally, denitrification happens because of the presence of denitrifying bacteria and conditions favourable to their metabolisms.

9.5.2 *How to Improve Nutrient Cycling?*

To conclude, nutrient cycling still needs to be improved in order to optimise plant growth in aquaponics. Several options are therefore currently explored in Chap. 8. To avoid losing the nutrients captured in the sludge, sludge remineralisation units have been developed (Chap. 10). The aim of these units is to extract the nutrients captured in solid form in the sludge and to reinject these into the system under a form which the plants can absorb (Delaide 2017). A further technique to reduce nutrient loss would be to foster plant uptake through the concentration of the aquaponic solution (i.e. the removal of a fraction of the water to keep the same amount of nutrients but in a lesser water volume). Such a concentration could be achieved via the addition of a desalination unit as part of the aquaponic system (Goddek and Körner 2019; Goddek and Keesman 2018). Finally, the use of decoupled/multi-loop systems enables optimal living and growing conditions for all fish, plants and microorganisms. While some research has been undertaken in this field, more research should be conducted to better understand nutrient cycling in aquaponics. Indeed, more information concerning the exact cycles of each macronutrient (what form, how it can be transformed or not by microorganisms, how it is taken up by plants in aquaponics) or the influence of the plant and fish species and water parameters on the nutrient cycles could greatly help the understanding of aquaponic systems.

References

Bartelme RP, Oyserman BO, Blom JE, Sepulveda-Villet OJ, Newton RJ (2018) Stripping away the soil: plant growth promoting microbiology opportunities in aquaponics. Front Microbiol 9(8). https://doi.org/10.3389/fmicb.2018.00008

Bittsanszky A, Uzinger N, Gyulai G, Mathis A, Junge R, Villarroel M, Kotzen B, Komives T (2016) Nutrient supply of plants in aquaponic systems. Ecocycles 2(1720). https://doi.org/10.19040/ecocycles.v2i2.57

Blancheton JP, Attramadal KJK, Michaud L, d'Orbcastel ER, Vadstein O (2013) Insight into bacterial population in aquaculture systems and its implication. Aquac Eng 53:30–39

Boyd CE (2015) Overview of aquaculture feeds: global impacts of ingredient use. Feed Feed Pract Aquac 3–25. https://doi.org/10.1016/B978-0-08-100506-4.00001-5

Bugbee B (2004) Nutrient management in recirculating hydroponic culture. Acta Hortic 648:99–112. https://doi.org/10.17660/ActaHortic.2004.648.12

Buzby KM, Lin LS (2014) Scaling aquaponic systems: balancing plant uptake with fish output. Aquac Eng 63:39–44. https://doi.org/10.1016/j.aquaeng.2014.09.002

Carsiotis M, Khanna S (1989) Genetic engineering of microbial nitrification. United States Environmental Protection Agency, Risk Reduction Engineering Laboratory, Cincinnati

Cerozi BS, Fitzsimmons K (2017) Phosphorus dynamics modeling and mass balance in an aquaponics system. Agric Syst 153:94–100. https://doi.org/10.1016/j.agsy.2017.01.020

Daims H, Lebedeva EV, Pjevac P, Han P, Herbold C, Albertsen M, Jehmlich N, Palatinszky M, Vierheilig J, Bulaev A, Kirkegaard RH, Von Bergen M, Rattei T, Bendinger B, Nielsen H, Wagner M (2015) Complete nitrification by Nitrospira bacteria. Nature 528:504. https://doi.org/10.1038/nature16461

Davidson J, Good C, Barrows FT, Welsh C, Kenney PB, Summerfelt ST (2013) Comparing the effects of feeding a grain- or a fish meal-based diet on water quality, waste production, and rainbow trout *Oncorhynchus mykiss* performance within low exchange water recirculating aquaculture systems. Aquac Eng 52:45–57. https://doi.org/10.1016/J.AQUAENG.2012.08.001

Delaide B(2017) A study on the mineral elements available in aquaponics, their impact on lettuce productivity and the potential improvement of their availability. PhD thesis. Gembloux Agro-Bio Tech, University of Liege

Delaide B, Goddek S, Gott J, Soyeurt H, Haissam Jijakli M, Lalman J, Junge R (2016) Lettuce (*Lactuca sativa* L. var. Sucrine) growth performance in complemented aquaponic solution outperforms hydroponics. Water 8. https://doi.org/10.3390/w8100467

Delaide B, Delhaye G, Dermience M, Gott J, Soyeurt H, Jijakli MH (2017) Plant and fish production performance, nutrient mass balances, energy and water use of the PAFF box, a small-scale aquaponic system. Aquac Eng 78:130–139. https://doi.org/10.1016/j.aquaeng.2017.06.002

Fitzgerald CM, Camejo P, Oshlag JZ, Noguera DR (2015) Ammonia-oxidizing microbial communities in reactors with efficient nitrification at low-dissolved oxygen. Water Res 70:38–51. https://doi.org/10.1016/J.WATRES.2014.11.041

Geay F, Ferraresso S, Zambonino-Infante JL, Bargelloni L, Quentel C, Vandeputte M, Kaushik S, Cahu CL, Mazurais D (2011) Effects of the total replacement of fish-based diet with plant-based diet on the hepatic transcriptome of two European sea bass (*Dicentrarchus labrax*) half-sibfamilies showing different growth rates with the plant-based diet. BMC Genomics 12:522. https://doi.org/10.1186/1471-2164-12-522

Gentile ME, Lynn Nyman J, Criddle CS (2007) Correlation of patterns of denitrification instability in replicated bioreactor communities with shifts in the relative abundance and the denitrification patterns of specific populations. ISME J 1:714–728. https://doi.org/10.1038/ismej.2007.87

Goddek S, Keesman KJ (2018) The necessity of desalination technology for designing and sizing multi-loop aquaponics systems. Desalination 428:76–85. https://doi.org/10.1016/J.DESAL.2017.11.024

Goddek S, Körner O (2019) A fully integrated simulation model of multi-loop aquaponics: a case study for system sizing in different environments. Agric Syst 171:143

Goddek S, Delaide B, Mankasingh U, Ragnarsdottir KV, Jijakli H, Thorarinsdottir R (2015) Challenges of sustainable and commercial aquaponics. Sustainability 7:4199–4224. https://doi.org/10.3390/su7044199

Goddek S, Schmautz Z, Scott B, Delaide B, Keesman K, Wuertz S, Junge R (2016) The effect of anaerobic and aerobic fish sludge supernatant on hydroponic lettuce. Agronomy 6:37. https://doi.org/10.3390/agronomy6020037

Goddek S, Delaide BPL, Joyce A, Wuertz S, Jijakli MH, Gross A, Eding EH, Bläser I, Reuter M, Keizer LCP et al (2018) Nutrient mineralization and organic matter reduction performance of RAS-based sludge in sequential UASB-EGSB reactors. Aquac Eng 83:10–19

Graber A, Junge R (2009) Aquaponic systems: nutrient recycling from fish wastewater by vegetable production. Desalination 246:147–156. https://doi.org/10.1016/j.desal.2008.03.048

Guangzhi G (2001) Mass balance and water quality in aquaculture tanks. The United Nations University, Fisheries Training Programme, Reyjavik

Hu B, Shen L, Xu X, Zheng P (2011) Anaerobic ammonium oxidation (anammox) in different natural ecosystems. Biochem Soc Trans 39:1811–1816. https://doi.org/10.1042/BST20110711

Hu Z, Lee JW, Chandran K, Kim S, Brotto AC, Khanal SK (2015) Effect of plant species on nitrogen recovery in aquaponics. Bioresources 188:92–98. https://doi.org/10.1016/j.biortech.2015.01.013

Hua K, Bureau DP (2012) Exploring the possibility of quantifying the effects of plant protein ingredients in fish feeds using meta-analysis and nutritional model simulation-based approaches. Aquaculture 356–357:284–301. https://doi.org/10.1016/J.AQUACULTURE.2012.05.003

Jorquera M, Martínez O, Maruyama F, Marschner P, de la Luz Mora M (2008) Current and future biotechnological applications of bacterial Phytases and phytase-producing Bacteria. Microbes Environ 23:182–191. https://doi.org/10.1264/jsme2.23.182

Körner O, Aaslyng JM, Andreassen AU, Holst N (2007) Modelling microclimate for dynamic greenhouse climate control. Hortscience 42:272–279

Körner O, Gutzmann E, Kledal PR (2017) A dynamic model simulating the symbiotic effects in aquaponic systems. Acta Hortic 1170:309–316

Kuhn DD, Drahos DD, Marsh L, Flick GJ (2010) Evaluation of nitrifying bacteria product to improve nitrification efficacy in recirculating aquaculture systems. Aquac Eng 43:78–82. https://doi.org/10.1016/J.AQUAENG.2010.07.001

Le Corre KS, Valsami-Jones E, Hobbs P, Parsons SA (2005) Impact of calcium on struvite crystal size, shape and purity. J Cryst Growth 283:514–522. https://doi.org/10.1016/J.JCRYSGRO.2005.06.012

Lennard WA, Leonard BV (2006) A comparison of three different hydroponic sub-systems (gravel bed, floating and nutrient film technique) in an Aquaponic test system. Aquac Int 14:539–550. https://doi.org/10.1007/s10499-006-9053-2

Letelier-Gordo CO, Dalsgaard J, Suhr KI, Ekmann KS, Pedersen PB (2015) Reducing the dietary protein:energy (P:E) ratio changes solubilization and fermentation of rainbow trout (*Oncorhynchus mykiss*) faeces. Aquac Eng 66:22–29

Licamele J (2009) Biomass production and nutrient dynamics in an aquaponics system. PhD thesis. Department of Agriculture and biosystems engineering, University of Arizona

Losordo TM, Masser MP, Rakocy J (1998) Recirculating aquaculture tank production systems: an overview of critical considerations. SRAC No. 451, pp 18–31

Madigan MT, Martinko JM (2007) Biologie des micro-organismes, 11th edn. Pearson Education France, Paris

Meriac A, Eding EH, Schrama J, Kamstra A, Verreth JAJ (2014) Dietary carbohydrate composition can change waste production and biofilter load in recirculating aquaculture systems. Aquaculture 420–421:254–261. https://doi.org/10.1016/j.aquaculture.2013.11.018

Michaud L, Blancheton JP, Bruni V, Piedrahita R (2006) Effect of particulate organic carbon on heterotrophic bacterial populations and nitrification efficiency in biological filters. Aquac Eng 34:224–233. https://doi.org/10.1016/j.aquaeng.2005.07.005

Munguia-Fragozo P, Alatorre-Jacome O, Rico-Garcia E, Torres-Pacheco I, Cruz-Hernandez A, Ocampo-Velazquez RV, Garcia-Trejo JF, Guevara-Gonzalez RG (2015) Perspective for aquaponic systems: "omic" technologies for microbial community analysis. Biomed Res Int 2015:1. https://doi.org/10.1155/2015/480386

Neto MR, Ostrensky A (2015) Nutrient load estimation in the waste of Nile tilapia *Oreochromis niloticus* (L.) reared in cages in tropical climate conditions. Aquac Res 46:1309–1322. https://doi.org/10.1111/are.12280

Orozco-Mosque MC, Rocha-Granados MC, Glick BR, Santoyo G (2018) Microbiome engineering to improve biocontrol and plant growth-promoting mechanism. Microbiol Res. In Press

Palm HW, Knaus U, Appelbaum S, Goddek S, Strauch SM, Vermeulen T, Haïssam Jijakli M, Kotzen B (2018) Towards commercial aquaponics: a review of systems, designs, scales and nomenclature. Aquac Int 26:813–842

Prabhu AS, Fageria NK, Berni RF, Rodrigues FA (2007) Phosphorus and plant disease. In: Datnoff LE, Elmer WH, Huber DM (eds) Mineral nutrition and plant disease. The American Phytopathological Society, St. Paul, pp 45–55

Rafiee G, Saad CR (2005) Nutrient cycle and sludge production during different stages of red tilapia (Oreochromis sp.) growth in a recirculating aquaculture system. Aquaculture 244:109–118. https://doi.org/10.1016/J.AQUACULTURE.2004.10.029

Rakocy JE, Shultz RC, Bailey DS, Thoman ES (2004) Aquaponic production of tilapia and basil: comparing a batch and staggered cropping system. Acta Hortic 648:63–69. https://doi.org/10.17660/ActaHortic.2004.648.8

Rakocy JE, Masser MP, Losordo TM (2006) Recirculating aquaculture tank production systems: aquaponics- integrating fish and plant culture. SRAC Publ South Reg Aquac Cent 16. https://doi.org/454

Randall D, Tsui TK (2002) Ammonia toxicity in fish. Mar Pollut Bull 45:17–23. https://doi.org/10.1016/S0025-326X(02)00227-8

Resh HM (2013) Hydroponic food production: a definitive guidebook for the advanced home gardener and the commercial hydroponic grower, 7th edn. CRC Press, Boca Raton

Roosta HR, Hamidpour M (2011) Effects of foliar application of some macro- and micro-nutrients on tomato plants in aquaponic and hydroponic systems. Sci Hortic 129:396–402. https://doi.org/10.1016/J.SCIENTA.2011.04.006

Roosta HR, Hamidpour M (2013) Mineral nutrient content of tomato plants in Aquaponic and hydroponic systems: effect of foliar application of some macro- and micro-nutrients. J Plant Nutr 36:2070–2083. https://doi.org/10.1080/01904167.2013.821707

Ru D, Liu J, Hu Z, Zou Y, Jiang L, Cheng X, Lv Z (2017) Improvement of aquaponic performance through micro- and macro-nutrient addition. Environ Sci Pollut Res 24:16328–16335. https://doi.org/10.1007/s11356-017-9273-1

Rurangwa E, Verdegem MCJ (2013) Microorganisms in recirculating aquaculture systems and their management. Rev Aquac 7:117–130. https://doi.org/10.1111/raq.12057

Schmautz Z, Graber A, Mathis A, Griessler Bulc T, Junge R (2015) Tomato production in aquaponic system: mass balance and nutrient recycling (abstract)

Schmautz Z, Loeu F, Liebisch F, Graber A, Mathis A, Bulc TG, Junge R (2016) Tomato productivity and quality in aquaponics: comparison of three hydroponic methods. Water 8:1–22. https://doi.org/10.3390/w8110533

Schmautz Z, Graber A, Jaenicke S, Goesmann A, Junge R, Smits THM (2017) Microbial diversity in different compartments of an aquaponics system. Arch Microbiol 199:613. https://doi.org/10.1007/s00203-016-1334-1

Schneider O, Sereti V, Eding EH, Verreth JAJ (2004) Analysis of nutrient flows in integrated intensive aquaculture systems. Aquac Eng 32:379–401. https://doi.org/10.1016/j.aquaeng.2004.09.001

Seawright DE, Stickney RR, Walker RB (1998) Nutrient dynamics in integrated aquaculture–hydroponics systems. Aquaculture 160:215–237

Shoda M (2014) Heterotrophic nitrification and aerobic denitrification by *Alcaligenes faecalis*. J Biosci Bioeng 117:737–741. https://doi.org/10.5772/68052

Somerville C, Stankus A, Lovatelli A (2014) Small-scale aquaponic food production. Integrated fish and plant farming. Food and Agriculture Organisation of the United Nations, Rome

Sonneveld C, Voogt W (2009) Plant nutrition of greenhouse crops. Springer, Dordrecht/Heidelberg/London/New York

Sugita H, Nakamura H, Shimada T (2005) Microbial communities associated with filter materials in recirculating aquaculture systems of freshwater fish. Aquaculture 243:403. https://doi.org/10.1016/j.aquaculture.2004.09.028

Suhl J, Dannehl D, Kloas W, Baganz D, Jobs S, Scheibe G, Schmidt U (2016) Advanced aquaponics: evaluation of intensive tomato production in aquaponics vs. conventional hydroponics. Agric Water Manag 178:335–344. https://doi.org/10.1016/j.agwat.2016.10.013

Tacon AGJ, Metian M (2008) Global overview on the use of fish meal and fish oil in industrially compounded aquafeeds: trends and future prospects. Aquaculture 285:146–158. https://doi.org/10.1016/j.aquaculture.2008.08.015

Timmons MB, Ebeling JM (2013) Recirculating aquaculture. Ithaca Publishing, New York

Turcios AE, Papenbrock J (2014) Sustainable treatment of aquaculture effluents—what can we learn from the past for the future? Sustainability 6:836–856. https://doi.org/10.3390/su6020836

van Lier JB, Mahmoud N, Zeeman G (2008) In: Henze M, van Loosdrecht MCM, Ekama GA, Brdjanovic D (eds) Anaerobic wastewater treatment, in: biological wastewater treatment: principles, modelling and design. IWA Publishing, London. ISBN: 9781843391883

van Rijn J (2013) Waste treatment in recirculating aquaculture systems. Aquac Eng 53:49–56. https://doi.org/10.1016/J.AQUAENG.2012.11.010

Wongkiew S, Hu Z, Chandran K, Lee JW, Khanal SK (2017) Nitrogen transformations in aquaponic systems: a review. Aquac Eng 76:9–19. https://doi.org/10.1016/j.aquaeng.2017.01.004

Xu G, Fan X, Miller AJ (2012) Plant nitrogen assimilation and use efficiency. Annu Rev Plant Biol 63:153–182. https://doi.org/10.1146/annurev-arplant-042811-105532

Yildiz HY, Robaina L, Pirhonen J, Mente E, Domínguez D, Parisi G (2017) Fish welfare in Aquaponic systems: its relation to water quality with an emphasis on feed and Faeces—a review. Water 9. https://doi.org/10.3390/w9010013

Yogev U, Barnes A, Gross A (2016) Nutrients and energy balance analysis for a conceptual model of a three loops off grid, aquaponics. Water 8. https://doi.org/10.3390/w8120589

Zekki H, Gauthier L, Gosselin A (1996) Growth, productivity, and mineral composition of hydroponically cultivated greenhouse tomatoes, with or without nutrient solution recycling. J Am Soc Hortic Sci 121:1082–1088

Zou Y, Hu Z, Zhang J, Xie H, Guimbaud C, Fang Y (2016) Effects of pH on nitrogen transformations in media-based aquaponics. Bioresour Technol 210:81–87. https://doi.org/10.1016/J.BIORTECH.2015.12.079

10

Sludge Treatment in Aquaponics: Aerobic and Anaerobic Methods

Boris Delaide, Hendrik Monsees, Amit Gross, and Simon Goddek

Abstract Recirculating aquaculture systems, as part of aquaponic units, are effective in producing aquatic animals with a minimal water consumption through effective treatment stages. Nevertheless, the concentrated sludge produced after the solid filtration stage, comprising organic matter and valuable nutrients, is most often discarded. One of the latest developments in aquaponic technology aims to reduce this potential negative environmental impact and to increase the nutrient recycling by treating the sludge on-site. For this purpose, microbial aerobic and anaerobic treatments, dealt with either individually or in a combined approach, provide very promising opportunities to simultaneously reduce the organic waste as well as to recover valuable nutrients such as phosphorus. Anaerobic sludge treatments additionally offer the possibility of energy production since a by-product of this process is biogas, i.e. mainly methane. By applying these additional treatment steps in aquaponic units, the water and nutrient recycling efficiency is improved and the dependency on external fertiliser can be reduced, thereby enhancing the sustainability of the system in terms of resource utilisation. Overall, this can pave the way for the economic improvement of aquaponic systems because costs for waste disposal and fertiliser acquisition are decreased.

B. Delaide (✉)
Developonics asbl, Brussels, Belgium
e-mail: boris@developonics.com; borisdelaide@hotmail.com

H. Monsees
Leibniz-Institute of Freshwater Ecology and Inland Fisheries, Berlin, Germany
e-mail: h.monsees@igb-berlin.de

A. Gross
Department of Environmental Hydrology and Microbiology, Zuckerberg Institute for Water Research, Blaustein Institutes for Desert Research, Ben-Gurion University of the Negev, Beersheba, Israel
e-mail: amgross@bgu.ac.il

S. Goddek
Mathematical and Statistical Methods (Biometris), Wageningen University, Wageningen, The Netherlands
e-mail: simon.goddek@wur.nl; simon@goddek.nl

Keywords Sludge recycling · Phosphorus · Microbial sludge conversion · Mass balance · Nutrient recycling

10.1 Introduction

The concept of aquaponics is associated with being a sustainable production system, as it re-utilises recirculating aquaculture system (RAS) wastewater enriched in macronutrients (i.e. nitrogen (N), phosphorus (P), potassium (K), calcium (Ca), magnesium (Mg) and sulphur (S)) and micronutrients (i.e. iron (Fe), manganese (Mn), zinc (Zn), copper (Cu), boron (B) and molybdenum (Mo)) to fertilise the plants (Graber and Junge 2009; Licamele 2009; Nichols and Savidov 2012; Turcios and Papenbrock 2014). A much debated question is whether this concept can match its own ambition of being a quasi-closed-loop system, as high amounts of the nutrients that enter the system are wasted by discharging the nutrient-rich fish sludge (Endut et al. 2010; Naylor et al. 1999; Neto and Ostrensky 2013). Indeed, to maintain a good water quality in a RAS and aquaponic systems, the water has to be constantly filtrated for solid removal. The two main techniques for solid filtration are to retain the particles in a mesh (i.e. mesh filtration as drum filters) and to allow the particles to decant in clarifiers. In most conventional plants, sludge is recovered out of these mechanical filtration devices and is discharged as sewage. In the best cases, the sludge is dried and applied as fertiliser on land fields (Brod et al. 2017). Notably, up to 50% (in dry matter) of the feed ingested is excreted as solids by fish (Chen et al. 1997), and most of the nutrients that enter aquaponic systems via fish feed accumulate in these solids and so in the sludge (Neto and Ostrensky 2013; Schneider et al. 2005). Hence, effective solid filtration removes, for example, more than 80% of the valuable P (Monsees et al. 2017) that could otherwise be used for plant production. Therefore, recycling these valuable nutrients for aquaponic applications is of major importance. Developing an appropriate sludge treatment able to mineralise the nutrients contained in sludge for re-using them in the hydroponic unit seems to be a necessary process for contributing to close the nutrient loop to a higher degree and thus lowering the environmental impact of aquaponic systems (Goddek et al. 2015; Goddek and Keesman 2018; Goddek and Körner 2019).

It has been shown in experimental studies that complemented aquaponic nutrient solution (i.e. after addition of lacking nutrients) promotes plant growth compared to hydroponics (Delaide et al. 2016; Ru et al. 2017; Saha et al. 2016). Hence, sludge mineralisation is also a promising way to improve the aquaponic system performance as the nutrients recovered are used to complement the aquaponic solution. In addition, on-site mineralisation units can also increase the self-sufficiency of aquaponic facilities, especially with respect to finite resources as P which is essential for plant growth. P is produced by mining activities, whereby the deposits are not equally distributed around the world. In addition, its price has risen by up to 800% within the last decade (McGill 2012). Thus, mineralisation units applied in aquaponic systems are also likely to increase its future economic success and stability.

Sludge treatment in aquaponics needs to be approached differently than has been done in the past. Indeed, in conventional wastewater treatment, the main objective is to obtain a decontaminated and clean effluent. The treatment performances are expressed in terms of removal of contaminants (e.g. solids, nitrogen (N), phosphorus (P), etc.) out of the wastewater and by quantifying the effluents with respect to the achieved quality (Techobanoglous et al. 2014). Using this conventional approach, several studies have provided quantitative evidence that a consistent proportion of chemical oxygen demand (COD) and total suspended solids (TSS) can be removed by digesting the RAS wastewater under aerobic, anaerobic and sequential aerobic–anaerobic conditions (Goddek et al. 2018; Chowdhury et al. 2010; Mirzoyan et al. 2010; Van Rijn 2013). However, in aquaponic systems, the wastewater from fish is considered to be a valuable fertiliser source. Within a closed-loop approach, the solid part discharged needs to be minimised (i.e. organic reduction maximised), and the nutrient content in the effluents needs to be maximised (i.e. nutrient mineralisation maximised). Therefore, the wastewater treatment performance in aquaponics no longer needs to be expressed in terms of contaminants removal but in terms of its contaminant reduction and nutrient mineralisation ability.

A few studies have demonstrated the functional capability of digesting fish sludge with aerobic and anaerobic treatments for organic reduction purposes (Goddek et al. 2018; van Rijn et al. 1995). With anaerobic treatment in bioreactor, high dry matter (i.e. TSS) reduction performance (e.g. higher than 90%) can be achieved while methane can also be produced (van Lier et al. 2008; Mirzoyan and Gross 2013; Yogev et al. 2016).

The aerobic treatment of sludge is also a very effective way to reduce organic matter, which is oxidised to CO_2 during respiration (see Eq. 10.1). For example, reduction rates of 90% (here determined as suspended solids, COD and BOD reduction) were reported from a water resource recovery facility (Seo et al. 2017). Aerobic processes are faster than anaerobic, but they can be more expensive (Chen et al. 1997) as a constant aeration of the sludge–water mixture requires energy-intensive pumps or motors. Moreover, significant fractions of the nutrients are converted to microbial biomass and do not stay dissolved in the water.

Although these studies have shown the organic reduction potential of fish sludge, only a few authors have examined the release of specific nutrients (e.g. for N and P) from fish sludge. Most of these studies were for short in vitro batch experiments (Conroy and Couturier 2010; Monsees et al. 2017; Stewart et al. 2006) and from an operating RAS (Yogev et al. 2016), rather than an aquaponic setup. While discussed to some extent in theory (Goddek et al. 2016; Yogev et al. 2017), the research has to start now to systematically investigate the organic reduction and nutrient mineralisation performance of fish sludge for both aerobic and anaerobic reactors and its effect on the water composition and plant growth. Therefore, this chapter aims to give an overview on the diverse fish sludge treatments that can be integrated into aquaponic setups to achieve organic reduction and nutrient mineralisation. Some design approaches will be highlighted. The nutrient mass balance approach in the context of aquaponic sludge treatment will be discussed, and specific methodology to quantify the sludge treatment performance will be developed.

10.2 Wastewater Treatment Implementation in Aquaponics

In aquaponics, the wastewater charged with solids (i.e. the sludge) is a valuable source of nutrients, and appropriate treatments need to be carried out. The treatment goals differ from conventional wastewater treatment because in aquaponics solids and water conservation is of interest. Moreover, regardless of the wastewater treatment applied, its aim should be to reduce solids and at the same time mineralise its nutrients. In other words, the aim is to obtain a solid-free effluent but rich in solubilised nutrients (i.e. anions and cations) that can be reinserted into the water loop in a coupled setup (Fig. 10.1a) or directly into the hydroponic grow beds in a decoupled setup (Fig. 10.1b). Fish sludge solids are mainly composed of degradable organic matter so that the solid reduction can be called organic reduction. Indeed, the complex organic molecules (e.g. proteins, lipids, carbohydrates, etc.) are principally composed of carbon and will be successively reduced to lower molecular weight compounds until the ultimate gaseous forms of CO_2 and CH_4 (in the case of anaerobic fermentation). During this degradation process, the macronutrients (i.e. N, P, K, Ca, Mg and S) and micronutrients (i.e. Fe, Mn, Zn, Cu, B and Mo) that were bound to the organic molecules are released into the water in their ionic forms. This phenomenon is called nutrient leaching or nutrient mineralisation. It can be assumed that when high organic reduction is achieved, high nutrient mineralisation would also be achieved. On the one hand, sludge contains a proportion of undissolved minerals, and on the other hand, some macro- and micronutrients are released during the mineralisation process. These can quickly precipitate together and form insoluble minerals. The state between ions and precipitated minerals of most of the macro- and micronutrients is pH dependent. The most well-known minerals that precipitate in bioreactors are calcium phosphate, calcium sulphate, calcium carbonate, pyrite and struvite (Peng et al. 2018; Zhang et al. 2016). Conroy and Couturier (2010) observed that Ca and P were released in anaerobic reactor when the pH dropped under 6. They showed that the release corresponded exactly to the mineralisation of calcium phosphate. Goddek et al. (2018) also observed the solubilisation of P, Ca and other macronutrients in upflow anaerobic sludge blanket reactor (UASB) that turned acidic. Jung and Lovitt (2011) reported a 90% nutrient mobilisation of aquaculture-derived sludge at a very low pH value of 4. In this condition, all the macro- and micronutrients were solubilised. There is thus an antagonism between organic reduction and nutrient mineralisation. Indeed, organic reduction is maximal when the microorganisms are active for degrading the organic compounds, and this happens at pH in a range of 6–8. Because nutrient leaching occurs at pH below 6, for optimal organic reduction and nutrient mineralisation, the most effective would be to divide the process in two steps, i.e. an organic reduction step at pH close to neutral and a nutrient leaching step under acidic conditions. To our knowledge, no operation using this two-step approach has been yet reported. This opens a new field in wastewater treatment and more research for implementation in aquaponics is needed.

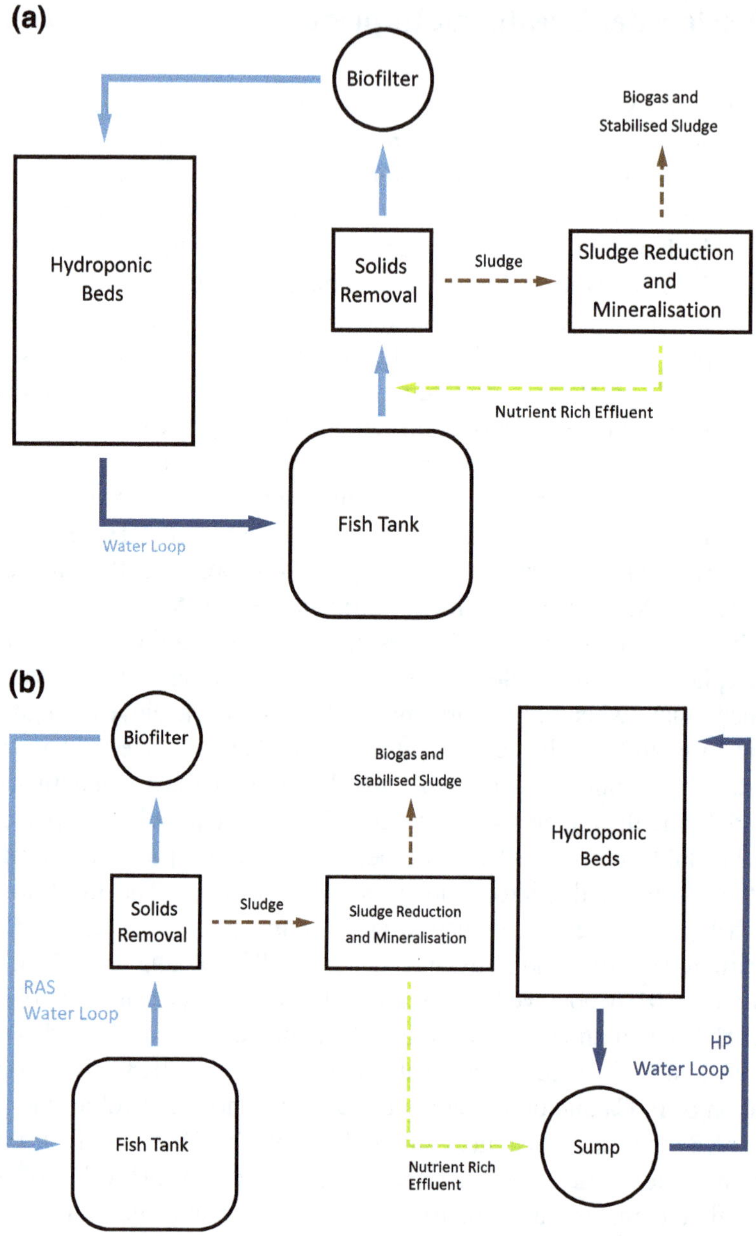

Fig. 10.1 Schematic implementation of sludge treatment in one loop aquaponic system (**a**) and in decoupled aquaponic system (**b**)

The choice of feed is also important in this context. In animal-based feeds where a major ingredient fraction is still based on animal sources (e.g. fishmeal, bone meal), bound phosphate, e.g. as apatite (derived from bone meal), is easily available under acidic conditions, whereas plant-based feeds contain phytate as a major phosphate source. Phytate in contrast to, e.g. apatite requires enzymatic (phytase) conversion (Kumar et al. 2012), and so the phosphate is not as easily available.

10.3 Aerobic Treatments

Aerobic treatment enhances the oxidation of the sludge by supporting its contact with oxygen. In this case, the oxidation of the organic matter is driven mainly by the respiration of heterotrophic microorganisms. CO_2, the end product of respiration, is released as is shown in Eq. (10.1).

$$C_6H_{12}O_6 + 6\,O_2 \rightarrow 6\,CO_2 + 6\,H_2O + \text{energy} \tag{10.1}$$

This process in aerobic reactors is mainly achieved by injecting air into the sludge–water mixture with air blowers connected to diffusers and propellers. Air injection also ensures a proper mixing of the sludge.

During this oxidative process, the macro- and micronutrients bound to the organic matter are released. This process is called aerobic mineralisation. Therefore, further nutrients can be recycled during the mineralisation process, whereas some nutrients, e.g. sodium and chloride, can also exceed their threshold for hydroponic application and must be monitored carefully before application (Rakocy et al. 2007). Aerobic mineralisation of organic matter, derived from the solid removal unit (e.g. clarifier or drum filter) in RAS, is an easy way to recycle nutrients for subsequent aquaponic application.

Moreover, during the aerobic digestion process, the pH drops and promotes the mineralisation of bound minerals trapped in the sludge. For example, Monsees et al. (2017) showed that P was released from RAS sludge due to this pH shift. This decrease in pH is mainly driven by respiration and to a lower extent probably by nitrification.

Due to a constant supply of oxygen via aeration of the mineralisation chamber and the abundance of organic matter, heterotrophic microorganisms find ideal conditions to grow. This results in an increase of respiration and the release of CO_2 that dissolves in water. CO_2 forms carbonic acid which dissociates and thereby lowers the pH of the process water as illustrated in the following equation:

$$CO_{2(g)} + 2H_2O \rightarrow H_3O^+ + HCO_3^- \tag{10.2}$$

RAS-derived wastewater often contains NH_4^+ and additionally is characterised by a neutral pH of around 7, because the pH in RAS needs to be kept at that level to ensure optimal microbial conversion of NH_4^+ to NO_3^- within the biofilter (i.e. nitrification). The nitrification process can contribute to the decrease in pH in aerobic reactors in the starting phase by releasing protons to the process water as can been seen in the following equation:

$$NH_4^+ + 2\,O_2 \rightarrow NO_3^- + 2\,H^+ + H_2O + \text{energy} \tag{10.3}$$

This is at least valid for the starting phase where the pH is still above 6. At a pH $\leq$ 6, nitrification might significantly slow down or even cease (Ebeling et al. 2006). However, this does not represent a problem for the mineralisation unit.

The general decrease of the pH in the aerobic mineralisation unit in the ongoing process is the main driver of the release of nutrients present under the form of precipitated minerals as calcium phosphates. Monsees et al. (2017) noted that around 50% of the phosphate in the sludge was acid soluble, derived from a tilapia RAS where a standard feed containing fishmeal was applied. Here, around 80% of the phosphate within the RAS was lost by the cleaning of the decanter and the discarding of the sludge–water mixture. Considering this fact, the big potential of mineralisation units for aquaponic applications becomes clears.

The advantages of aerobic mineralisation are the low maintenance with no need for skilled personnel and no subsequent reoxygenation. The enriched water can be used directly for plant fertilisation, ideally managed by an online system for the adequate preparation of the nutrient solution. A disadvantage compared to anaerobic mineralisation is that no methane is produced (Chen et al. 1997) and, as already mentioned, the higher energy demand due to the need for constant aeration.

10.3.1 Aerobic Mineralisation Units

A design example of an aerobic mineralisation unit is presented in Fig. 10.2. The inlet is connected to the solid removal unit via a valve, which allows discontinuous refilling of the mineralisation chamber with a mixture of sludge and water. The

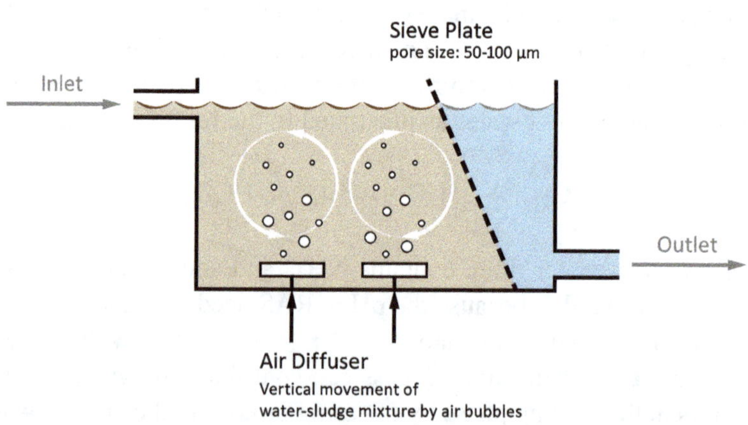

Fig. 10.2 Schematic example of an aerobic mineralisation unit operated in a batch mode. Mineralisation chamber (brown) is separated from the outlet chamber (blue) by a sieve plate that is covered by a solid cover plate during the mineralisation process (strong aeration) to prevent clogging and formation of fine particles. Organic-rich water from a clarifier or drum filter enters the mineralisation unit via the inlet. After a mineralisation cycle is completed, nutrient-rich, solid-free water exits the mineralisation unit via the outlet and is either directly transferred to the hydroponic unit or kept in a storage tank until needed

mineralisation chamber is aerated via compressed air to promote the respiration of heterotrophic bacteria and to keep anaerobic denitrification processes as minimal as possible. To prevent organic material from leaving the mineralisation chamber, a sieve plate could serve as a barrier. Ideally, a second, impermeable cover plate should be used to cover the sieve during the mineralisation process (during aeration). This should prevent the sieve plate from clogging as during the heavy aeration the organic material would be constantly moved against the sieve plate. Before transferring the nutrient-rich water from the mineralisation chamber to the hydroponic unit, aeration is stopped to allow the particles to settle. Subsequently, the cover plate is removed, and the nutrient-enriched water can pass through the sieve plate and leave the mineralisation chamber via the outlet as suggested in Fig. 10.2. Finally, the cover plate is put in place again, mineralisation chamber is refilled with RAS-derived sludge–water mixture, and the mineralisation process starts again (i.e. batch process).

The mineralisation unit should have at least twice the volume of the clarifier to allow for a continuous mineralisation. One mineralisation cycle can last for up to 5–30 days depending on the system, organic load and required nutrient profile and has to be elaborated for each individual system. For systems including a drum filter, as it is the case in most modern RAS, the mineralisation unit size has to be adjusted according to the daily or weekly sludge outflow of the drum filter. Since that has not been tested in an experimental setup so far, specific recommendations are not currently possible.

10.3.2 Implementation

An example of the implementation of an aerobic mineralisation unit into a decoupled aquaponic system is presented in Fig. 10.3. Since no pre- and post-treatment (e.g. re-oxygenation) is required, the mineralisation unit can be directly placed between the solid removal unit and the hydroponic beds. By installing a valve before and after the mineralisation unit, a discontinuous operation and nutrient delivery to the hydroponic unit on-demand are possible, but in many cases, an additional storage tank would be required. Ideally, after directing nutrient-rich water to the hydroponic unit, the displaced water is replaced with new sludge and water from the solid removal unit. Depending on the volume of the mineralisation unit, it is important to note that refilling with new sludge–water mixture can lead to an increase in pH again, and thus the mineralisation process could be interrupted. By increasing the size of the mineralisation unit, this effect would be buffered. In the study by Rakocy et al. (2007) investigating liquid organic waste from two aquaculture systems, a retention time of 29 days for aerobic mineralisation resulted in a substantial mineralisation success. Nevertheless, this also depends on the TS content within the mineralisation chamber, on the feed applied to the RAS, on the temperature and on the nutrient requirements of the plants that are produced within the hydroponic unit.

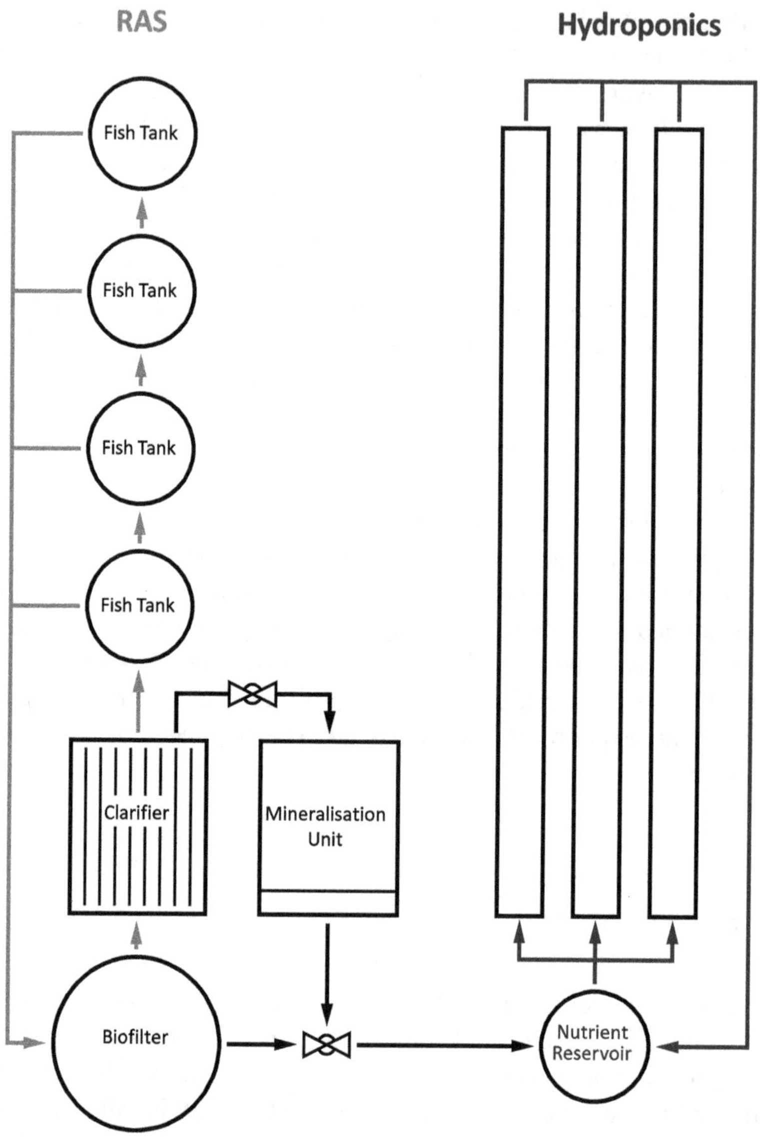

Fig. 10.3 Schematic picture of a decoupled aquaponic system including an aerobic mineralisation unit. Water can be transferred to the nutrient reservoir either from the RAS water loop or directly from the mineralisation unit

10.4 Anaerobic Treatments

Anaerobic digestion (AD) has long been used for the stabilisation and reduction of sludge mass process, mainly because of the simplicity of operation, relatively low costs and production of biogas as potential energy source. General stoichiometric representation of anaerobic digestion can be described as follows:

$$\begin{aligned} \mathrm{CnHaOb} + (\mathrm{n} - \mathrm{a}/4 - \mathrm{b}/2) \cdot \mathrm{H_2O} \rightarrow & (\mathrm{n}/2 - \mathrm{a}/8 + \mathrm{b}/4) \cdot \mathrm{CO_2} \\ & + (\mathrm{n}/2 + \mathrm{a}/8 - \mathrm{b}/4) \cdot \mathrm{CH4} \end{aligned} \tag{10.4}$$

Equation 10.4 Biogas general mass balance (Marchaim 1992).

And the theoretical methane concentration can be calculated as follows:

$$[\mathrm{CH_4}] = 0.5 + (\ \mathrm{a}/4 + \mathrm{b}/2\ )/2\mathrm{n} \tag{10.5}$$

Equation 10.5 Theoretical expected methane concentration in the biogas (Marchaim 1992).

The ultimate products from AD are mostly inorganic material (e.g. minerals), slightly degraded organic compounds and biogas which is typically composed of >55% methane (CH_4) and carbon dioxide (CO_2), with only small levels (<1%) of hydrogen sulphide (H_2S) and total ammonia nitrogen (NH_3^+/$NH4^+$) (Appels et al. 2008).

During the process of AD, the organic sludge undergoes considerable changes in its physical, chemical and biological properties and schematically can be divided into four stages (Fig. 10.4). The first stage is hydrolysis, where complex organic matter such as lipids, polysaccharides, proteins and nucleic acids degrade into soluble organic substances (sugars, amino acids and fatty acids). This step is generally considered rate-limiting (Deublein and Steinhauser 2010). In the second acidogenesis step, the monomers formed in the first step split further, and volatile fatty acids (VFA) are produced by acidogenic (fermentative) bacteria along with ammonia, CO_2, H_2S and other by-products. The third step is acetogenesis, where the VFA and alcohols are further digested by acetogens to produce mainly acetic acid as well as CO_2 and H_2. This conversion is controlled to a large extent by the partial pressure of H_2 in the mixture. The last step is methanogenesis where methane is mainly produced by two groups of methanogenic bacteria: acetotrophic archaea, which split acetate into methane and CO_2, and hydrogenotrophic archaea, which use hydrogen as an electron donor and carbon dioxide as electron acceptor to produce methane (Appels et al. 2008).

Various factors such as sludge pH, salinity, mineral composition, temperature, loading rate, hydraulic retention time (HRT), carbon-to-nitrogen (C/N) ratio and volatile fatty acid content influence the digestibility of the sludge and the biogas production (Khalid et al. 2011).

Anaerobic sludge treatment from RAS began about 30 years ago with reports on sludge from freshwater RAS (Lanari and Franci 1998) followed by reports on marine (Arbiv and van Rijn 1995; Klas et al. 2006; McDermott et al. 2001) and brackish water operations (Gebauer and Eikebrokk 2006; Mirzoyan et al. 2008). Recently, the use of UASB (Fig. 10.5) for AD of RAS sludge followed by biogas production as an alternative source of energy was suggested (Mirzoyan et al. 2010). The reactor is made of a tank, part of which is filled with an anaerobic granular sludge blanket containing the active microorganism species. Sludge flows upwards through a 'microbial blanket' where it is degraded by the anaerobic microorganisms and

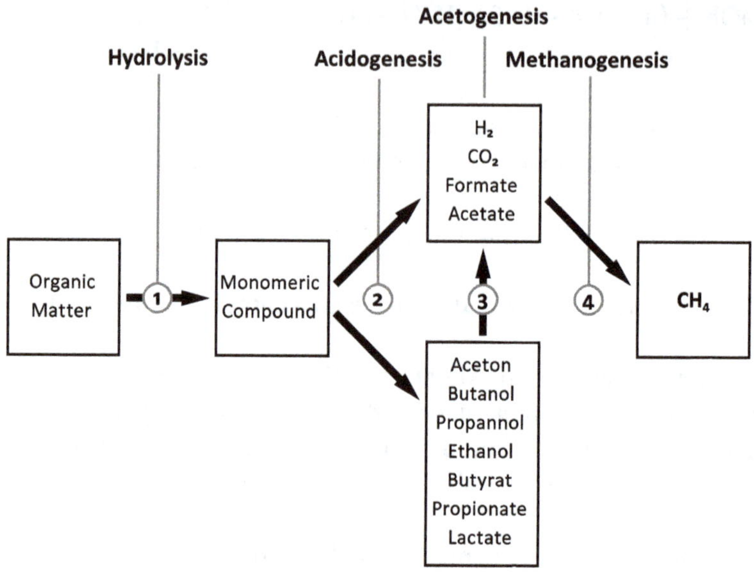

Fig. 10.4 Schematic diagram showing anaerobic degradation of organic matter based on Garcia et al. (2000)

Fig. 10.5 Scheme of an upflow anaerobic sludge blanket reactor (UASB)

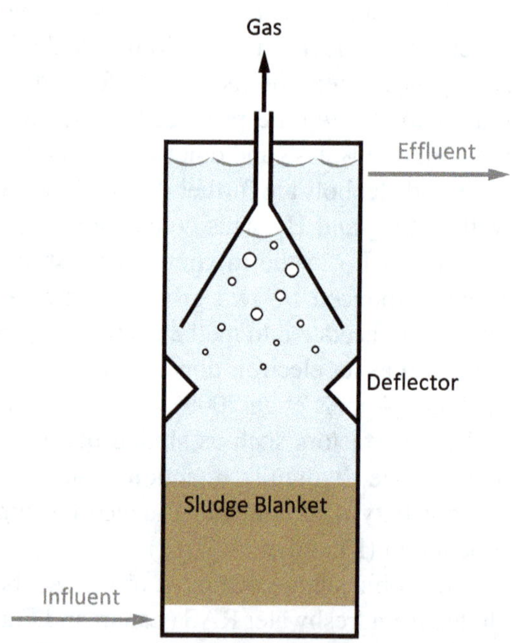

biogas is produced. An inverted cone settler at the top of the digester allows gas–liquid separation. When the biogas is released from the floc, it is oriented into the cone by the deflectors to be collected. A slow mixing in the reactor results from the upwards flow coupled with the natural movement of the microbial flocs that are attached to biogas bubbles. At some point, the floc leaves the gas bubble and settles back down allowing for the effluent to be free from TSS, which can then be recycled back to the system or released. The main advantages of the UASB are the low operational costs and simplicity of operation while providing high (>92%) solid-removal efficiency for wastes with low (1–3%) TSS content (Marchaim 1992; Yogev et al. 2017).

Two recent case studies demonstrated the use of UASB as a treatment for solids in pilot scale marine and saline RAS, which provide an example of the potential advantages of this unit in aquaponics (Tal et al. 2009; Yogev et al. 2017). A detailed look at the carbon balance suggested that about 50% of the introduced carbon (from feed) was removed by fish assimilation and respiration, 10% was removed by aerobic biodegradation in the nitrification bioreactor and 10% was removed in the denitrification reactor (Yogev et al. 2017). Therefore, overall about 25% carbon was introduced into the UASB reactor of which 12.5% was converted to methane, 7.5% to CO_2 and the rest (~5%) remained as nondegradable carbon in the UASB. In summary, it was demonstrated that the use of UASB allowed better water recirculation (>99%), smaller (<8%) production of sludge when compared with typical RAS that do not have on-site solid treatment, and recovery of energy that can account for 12% of the overall energy demand of the RAS. It should be noted that using UASB in aquaponics will also allow significant recovery of up to 50% more nutrients such as nitrogen, phosphorus and potassium since they are released into the water as a result of solid biodegradation (Goddek et al. 2018).

The anaerobic membrane bioreactor (AnMBR) is a more advanced technology. The main process consists of using a special membrane to separate the solids from the liquid instead of using a decanting process as in UASB. The sludge fermentation occurs in a simple anaerobic tank and the effluents leave it through the membrane. Depending on the membrane pore size (going down to 0.1–0.5 μm) even microorganisms can be retained. There are two types of membrane bioreactor design: one uses a side-stream mode outside the tank, and the other has the membrane unit submerged into the tank (Fig. 10.6), the latter being more favourable in AnMBR application due to its more compact configuration and lower energy consumption (Chang 2014). Membranes of different materials such as ceramic or polymeric (e.g. polyvinylidene fluoride (PVDF), polyethylene, polyethersulfone (PES), polyvinyl chloride (PVC)) may be configured as plate and frame, hollow fibre or tubular units (Gander et al. 2000; Huang et al. 2010). AnMBR has several significant advantages over typical biological reactors such as the UASB, namely, decoupling of (long) sludge retention time (SRT) and (short) hydraulic residence time (HRT), hence enabling the problem of the AD process's slow kinetics to be overcome; very high effluent quality in which most nutrients remain; and removal of pathogens and a small footprint (Judd and Judd 2008). In addition, efficient biogas production in the AnMBR can possibly result in a net energy balance.

While this technology deserves a lot of attention and research, it should be noted that since it is a fairly new technology, there are still several significant drawbacks that must be addressed before AnMBR would be adopted by the aquaculture industry. These are the high operational costs due to membrane maintenance to prevent biofouling, regular membrane exchange and high CO_2 fraction (30–50%) in the biogas which limits its utilisation and contributes to greenhouse gas (GHG)

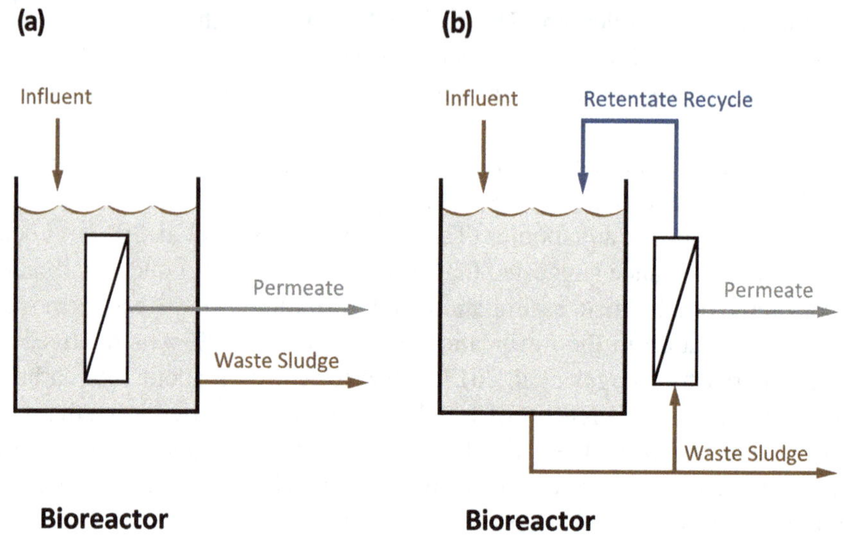

Fig. 10.6 (**a**) Side-stream MBR with a separate filtration unit with the retained fluid recycled back to the bioreactor; (**b**) submerged MBR: filtration unit integrated into the bioreactor. (Gander et al. 2000)

emission (Cui et al. 2003). On a positive note, in the near future, new biofouling prevention techniques will be developed while the membrane cost will certainly drop with the broader use of this technology. The combination of a UASB with a membrane reactor to filter the UASB effluent has been successfully studied to remove organic carbon and nitrogen (An et al. 2009). This combination seems a promising option for aquaponics for safe and sanitary use of UASB effluents.

10.4.1 Implementation

One possible solution of implementing anaerobic reactors is in a sequential manner (see also Chap. 8). A 'high pH–low pH' combination allows for harvesting methane (and thus reducing carbon) in the first high pH step and mobilising nutrients in the decarbonised sludge in a subsequent low pH environment. The advantage of this method is that the carbon reduction under high pH conditions results in less VFAs, which can occur during the low pH second step (Fig. 10.7). This approach also allows for co-digestion of green vegetative matter (i.e. from any harvesting of plants, there will be waste vegetative matter which could be put through such a digester) to increase both biogas production and nutrient recovery from the overall scheme.

Another technical integration possibility has been presented by Ayre et al. (2017). They propose to discharge the effluent of a high-pH anaerobic digester to an algal culture pond. Within that pond, algae are grown, whose biomass can be used for animal–aquaculture feed or biofertilisation (Fig. 10.8). More detailed information on this approach can be found in Chap. 11.

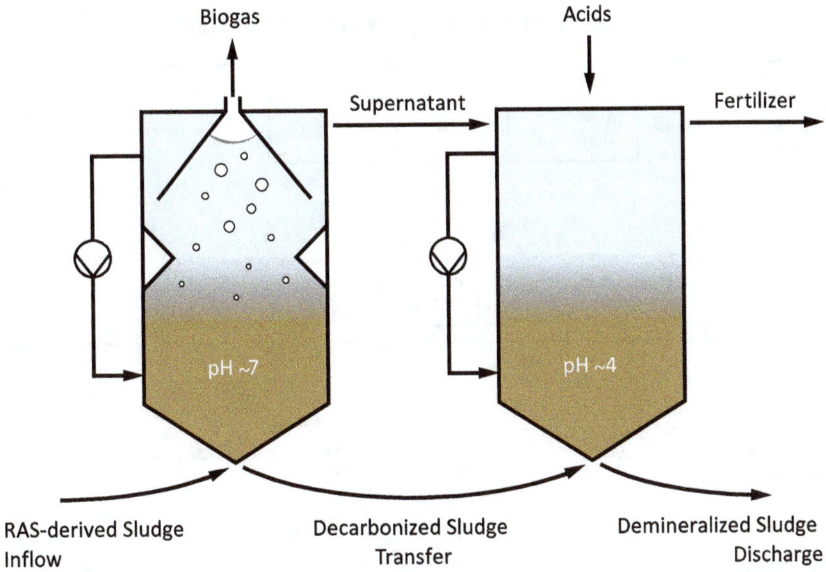

Fig. 10.7 Two-stage anaerobic system. In the first stage (high pH), the carbon will be removed from the sludge as biogas, whereas the low pH in the second stage allows nutrients that are trapped in the sludge do dissolve in the water. Usually, volatile fatty acids (VFA) would form in low pH environments. The removal of the carbon source in the first stage, however, limits VFA production in such a sequential setup

10.5 Methodology to Quantify the Sludge Reduction and Mineralisation Performance

To determine the digestion of aquaponic sludge treatment in aerobic and anaerobic bioreactors, a specific methodology needs to be followed. A methodology adapted for aquaponic sludge treatment purposes is presented in this chapter. Specific equations have been developed to precisely quantify their performance (Delaide et al. 2018), and these should be used to evaluate the performance of the treatment applied in a specific aquaponic plant.

In order to evaluate the treatment's performance, a mass balance approach needs to be achieved. It requires that TSS, COD and nutrient masses are determined for the all reactor inputs (i.e. fresh sludge) and outputs (i.e. effluents). The reactor content also needs to be sampled at the beginning and at the end of the studied period. The input, output and content of the reactors have to be perfectly mixed for sampling. Reactor input and output should basically be sampled every time the reactors are fed with fresh sludge.

Then, reactor sludge reduction performance (*η*) can be formulated as follows:

$$\eta_S = 100\%(1 - (\Delta S + S_{out})/S_{in})) \tag{10.6}$$

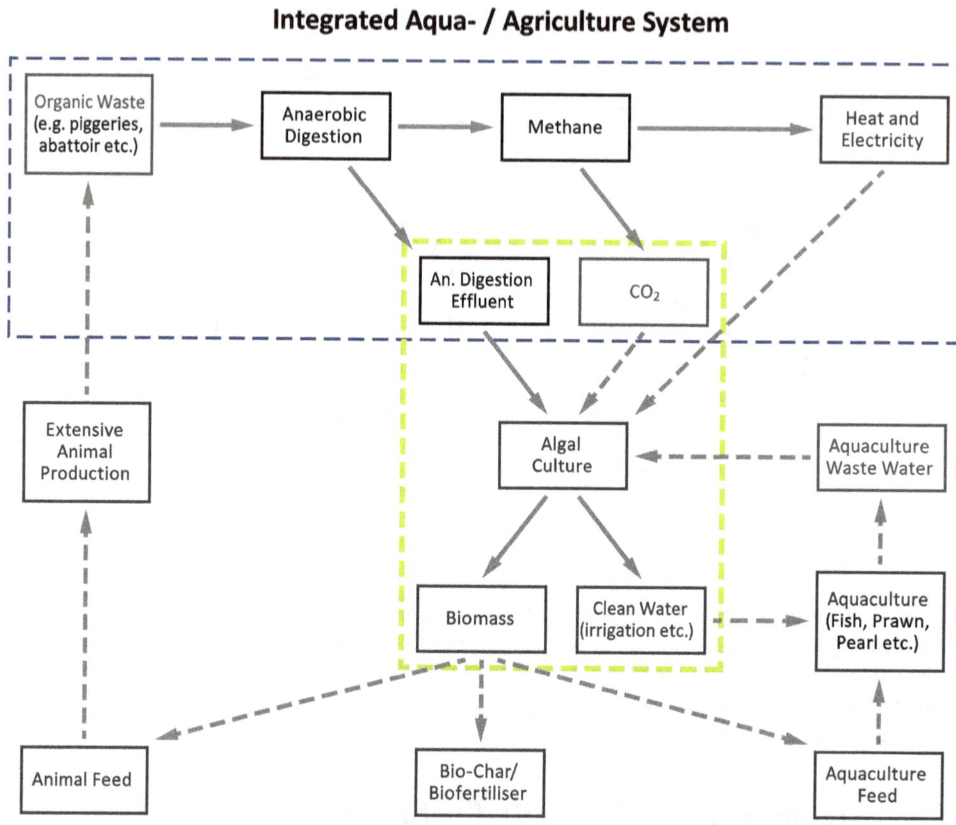

Fig. 10.8 Anaerobic digestion system integrated with aquaculture and algal culture based on Ayre et al. (2017)

where ΔS is the sludge inside the reactor at the end of the studied period minus the one at the beginning of the period, S_{out} is the total sludge that left the reactor in the outflow, and S_{in} is the total sludge that entered the reactor via inflow.

For organic reduction, the sludge (i.e. the term S) can be characterised by the dry mass of sludge (i.e. TSS) or the mass of oxygen needed to oxidise the sludge (i.e. COD). Thus, for COD and TSS reduction performances, the smaller the accumulation and the smaller the quantity in the outflow, the higher the reduction performance (i.e. high percentage) and so the less solids discharged out of the loop.

Based on the same mass balance, the nutrient mineralisation performance of the treatment (ζ), i.e. the conversion into soluble ions of the macro- and micronutrients present in the sludge under undissolved forms, the following formula can be used:

$$\zeta_N = 100\%((DN_{out} - DN_{in})/(TN_{in} - DN_{in})) \tag{10.7}$$

where ζ is the recovery of N nutrient at the end of the studied period in percent, DN_{out} is the total mass of dissolved nutrient in the outflow, DN_{in} is the total mass of dissolved nutrient in the inflow and TN_{in} is the total mass of dissolved plus undissolved nutrients in the inflow.

Thus, similar to the organic reduction performances, the smaller the accumulation inside the reactor and undissolved nutrient content in the outflow, the higher the mineralisation performance (i.e. high percentage) and so the dissolved nutrient recovered in the effluent (or outflow) for aquaponic crop fertilisation (see Example 10.1). The presented mass balance equations are used in the example box.

Example 10.1

The digestion performance of a 250-L anaerobic bioreactor has been evaluated for an 8-week period. It was fed once a day with 25 L of fresh sludge coming from a tilapia RAS system, and the equivalent supernatant volume (or output) was removed from the bioreactor. The fresh sludge (input) had a TSS of 10 g dry mass (DM) per litre or 1%, and the supernatant (output) had a TSS of 1 gDM/L or 0.1%. The TSS inside the bioreactor at the beginning and at the end of the period was 20 gDM/L. Consequently, the total DM inputs, outputs and inside the bioreactor during the evaluated period are calculated as follows:

DM in = 0.01 kg/Ld × 25 L × 7 days × 8 weeks = 14 kg
DM out = 0.001 kg/Ld × 25 L × 7 days × 8 weeks = 1.4 kg
DM to = DM tf = 250 L × 0.02 kg/L = 5 kg

The TSS reduction performance (η_{TSS}) of the bioreactor can then be calculated as follows:

$$\eta_{TSS} = 100\%(1 - ((5 - 5) + 1.4)/14)) = 90\%$$

The bioreactor P mineralisation performance can be evaluated knowing that the fresh sludge (input) had a concentration of dissolved P of 15 mg/L and a total P content of 90 mg/L. The concentration of dissolved P in the supernatant (output) was 20 mg/L. Consequently, the total P content in the input, the total dissolved P in the inputs and outputs during the evaluated period are calculated as follows:

TP in = 0.090 g/Ld × 25 L × 7 days × 8 weeks = 126 g
DP in = 0.015 g/Ld × 25 L × 7 days × 8 weeks = 21 g
DP out = 0.020 g/Ld × 25 L × 7 days × 8 weeks = 28 g

The P mineralisation performance (ζ_P) of the bioreactor can then be calculated as follows:

$$\zeta_P = 100\%((28 - 21)/(126 - 21)) = 6.67\%$$

10.6 Conclusions

Fish sludge treatment for reduction and nutrient recovery is in an early phase of implementation. Further research and improvements are needed and will see the day with the increased concern of circular economy. Indeed, fish sludge needs to be considered more as a valuable source instead of a disposable waste.

References

An Y, Yang F, Wong FS, Chua HC (2009) Effect of recirculation ratio on simultaneous methanogenesis and nitrogen removal using a combined up-flow anaerobic sludge blanket–membrane bioreactor. Environ Eng Sci 26:1047–1053. https://doi.org/10.1089/ees.2007.0317

Appels L, Baeyens J, Degrève J, Dewil R (2008) Principles and potential of the anaerobic digestion of waste-activated sludge. Prog Energy Combust Sci 34:755. https://doi.org/10.1016/j.pecs.2008.06.002

Arbiv R, van Rijn J (1995) Performance of a treatment system for inorganic nitrogen removal in intensive aquaculture systems. Aquac Eng 14:189. https://doi.org/10.1016/0144-8609(94)P4435-E

Ayre JM, Moheimani NR, Borowitzka MA (2017) Growth of microalgae on undiluted anaerobic digestate of piggery effluent with high ammonium concentrations. Algal Res 24:218–226. https://doi.org/10.1016/j.algal.2017.03.023

Brod E, Oppen J, Kristoffersen AØ, Haraldsen TK, Krogstad T (2017) Drying or anaerobic digestion of fish sludge: nitrogen fertilisation effects and logistics. Ambio 46:852. https://doi.org/10.1007/s13280-017-0927-5

Chang S (2014) Anaerobic membrane bioreactors (AnMBR) for wastewater treatment. Adv Chem Eng Sci 4:56. https://doi.org/10.4236/aces.2014.41008

Chen SL, Coffin DE, Malone RF (1997) Sludge production and management for recirculating aquacultural systems. J World Aquac Soc 28:303–315. https://doi.org/10.1111/j.1749-7345.1997.tb00278.x

Chowdhury P, Viraraghavan T, Srinivasan A (2010) Biological treatment processes for fish processing wastewater–a review. Bioresour Technol 101:439–449. https://doi.org/10.1016/j.biortech.2009.08.065

Conroy J, Couturier M (2010) Dissolution of minerals during hydrolysis of fish waste solids. Aquaculture 298:220–225. https://doi.org/10.1016/j.aquaculture.2009.11.013

Cui ZF, Chang S, Fane AG (2003) The use of gas bubbling to enhance membrane processes. J Memb Sci. https://doi.org/10.1016/S0376-7388(03)00246-1

Delaide B, Goddek S, Gott J, Soyeurt H, Jijakli MH (2016) Lettuce (*Lactuca sativa* L. var. Sucrine) growth performance in complemented aquaponic solution outperforms hydroponics. Water (Switzerland) 8. https://doi.org/10.3390/w8100467

Delaide B, Goddek S, Keesman KJ, Jijakli MH (2018) A methodology to quantify the aerobic and anaerobic sludge digestion performance for nutrient recycling in aquaponics. Biotechnol Agron Soc Environ 22

Deublein D, Steinhauser A (2010) Biogas from waste and renewable resources: an introduction. In: Biogas from waste and renewable resources: an introduction, 2nd edn. https://doi.org/10.1002/9783527632794

Ebeling JM, Timmons MB, Bisogni JJ (2006) Engineering analysis of the stoichiometry of photoautotrophic, autotrophic, and heterotrophic removal of ammonia-nitrogen in aquaculture systems. Aquaculture 257:346. https://doi.org/10.1016/j.aquaculture.2006.03.019

Endut A, Jusoh A, Ali N, Wan Nik WB, Hassan A (2010) A study on the optimal hydraulic loading rate and plant ratios in recirculation aquaponic system. Bioresour Technol 101:1511–1517. https://doi.org/10.1016/j.biortech.2009.09.040

Gander M, Je B, Judd S (2000) Aerobic MBRs for domestic wastewater treatment: a review with cost considerations. Sep Purif Technol 18:119–130

Garcia J-L, Patel BKC, Ollivier B (2000) Taxonomic, phylogenetic, and ecological diversity of methanogenic archaea. Anaerobe 6:205. https://doi.org/10.1006/anae.2000.0345

Gebauer R, Eikebrokk B (2006) Mesophilic anaerobic treatment of sludge from salmon smolt hatching. Bioresour Technol 97:2389–2401. https://doi.org/10.1016/j.biortech.2005.10.008

Goddek S, Keesman KJ (2018) The necessity of desalination technology for designing and sizing multi-loop aquaponics systems. Desalination 428:76–85. https://doi.org/10.1016/j.desal.2017.11.024

Goddek S, Körner O (2019) A fully integrated simulation model of multi-loop aquaponics: a case study for system sizing in different environments. Agric Syst 171:143

Goddek S, Delaide B, Mankasingh U, Ragnarsdottir K, Jijakli H, Thorarinsdottir R (2015) Challenges of sustainable and commercial aquaponics. Sustainability 7:4199–4224. https://doi.org/10.3390/su7044199

Goddek S, Espinal CA, Delaide B, Jijakli MH, Schmautz Z, Wuertz S, Keesman KJ (2016) Navigating towards decoupled aquaponic systems: a system dynamics design approach. Water (Switzerland) 8. https://doi.org/10.3390/W8070303

Goddek S, Delaide B, Oyce A, Wuertz S, Jijakli MH, Gross A, Eding EH, Bläser I, Keizer LCP, Morgenstern R, Körner O, Verreth J, Keesman KJ (2018) Nutrient mineralisation and organic matter reduction performance of RAS-based sludge in sequential UASB-EGSB reactors. Aquac Eng 83:10. https://doi.org/10.1016/J.AQUAENG.2018.07.003

Graber A, Junge R (2009) Aquaponic systems: nutrient recycling from fish wastewater by vegetable production. Desalination 246:147–156

Huang X, Xiao K, Shen Y (2010) Recent advances in membrane bioreactor technology for wastewater treatment in China. Front Environ Sci Eng China 4:245. https://doi.org/10.1007/s11783-010-0240-z

Judd S, Judd C (2008) The MBR book: principles and applications of membrane bioreactors in water and wastewater treatment. Elsevier. https://doi.org/10.1016/B978-185617481-7/50005-2

Jung IS, Lovitt RW (2011) Leaching techniques to remove metals and potentially hazardous nutrients from trout farm sludge. Water Res 45:5977–5986. https://doi.org/10.1016/j.watres.2011.08.062

Khalid A, Arshad M, Anjum M, Mahmood T, Dawson L (2011) The anaerobic digestion of solid organic waste. Waste Manag 31:1737. https://doi.org/10.1016/j.wasman.2011.03.021

Klas S, Mozes N, Lahav O (2006) Development of a single-sludge denitrification method for nitrate removal from RAS effluents: lab-scale results vs. model prediction. Aquaculture 259:342. https://doi.org/10.1016/j.aquaculture.2006.05.049

Kumar V, Sinha AK, Makkar HPS, De Boeck G, Becker K (2012) Phytate and phytase in fish nutrition. J Anim Physiol Anim Nutr (Berl) 96:335. https://doi.org/10.1111/j.1439-0396.2011.01169.x

Lanari D, Franci C (1998) Biogas production from solid wastes removed from fish farm effluents. Aquat Living Resour 11:289–295. https://doi.org/10.1016/S0990-7440(98)80014-4

Licamele JD (2009) Biomass production and nutrient dynamics in an aquaponics system. The University of Arizona

Marchaim U (1992) Biogas processes for sustainable development. FAO Agricultural Services Bulletin 95. Food and Agriculture Organization of the United Nations

McDermott BL, Chalmers AD, Goodwin JAS (2001) Ultrasonication as a pre-treatment method for the enhancement of the psychrophilic anaerobic digestion of aquaculture effluents. Environ Technol (United Kingdom) 22:823. https://doi.org/10.1080/095933322086180317

McGill SM (2012) 'Peak' phosphorus? The implications of phosphate scarcity for sustainable investors. J Sustain Financ Invest. https://doi.org/10.1080/20430795.2012.742635

Mirzoyan N, Gross A (2013) Use of UASB reactors for brackish aquaculture sludge digestion under different conditions. Water Res 47:2843–2850. https://doi.org/10.1016/j.watres.2013.02.050

Mirzoyan N, Parnes S, Singer A, Tal Y, Sowers K, Gross A (2008) Quality of brackish aquaculture sludge and its suitability for anaerobic digestion and methane production in an upflow anaerobic sludge blanket (UASB) reactor. Aquaculture 279:35–41. https://doi.org/10.1016/j.aquaculture.2008.04.008

Mirzoyan N, Tal Y, Gross A (2010) Anaerobic digestion of sludge from intensive recirculating aquaculture systems: review. Aquaculture 306:1–6. https://doi.org/10.1016/j.aquaculture.2010.05.028

Monsees H, Keitel J, Paul M, Kloas W, Wuertz S (2017) Potential of aquacultural sludge treatment for aquaponics: evaluation of nutrient mobilization under aerobic and anaerobic conditions. Aquac Environ Interact 9:9–18. https://doi.org/10.3354/aei00205

Naylor SJ, Moccia RD, Durant GM (1999) The chemical composition of settleable solid fish waste (manure) from commercial rainbow trout farms in Ontario, Canada. North Am J Aquac 61:21–26

Neto RM, Ostrensky A (2013) Nutrient load estimation in the waste of Nile tilapia *Oreochromisniloticus* (L.) reared in cages in tropical climate conditions. Aquac Res 46:1309–1322. https://doi.org/10.1111/are.12280

Nichols MA, Savidov NA (2012) Aquaponics: a nutrient and water efficient production system. Acta Hortic:129–132

Peng L, Dai H, Wu Y, Peng Y, Lu X (2018) A comprehensive review of phosphorus recovery from wastewater by crystallization processes. Chemosphere 197:768. https://doi.org/10.1016/j.chemosphere.2018.01.098

Rakocy JE, Bailey DS, Shultz RC, Danaher JJ (2007) Preliminary evaluation of organic waste from two aquaculture systems as a source of inorganic nutrients for hydroponics. Acta Hortic 742:201–208

Ru D, Liu J, Hu Z, Zou Y, Jiang L, Cheng X, Lv Z (2017) Improvement of aquaponic performance through micro- and macro-nutrient addition. Environ Sci Pollut Res 24:16328. https://doi.org/10.1007/s11356-017-9273-1

Saha S, Monroe A, Day MR (2016) Growth, yield, plant quality and nutrition of basil (*Ocimumbasilicum* L.) under soilless agricultural systems. Ann Agric Sci 61:181–186. https://doi.org/10.1016/j.aoas.2016.10.001

Schneider O, Sereti V, Eding EH, Verreth JAJ (2005) Analysis of nutrient flows in integrated intensive aquaculture systems. Aquac Eng 32:379–401. https://doi.org/10.1016/j.aquaeng.2004.09.001

Seo KW, Choi YS, Gu MB, Kwon EE, Tsang YF, Rinklebe J, Park C (2017) Pilot-scale investigation of sludge reduction in aerobic digestion system with endospore-forming bacteria. Chemosphere 186:202–208. https://doi.org/10.1016/j.chemosphere.2017.07.150

Stewart NT, Boardman GD, Helfrich LA (2006) Characterization of nutrient leaching rates from settled rainbow trout (*Oncorhynchus mykiss*) sludge. Aquac Eng 35:191–198. https://doi.org/10.1016/j.aquaeng.2006.01.004

Tal Y, Schreier HJ, Sowers KR, Stubblefield JD, Place AR, Zohar Y (2009) Environmentally sustainable land-based marine aquaculture. Aquaculture 286:28–35. https://doi.org/10.1016/j.aquaculture.2008.08.043

Techobanoglous G, Burton FL, Stensel HD (2014) Wastewater engineering: treatment and reuse, 5th edn. Metcalf and Eddy. https://doi.org/10.1016/0309-1708(80)90067-6

Turcios AE, Papenbrock J (2014) Sustainable treatment of aquaculture effluents-what can we learn from the past for the future? Sustain 6:836–856

Van Lier JB, Mahmoud N, Zeeman G (2008) Anaerobic wastewater treatment, Biological wastewater treatment: principles, modelling and design. https://doi.org/10.1021/es00154a002

Van Rijn J (2013) Waste treatment in recirculating aquaculture systems. Aquac Eng 53:49–56. https://doi.org/10.1016/j.aquaeng.2012.11.010

van Rijn J, Fonarev N, Berkowitz B (1995) Anaerobic treatment of intensive fish culture effluents: digestion of fish feed and release of volatile fatty acids. Aquaculture 133:9–20. https://doi.org/10.1016/0044-8486(94)00385-2

Yogev U, Barnes A, Gross A (2016) Nutrients and energy balance analysis for a conceptual model of a three loops off grid, Aquaponics. Water 8:589. https://doi.org/10.3390/W8120589

Yogev U, Sowers KR, Mozes N, Gross A (2017) Nitrogen and carbon balance in a novel near-zero water exchange saline recirculating aquaculture system. Aquaculture 467:118–126. https://doi.org/10.1016/j.aquaculture.2016.04.029

Zhang X, Hu J, Spanjers H, van Lier JB (2016) Struvite crystallization under a marine/brackish aquaculture condition. Bioresour Technol 218:1151. https://doi.org/10.1016/j.biortech.2016.07.088

11

Aquaponic Systems and their Mathematical Modelling

Karel J. Keesman, Oliver Körner, Kai Wagner, Jan Urban, Divas Karimanzira, Thomas Rauschenbach, and Simon Goddek

Abstract Mathematical models can take very different forms and very different levels of complexity. A systematic way to postulate, calibrate and validate, as provided by systems theory, can therefore be very helpful. In this chapter, dynamic systems modelling of aquaponic (AP) systems, from a systems theoretical perspective, is considered and demonstrated to each of the subsystems of the AP system, such as fish tanks, anaerobic digester and hydroponic (HP) greenhouse. It further shows the links between the subsystems, so that in principle a complete AP systems model can be built and integrated into daily practice with respect to management and control of AP systems. The main challenge is to choose an appropriate model complexity that meets the experimental data for estimation of parameters and states and allows us to answer questions related to the modelling objective, such as simulation, experiment design, prediction and control.

K. J. Keesman (✉) · S. Goddek
Mathematical and Statistical Methods (Biometris), Wageningen University, Wageningen, The Netherlands
e-mail: karel.keesman@wur.nl; simon.goddek@wur.nl; simon@goddek.nl

O. Körner
Leibniz-Institute of Vegetable and Ornamental Crops (IGZ), Grossbeeren, Germany
e-mail: koerner@igzev.de

K. Wagner
Institut für physikalische Prozesstechnik, University of Applied Sciences Saarbrücken, Saarbrücken, Germany
e-mail: kai.wagner@htwsaar.de

J. Urban
Laboratory of Signal and Image Processing, Institute of Complex Systems, South Bohemian Research Center of Aquaculture and Biodiversity of Hydrocenoses, Faculty of Fisheries and Protection of Waters, University of South Bohemia in Ceske Budejovice, Nove Hrady, Czech Republic
e-mail: urbanj@frov.jcu.cz

D. Karimanzira · T. Rauschenbach
Fraunhofer IOSB-AST, Ilmenau, Germany
e-mail: divas.karimanzira@iosb-ast.fraunhofer.de; Thomas.Rauschenbach@iosb-ast.fraunhofer.de

Keywords Modelling · Recirculating aquaculture system · Anaerobic digestion · Hydroponic greenhouse · Multi-loop aquaponic system · Tools

11.1 Introduction

In general, mathematical models can take very different forms depending on the system under study, which may range from social, economic and environmental to mechanical and electrical systems. Typically, the internal mechanisms of social, economic or environmental systems are not very well known or understood and often only small data sets are available, while the prior knowledge of mechanical and electrical systems is at a high level, and experiments can easily be done. Apart from this, the model form also strongly depends on the final objective of the modelling procedure. For instance, a model for process design or simulation should contain much more detail than a model used for studying different long-term scenarios.

In particular, for a wide range of applications (e.g. Keesman 2011), models are developed to:

- Obtain or enlarge insight in different phenomena, for example, recovering physical or economic relationships.
- Analyse process behaviour using simulation tools, for example, process training of operators or weather forecasts.
- Estimate state variables that cannot be easily measured in real time on the basis of available measurements, for instance, online process information.
- Control, for instance, in the internal model control or model-based predictive control concept or to manage processes.

A critical step in the modelling of any system is to find a mathematical model which adequately describes the actual situation or state. Firstly, the system boundaries and the system variables have to be specified. Then relationships between these variables have to be specified on the basis of prior knowledge, and assumptions about the uncertainties in the model have to be made. Combining this information defines the model structure. Still the model may contain some unknown or incompletely known coefficients, the model parameters, which in case of time-varying behaviour define an additional set of system variables. For a general introduction to mathematical modelling we refer to, for instance, Sinha and Kuszta (1983), Willems and Polderman (1998) and Zeigler et al. (2000).

In this chapter, the modelling of an aquaponic (food) production (AP) system will be described. Figure 11.1 shows a typical example of an AP system, i.e. the so-called decoupled three-loop aquaponic system. As a result of basic principles modelling, using conservation laws and constitutive relationships, mathematical models of all kinds of AP systems are usually represented as a set of ordinary or partial differential equations. These mathematical models are commonly used for design, estimation and control. In each of these specific modelling objectives, we distinguish between analysis and synthesis.

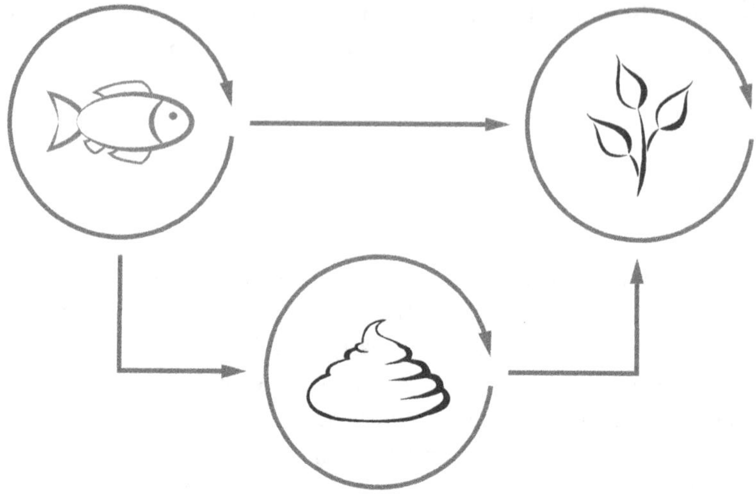

Fig. 11.1 Decoupled, three-loop aquaponic system with RAS, hydroponic and remineralization subsystems. (Goddek, 2017)

The outline of the chapter is as follows. In Sect. 11.1 some background on mathematical systems modelling is presented. Sections 11.2, 11.3, 11.4 and 11.5 describe the modelling of a recirculating aquaculture system (RAS), anaerobic digestion, hydroponic (HP) greenhouse and a multi-loop AP system, respectively. In Sect. 11.6 modelling tools are introduced and illustrated with some examples. The chapter concludes with a Discussion and Conclusions section.

11.2 Background

Many definitions of a system are available, ranging from loose descriptions to strict mathematical formulations. In what follows, a *system* is considered to be an object in which different variables interact at all kinds of time and space scales and that produces observable signals. These types of systems are also called *open systems*. A graphical representation of a general open system (*S*) with vector-valued input and output signals is represented in Fig. 11.2. Thus, multiple inputs or outputs are combined in one single arrow. So, the system variables may be scalars or vectors. In addition, they can be continuous or discrete functions of time. It is important to stress that the arrows in Fig. 11.2 represent signal flows and thus not necessarily physical flows.

It is also possible to connect systems into a network, as in an AP system, with parallel, feedback and feedforward paths. Figure 11.3 presents an example of such a network.

For controller/management analysis and synthesis, it is often convenient to connect the system (*S*) to the controller or management strategy (*C*), as in Fig. 11.4. Most often the input to the controller or management strategy is the external steering signal of the controlled system, and the output of the system is the observed system's behaviour.

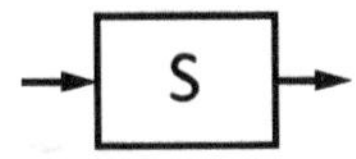

Fig. 11.2 General open system representation

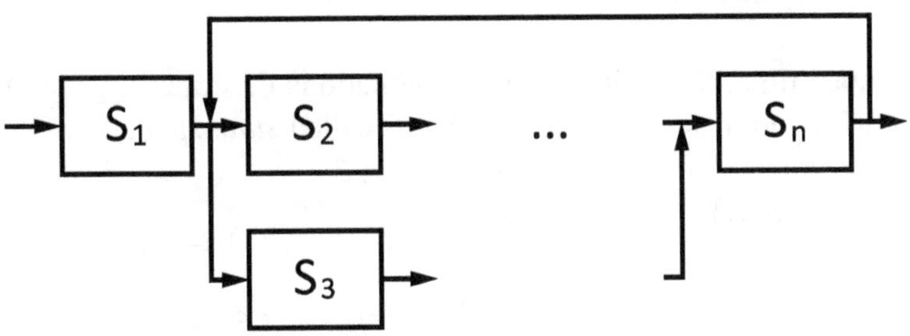

Fig. 11.3 Open system network representation

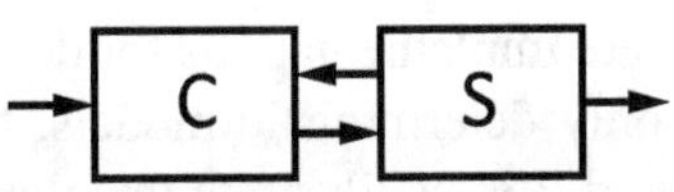

Fig. 11.4 Controlled system

Fig. 11.5 Model-based controlled system

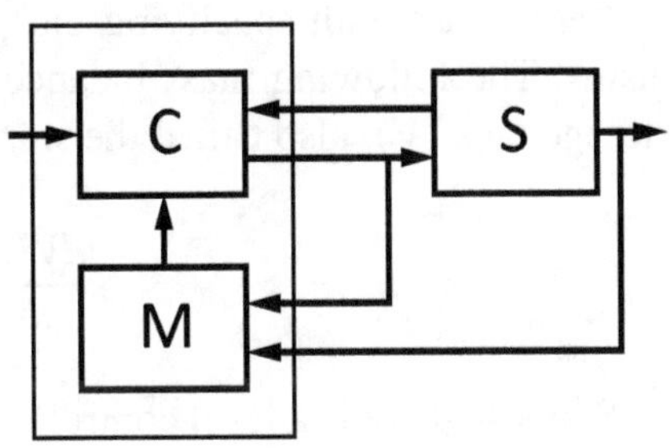

Finally, to emphasize the incorporation of a mathematical model (M) into the controller structure or management strategy, the following model-based controlled system representation is introduced (Fig. 11.5).

For now, it suffices to present the block diagram representation. In subsequent sections, the modelling of AP systems will be worked out in more detail.

In systems theory the basic structure of a mathematical model (M) is schematically represented as in Fig. 11.6. In Fig. 11.6, x is the so-called state of the system, u the control input, y the output, w the disturbance input and v the output noise. In general, each of these variables is vector-valued.

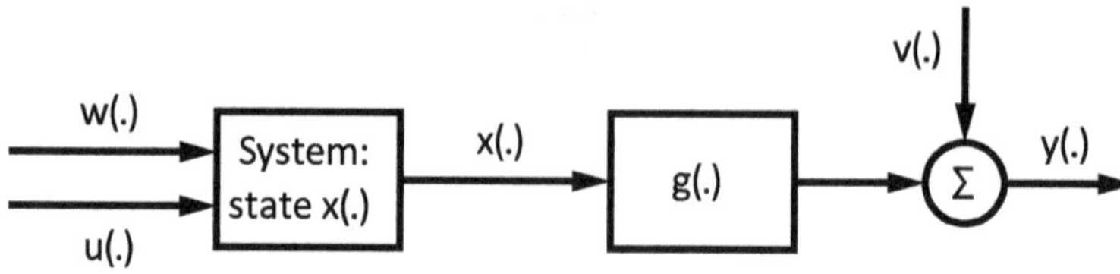

Fig. 11.6 Basic structure of mathematical model (M)

In continuous time, the following set of equations describes a general dynamic model (M), with parameter vector p, in what is called *state-space* form:

$$\begin{aligned} \frac{dx(t)}{dt} &= f(t, x(t), u(t), w(t); p), \qquad x(0) = x_0 \\ y(t) &= g(t, x(t), u(t); p) + v(t), \qquad t \in \mathfrak{R}^+ \end{aligned} \tag{11.1}$$

where the first equation describes the nonlinear and time-varying dynamics of the system in terms of state variables (x) and the second one expresses the algebraic relationship between u, x and y. This state-space model representation has been a starting point for many software implementations for design, control and estimation. In what follows, however, only deterministic models, thus without the stochastic vectors v and w, are considered. Let us illustrate this theory on a fish tank system.

Example: Fish Tank System

Consider the following fish tank, which is a typical example of the general system presented in Fig. 11.7.

Let us start with specifying our prior knowledge of the internal system mechanisms. The following mass balance can be defined in terms of the volume of the storage tank (V), also called the *state* of the system, inflows $u(t)$ and outflows $y(t)$:

$$\frac{dV(t)}{dt} = u(t) - y(t) \tag{11.2}$$

Suppose there is a level controller (LC) that keeps the outflow proportional to the volume in the tank. This can be enforced by implementing the following proportional control law,

$$Y(t) = KV(t) \tag{11.3}$$

with K a real, positive constant. Hence, after substituting Eq. (11.3) into (11.2), we obtain the following differential equation

$$\frac{dV(t)}{dt} + KV(t) = u(t) \tag{11.4}$$

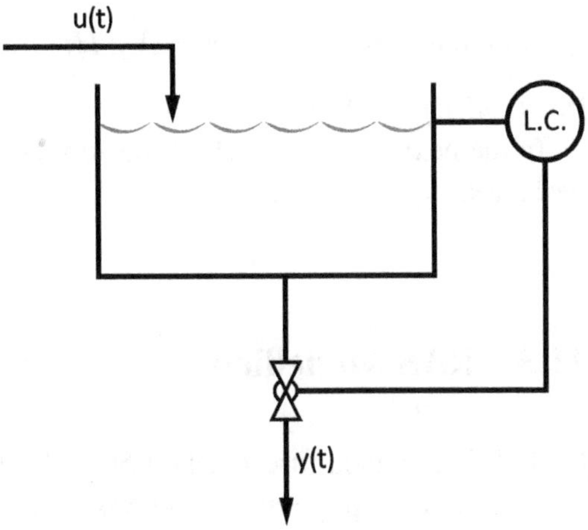

Fig. 11.7 Fish tank with volume-controlled flow using level controller (LC)

For this specific linear differential equation with constant coefficients, an analytical solution exists and is given by

$$y(t) = y(0)e^{-Kt} + \int_0^t Ke^{-K(t-s)}u(s)ds \tag{11.5}$$

under the assumption that $u(t) = 0$ for $t < 0$. From this example it is clear that applying first principles – mass conservation in this case – directly leads to an ordinary differential equation. In state-space format, the model can be represented as

$$\begin{aligned} \frac{dx(t)}{dt} &= -Kx(t) + u(t) \\ y(t) &= Kx(t) \end{aligned} \tag{11.6}$$

With x volume, u flow input and K controller gain. Thus, in terms of the general state-space Eq. (11.1), $f(t, x(t), u(t); p) \equiv -Kx(t) + u(t)$ and $g(t, x(t), u(t); p) \equiv Kx(t)$.

For two volume-controlled fish tanks in series with volume V_1 and V_2, and controller gain K_1 and K_2, respectively, two mass balances can be formulated, i.e.

$$\begin{aligned} \frac{dV_1(t)}{dt} &= -K_1V_1(t) + u(t) \\ \frac{dV_2(t)}{dt} &= K_1V_1(t) - K_2V_2(t) \end{aligned} \tag{11.7}$$

In vector-matrix form, and for physical outflow $y(t)$, we can write:

$$\begin{aligned} \frac{d}{dt}\begin{bmatrix} V_1(t) \\ V_2(t) \end{bmatrix} &= \begin{bmatrix} -K_1 & 0 \\ K_1 & -K_2 \end{bmatrix}\begin{bmatrix} V_1(t) \\ V_2(t) \end{bmatrix} + \begin{bmatrix} 1 \\ 0 \end{bmatrix}u(t) \\ y(t) &= K_2V_2(t) \end{aligned} \tag{11.8}$$

And thus, with $x_1 = V_1, x_2 = V_2$: $f(t, x(t), u(t); p) \equiv \begin{bmatrix} -K_1 x_1(t) + u(t) \\ K_1 x_1(t) - K_2 x_2(t) \end{bmatrix}$ and $g(t, x(t), u(t); p) \equiv K_2 x_2(t)$.

In the next sections, each of the subsystems of the AP system (Fig. 11.1) will be described in more detail.

11.3 RAS Modelling

Global fish aquaculture reached 50 million tons in 2014 (FAO 2016). Given the growing human population, there is a growing demand for fish proteins. Sustainable growth of aquaculture requires novel (bio)technologies such as recirculating aquaculture systems (RAS). RAS have a low water consumption (Orellana 2014) and allow for a recycling of excretory products (Waller et al. 2015). RAS provide suitable living conditions for fish, as a result of a multistep water treatment, such as particle separation, nitrification (biofiltration), gas exchange and temperature control. Dissolved and particulate excretory products can be transferred to secondary treatment such as plant (Waller et al. 2015) or algae production in integrated aqua-agriculture (IAAC) systems. IAAC systems are sustainable alternatives to conventional aquaculture systems and in particular are a promising expansion to RAS. In RAS it would be necessary to circulate the process water which has special implications for the process technology in both, the RAS and the algae/plant system. To combine RAS and algae/plant system, a deep understanding of the interaction between fish and water treatment is prerequisite and can be derived from dynamic modelling. The metabolism in fish follows a daily pattern which is well represented by the gastric evacuation rate (Richie et al. 2004). Particle separation, biofiltration and gas exchange are subjected to the same pattern. For design purposes the characterization of the basic components of a RAS treatment system should be investigated through simulation models. These simulation models are highly complex. Available numerical models for RAS capture only a small part of the complexity and consider only a part of the components with corresponding mechanisms. Hence, in this chapter, only a small part of a dynamic RAS model will be presented, i.e. nitrification-based biofiltration. The conversion of toxic ammonia into nitrate is a central process in the water treatment process in RAS. In the following, the dynamic modelling of the mass balance of ammonia excretion of fish and the conversion of ammonia into nitrate will be demonstrated as well as the transfer of the nutrient into an aquaponic system. With this it is possible not only to engineer a RAS but also to integrate fish production into an IAAC system based on valid parameters.

11.3.1 Dynamic Model of Nitrification-Based Biofiltration in RAS

The model is subdivided into a fish model for European seabass, *Dicentrarchus labrax*, a model describing the time- dependenting excretion of ammonia, and a nitrification model (Fig. 11.8). The fish excretion pattern is introduced into the model through the input vector u (Eq. 11.15), similar to the approach used by Wik et al. (2009). The complexity of the fish model is kept low to be able to explain its method of implementation. Nonetheless, a short introduction into modelling fish is presented in Sect. 11.3.2. Four basic aspects important to describe the nutrient flow in RAS (Badiola et al. 2012) are:

1. The flow Q, which is the total process water flow per unit time through the RAS, determines the mass transfer of all dissolved and particulate matter, including ammonia and nitrate.
2. The excretion of the fish input ammonia to the RAS process water and is depicted by the product of matrix B and vector u (Eq. 11.15).
3. The ammonia conversion into nitrate, taking place in the nitrification, is depicted in the nitrification vector n (Eq. 11.15).
4. The nutrient transfer from the RAS to a connected HP system is depicted in vector u (Eq. 11.15). Other important aspects of the RAS process chain such as solid removal, dissolved oxygen concentration and carbon dioxide concentration are not considered here. Hints for modelling these can be found in Sects. 3.1.1 and 3.2.2 of this book.

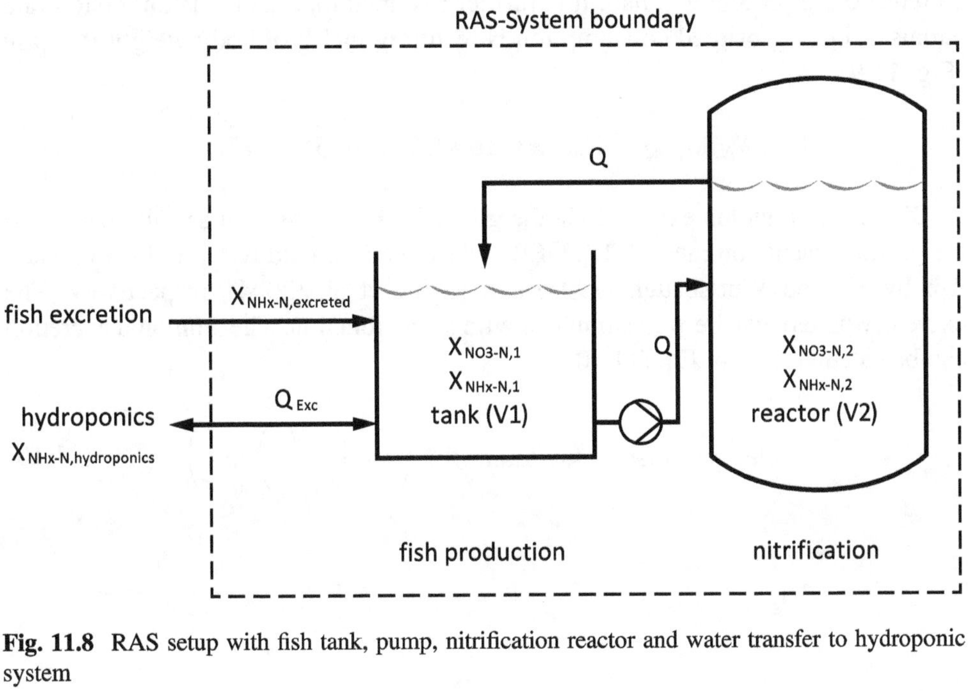

Fig. 11.8 RAS setup with fish tank, pump, nitrification reactor and water transfer to hydroponic system

11.3.2 Fish

A variety of models in the scientific literature predict growth and feed intake of different aquatic species. The models describe growth as weight gain per day, as percentage growth increment or as specific growth rate based on an exponential growth model. Models are often valid for specific life stages. Feed consumption, biomass and gender are influencing the model output as well as the environmental conditions such as temperature, oxygen level and nutrient concentration (Lugert et al. 2014). Careful research is needed to identify the correct model used for the specific application. Commercial RAS that consists of several cohorts of fish in different life stages require the modelling to incorporate cohorts into the model (Fig. 8.6) (Halamachi and Simon 2005). The excretory mass flow for the European seabass (*Dicentrarchus labrax*) can be estimated with algorithms published by Lupatsch and Kissil (1998).

Here the net nitrogen mass flow into the process water is estimated from the feed composition (protein content), the amount of given feed and the nitrogen retained in body tissues through the growth (weight increment) of fish. The faecal nitrogen losses are not included in the model, but excretion rate is corrected assuming a share of 0.25 and 0.75 of nitrogen excretion for faecal loss and ammonia excretion, respectively. The nitrogen input through the feeding of fish is estimated from the protein content and the average relative nitrogen content of proteins which is assumed to be 0.16. The protein content of seabass tissue is reported at around 0.17 g proteins g^{-1} seabass (Lupatsch et al. 2003). For a fish gaining body weight by consuming a given amount of feed, the nitrogen excretion ($X_{N,excreted}$, g) can be calculated from Eq. (11.9). It is assumed that the feed (X_{feed}) contains 0.5 g protein g^{-1} fish. It is further assumed that the feed conversion rate equals 1, i.e. 1 g of feed consumption is resulting in 1 g of body weight increase (Fig. 11.9):

$$X_{N,\text{excreted}} = X_{\text{feed}} * 0.16 * 0.75 * (0.5 - 0.17) \quad (11.9)$$

Dissolved ammonia excreted via the gills of fish follows similar daily pattern as the gastric evacuation rate (GER). GER is described for cold water and warm water fish by He and Wurtsbaugh (1993) and Richie et al. (2004), respectively. The excretory pattern can be well simulated with a sine function. The ammonia excretion can be calculated from Eq. (11.10):

$$X_{\text{NH}_x-N,\text{excreted}} = X_{N,\text{excreted}}\,[g] * \left(\sin\left(\frac{2\pi}{1440}\right) + 1\right) \quad (11.10)$$

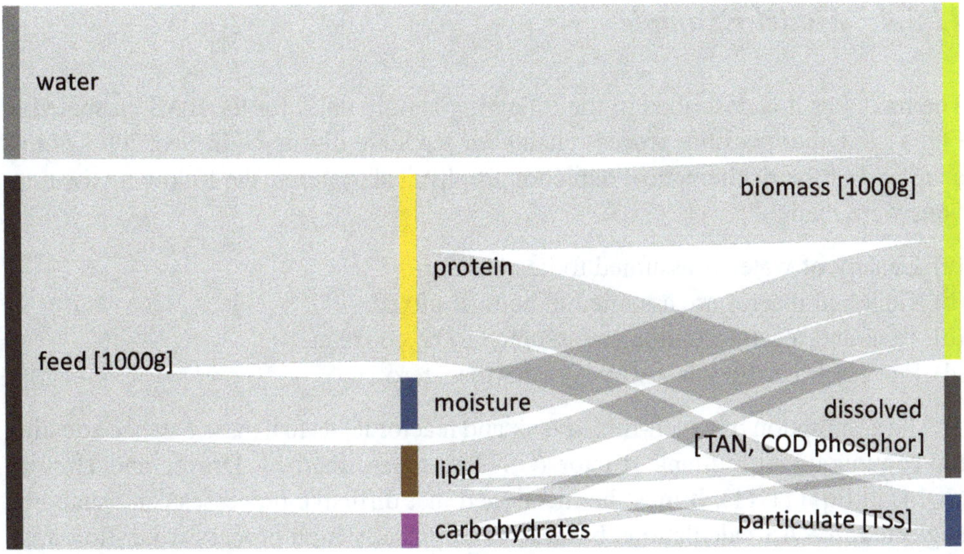

Fig. 11.9 Representation of the mass flows (Sankey chart) of feed ingredients and excretory products for a fish consuming 1000 g of feed assuming an FCR of 1

11.3.3 RAS

A variety of models describing RAS having different levels of complexity can be found in the literature. Very complex models are available for specific aspects, such as the interaction of soluble gases and alkalinity (Colt 2013) or the description of the microbial community (Henze et al. 2002). More practical models for the mass balance of RAS are published by Sánchez-Romero et al. (2016), Pagand et al. (2000), Wik et al. (2009) and Weatherley et al. (1993). All models provide information on excretory mass flows and/or nutrient flows in dependence of time and location in the process chain. Such models provide a base for the simulation of the coupling of RAS and HP. The most important dissolved matter in RAS modelling is total ammonia nitrogen (TAN). Besides TAN the chemical (COD) and biological (BOD) oxygen demand, the total suspended solids (TSS) and dissolved oxygen concentration need to be considered. However, different notations in the scientific literature make it sometimes hard to read, to convert and to implement the information into models. In the following, notations as recommended by Corominas et al. (2010) will be used. TAN will be rewritten as $X_{NHx\text{-}N}$ and nitrate nitrogen will be expressed as $X_{NO3\text{-}N}$.

11.3.4 Model Example

The model as it is described in the following is only valid for the RAS presented in Fig. 11.8. Other possible process chains for RAS are discussed in Sect. 11.3 of this chapter. For the mathematical depiction of physical systems, the following assumptions were made:

(a) Density of water is assumed to be constant.
(b) Tank and reactor are assumed to be well mixed.
(c) Tank and reactor volume are assumed to be constant.
(d) Process water flow is always greater than zero.

The assumption of a well-mixed tank and reactor leads to a mass balance equation for continuous stirred-tank reactor (CSTR) as described by Drayer and Howard (2014) in Eq. (11.11). It must be mentioned that diffusive processes can usually be neglected in RAS calculations because of a typically high process water flow rate. For a multi-tank RAS, the following holds:

$$\begin{gathered}\text{Accumulation} = \text{Inflow} - \text{outflow} + \text{generation} - \text{reduction} \\ V_i \dot{x}_i = Q_{\text{in}} x_{i,\text{in}} - Q_{\text{out}} x_{i,\text{out}} + x_{i,\text{gen}} - x_{i,\text{red}} \\ j = \begin{cases} n, & i = 1 \\ i-1, & i \neq 1 \end{cases}\end{gathered} \tag{11.11}$$

In the above given equation n represents the number of tanks in the System, $\dot{x}_i$ is the change of concentration of a given substrate x in a volume given by $V_{i.}$. The process water flow into the tank or reactor is represented by Q_{in}. V_i is the volume of the component where the process water flow Q_{in} is entering in. The process water flow Q_{in} came from a component having the volume V_j.

The conversion of $X_{\text{NHx-N}}$ into $X_{\text{NO3-N}}$ in nitrifying biofilters takes place on the surface area A [m^2] available on the bio-carriers in the nitrification reactor (Rusten 2006). The available bioactive surface in the nitrification is calculated by multiplying the volume of the reactor with the volume-specific active surface of the bio-carriers A_S [$m^2 \cdot m^{-3}$]. The total bioactive surface is calculated (Eq. 11.12) from the relative filling f_{bc} of the nitrification reactor which usually is 0.6 (for details, see Rusten 2006).

$$A = V_{\text{nitrification}} * A_S * f_{bc} \tag{11.12}$$

The total daily TAN microbial conversion μ_{max} [$g\ d^{-1}$] (nitrification) was calculated by multiplying the specific TAN conversion (nitrification) rate, $NHx_{\text{conversion-rate}}$ [$g\ m^{-2}\ d^{-1}$], with the total active surface area, A [m^2], of the bio-carriers. Values for TAN conversion in different types of nitrifying biofilters can be found in literature. For moving bed biofilm reactors (MBBR), values are reported by Rusten (2006). This rate is valid for certain process conditions, and it is assumed that the bacteria biofilm is fully developed over the whole.

$$\mu_{mm} = A^{*}NHx_{Conversion-rate} \tag{11.13}$$

The total mass of NH_x converted into NO_3-N can subsequently be calculated with a Monod kinetic (Eq. 11.14). For this the $NH_{x\text{-}N}$ concentration, $X_{NHx\text{-}N,2}$ [$g \cdot l^{-1}$], in the volume of the nitrification reactor (MBBR) V_2, is needed.

$$\begin{aligned} \frac{d}{dt}X_{NH_x-N,2} &= -\mu_{max} * \left(\frac{X_{NH_x-N,2}}{K_s + X_{NH_x-N,2}}\right) * \frac{1}{V_2} \quad \text{with} \quad K_s = \frac{\mu_{max}}{2} \\ \frac{d}{dt}X_{NO_s-N,2} &= +\mu_{max} * \left(\frac{X_{NH_x-N,2}}{K_s + X_{NH_x-N,2}}\right) * \frac{1}{V_2} \quad \text{with} \quad K_s = \frac{\mu_{max}}{2} \end{aligned} \tag{11.14}$$

Given Eqs. (11.9, 11.10, 11.11, 11.12, 11.13 and 11.14), the following state-space model (combining fish-nitrification) results

$$\frac{dx(t)}{dt} = \boldsymbol{A}^{*}\boldsymbol{X} + \boldsymbol{B}^{*}\boldsymbol{u} + \boldsymbol{n}$$

$$\boldsymbol{X} = \begin{bmatrix} X_{NH_x-N,1} \\ X_{NH_x-N,2} \\ X_{NO_3-N,1} \\ X_{NO_3-N,2} \end{bmatrix} \quad \boldsymbol{u} = \begin{bmatrix} X_{NH_x-N,\,excreted} \\ \boldsymbol{0} \\ \boldsymbol{Q}_{\mathbf{Exc}}{}^{*}X_{NH_x-N,\,hydroponics} \\ \boldsymbol{0} \end{bmatrix} \quad \boldsymbol{n} = \begin{bmatrix} \boldsymbol{0} \\ -\dfrac{\boldsymbol{\mu_{max}} * [\boldsymbol{X}]_2}{\boldsymbol{K_s} + [\boldsymbol{X}]_2} * \dfrac{1}{\boldsymbol{V_2}} \\ \boldsymbol{0} \\ +\dfrac{\boldsymbol{\mu_{max}} * [\boldsymbol{X}]_2}{\boldsymbol{K_s} + [\boldsymbol{X}]_2} * \dfrac{1}{\boldsymbol{V_2}} \end{bmatrix}$$

$$A = \begin{bmatrix} -\dfrac{Q}{V_1} - \dfrac{Q_{Exc}}{V_1} & \dfrac{Q}{V_1} & 0 & 0 \\ \dfrac{Q}{V_2} & -\dfrac{Q}{V_2} & 0 & 0 \\ 0 & 0 & -\dfrac{Q}{V_1} - \dfrac{Q_{Exc}}{V_1} & \dfrac{Q}{V_1} \\ 0 & 0 & \dfrac{Q}{V_2} & -\dfrac{Q}{V_1} \end{bmatrix}$$

$$\times B = \begin{bmatrix} \dfrac{1}{V_1} & 0 & 0 & 0 \\ 0 & \dfrac{1}{V_2} & 0 & 0 \\ 0 & 0 & \dfrac{1}{V_1} & 0 \\ 0 & 0 & 0 & \dfrac{1}{V_2} \end{bmatrix} \tag{11.15}$$

Example

In this example, a theoretical RAS with V_reactor = 1300 l and V_tank = 6000 l is simulated.

All simulations had a daily feed input of 2000 g/day with 500 g protein/kg feed (Eq. 11.8). The daily TAN excretion was assumed to be a sine curve (Eq. 11.9). Active surface of the bio-carriers A_S is 300 [m^2 m^{-3}], and the relative filling of the reactor f_{bc} is 0.6. Specific TAN conversion rate, $NHx_{conversion\text{-}rate}$, is 1.2 [g m^{-2} $^{-}$d], and the biofilm is supposed to be fully developed (Eqs. 11.11 and 11.12). The state-space representation (Eq. 11.14) was implemented in MATLAB Simulink. The Example showcases the importance of mass flow for nutrient concentrations in coupled systems (Fig. 11.10 and 11.11).

11.4 Modelling Anaerobic Digestion

Anaerobic digestion (AD) of organic material is a process that involves the sequential steps of hydrolysis, acidogenesis, acetogenesis and methanogenesis (Batstone et al. 2002). The anaerobic digestion of a mixture of proteins, carbohydrates and lipids is visualized in Figure 11.11. Most often, hydrolysis is considered as the rate-limiting step in the anaerobic digestion of complex organic matter (Pavlostathis and Giraldo-Gomez 1991). Thus, increasing the hydrolysis reaction rate will most likely lead to a

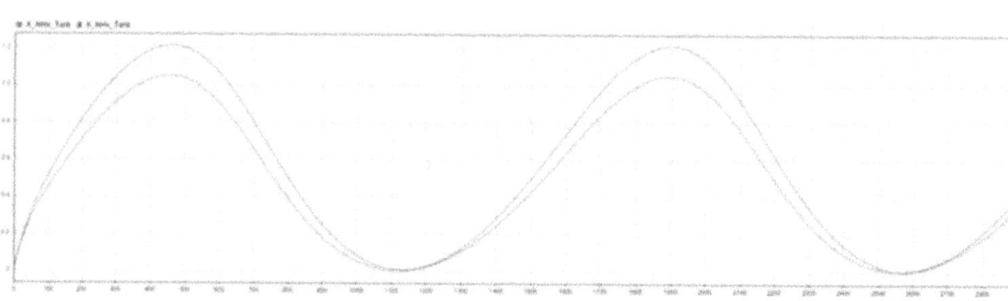

Fig. 11.10 Simulation of TAN ($X_{NHx\text{-}N,1}$) in [mg/l] over 2 days = 2880 min with Q = 300 l/min (blue) and Q = 200 l/min (orange)

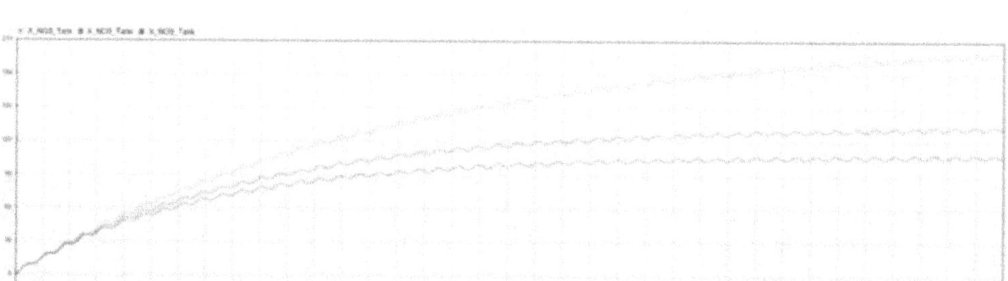

Fig. 11.11 Simulation of nitrate-N ($X_{NO3\text{-}N,1}$) in [mg/l] over 50 days = 72,000 min with Q_{Exc} = 300 l/day (yellow), Q_{Exc} = 480 l/day (orange) and Q_{Exc} = 600 l/day (blue)

higher anaerobic digestion reaction rate. However, increasing the reaction rates needs further understanding of the related process. Further understanding can be obtained via experimentation and/or mathematical modelling. As there are many factors influencing, for instance, the hydrolysis process, such as ammonia concentration; temperature; substrate composition; particle size; pH; intermediates; degree of hydrolysis; i.e. the potential of hydrolysable content; and residence time, it is almost impossible to evaluate the total effect of the factors on the hydrolysis reaction rate through experimentation. Mathematical modelling could therefore be an alternative, but as a result of all the uncertainties in model formulation, rate coefficients and initial conditions, no unique answers can be expected. But, a mathematical modelling framework would allow sensitivity and uncertainty analyses to facilitate the modelling process. As mentioned before, hydrolysis is just one of the steps in anaerobic digestion. Consequently, understanding and optimization of the full anaerobic digestion process needs connections from hydrolysis to the other processes taking place during anaerobic digestion and interactions between all these steps.

The well-known and widely used ADM1 (anaerobic digestion model #1) is a structured model including disintegration and hydrolysis, acidogenesis, acetogenesis and methanogenesis steps. Disintegration and hydrolysis are two extracellular steps. In the disintegration step, composite particulate substrates are converted into inert material, particulate carbohydrates, protein and lipids. Subsequently, the enzymatic hydrolysis step decomposes particulate carbohydrates, protein and lipids to monosaccharides, amino acids and long-chain fatty acids (LCFA), respectively (Batstone et al. 2002) (see Fig. 11.12).

ADM1 is a mathematical model that describes the biological processes and physicochemical processes of anaerobic digestion as a set of differential and algebraic equations (DAE). The model contains 26 dynamic state variables in terms of concentrations, 19 biochemical kinetic processes, 3 gas-liquid transfer kinetic processes and 8 implicit algebraic variables for each process unit. As an alternative, Galí et al. (2009) described the anaerobic process as a set of differential equations with 32 dynamic state variables in terms of concentrations and an additional 6 acid-base kinetic processes per process unit. For an overview of the modelling of anaerobic digestion processes, we refer to Ficara et al. (2012). However, in what follows and for some first insights into the AD process, we will present a simple nutrient-balance model of AD in a sequencing batch reactor (SBR).

11.4.1 Nutrient Mineralization

The nutrient mineralization can be calculated using the following equation (Delaide et al. 2018):

$$NR = 100\% \times \left(\frac{DN_{out} - DN_{in}}{TN_{in} - DN_{in}} \right) \tag{11.15a}$$

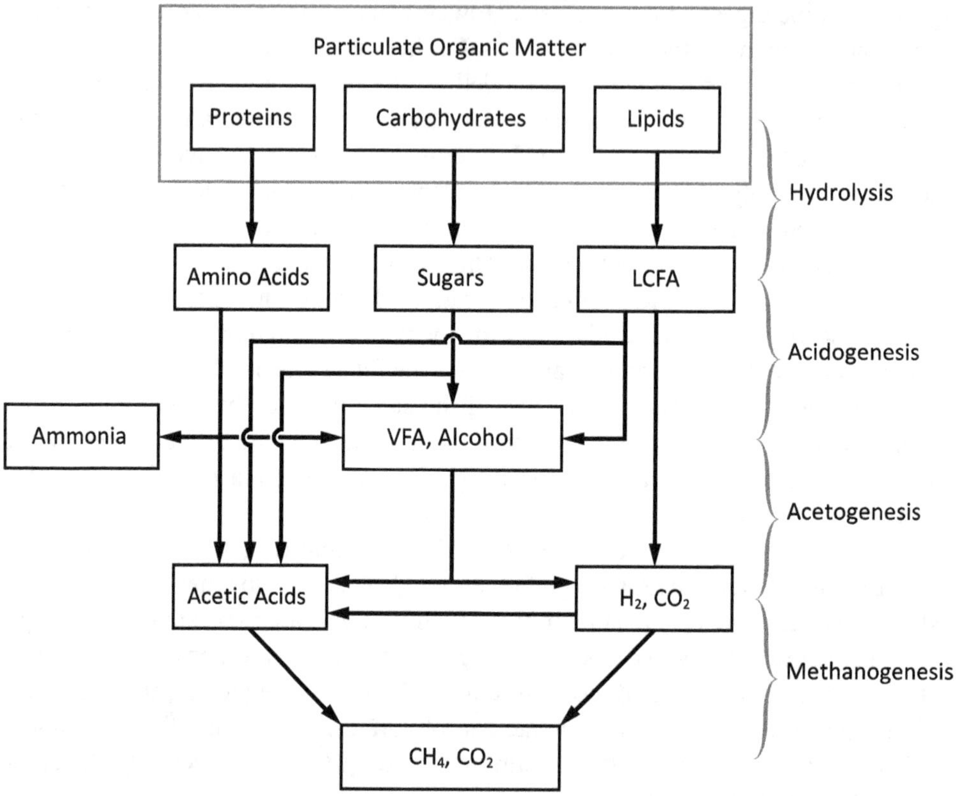

Fig. 11.12 A simplified scheme for the anaerobic digestion of complex particulate organic matter. (based on El-Mashad 2003)

where NR is the nutrient recovery at the end of the experiment in percent, DN_{out} is the total mass of dissolved nutrient in the outflow, DN_{in} is the total mass of dissolved nutrient in the inflow and TN_{in} is the total mass of dissolved plus undissolved nutrients in the inflow (see also Fig. 11.13).

11.4.2 Organic Reduction

The organic reduction performance of the reactor can be calculated using the following equation:

$$\eta_{OM} = 1 - \frac{\Delta OM + T_{OM\ out}}{T_{OM\ in}} \tag{11.15b}$$

where ΔOM is the organic matter (i.e. COD, TS, TSS, etc.) inside the reactor at the end of the experiment minus the one at the beginning of the experiment, $T_{OM\ out}$ is the total OM outflow and $T_{OM\ in}$ is the total OM inflow (see also Fig. 11.14).

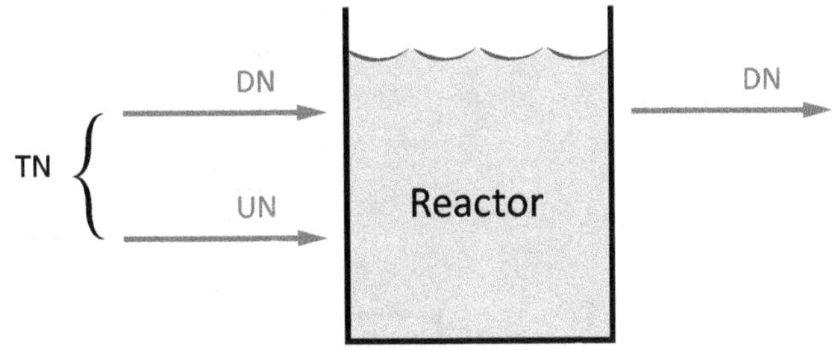

Fig. 11.13 Overall reactor scheme for determining the mineralization potential, where DN are the dissolved nutrients in the water, UN the undissolved nutrients in the sludge (i.e. TN-DN) and TN the total nutrients

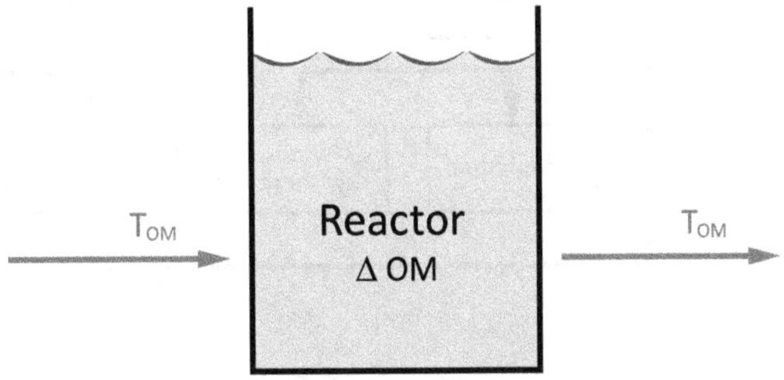

Fig. 11.14 Overall reactor scheme for determining the organic material reduction potential, where T_{OM} is the total organic matter and ΔOM the change of organic matter inside the reactor

11.5 HP Greenhouse Modelling

The crop water use and nutrient uptake is a central subsystem of aquaponics. The HP part is complex, as pure uptake of water and dissolved nutrients do not simply follow a rather simple linear relationship as, e.g. fish growth. To create a full-functional model, a complete greenhouse simulator is needed. This involves sub-model systems of greenhouse physics including climate controllers and crop biology covering interactive processes with biological and physical stressors.

However, from the HP point of view, greenhouse climate is the main driver for the complete aquaponic system, including, next to the nutrient balances, feedback loops of heat produced by the fish and additional CO_2 supplied to the plants as reported by Körner et al. (2017) (Fig. 11.15).

In this model, the fish culture produces heat through metabolic processes. The amount of heat produced by the fish is directly calculated from oxygen consumption that is a function of temperature and a constant for heat production for one unit oxygen consumed (i.e. 13608 J g^{-1} fish). Heat from breakdown of organic matter (Q_{bio}), e.g. faeces and feed remain, is also contributing to the heat balance. Energy supply to the water system can then be calculated by heat production through the fish calculated from an average oxygen consumption rate ($f_{O2,Twb}$). Additional heat production can then be calculated by biological breakdown of faeces (Fig. 11.16).

CO_2 production from the aquatic subsystem (d_{CO_2}, g h^{-1}), i.e. delivery to the aerial environment (d, g h^{-1}), can be calculated for the given water temperature (T_{H2O}, K) from oxygen delivery to the system (d_{O_2}, g h^{-1}) at water base temperature ($T_{H2O,b}$, K) and the Q_{10} value of fish respiration ($Q_{10,R}$). The following relationships are used:

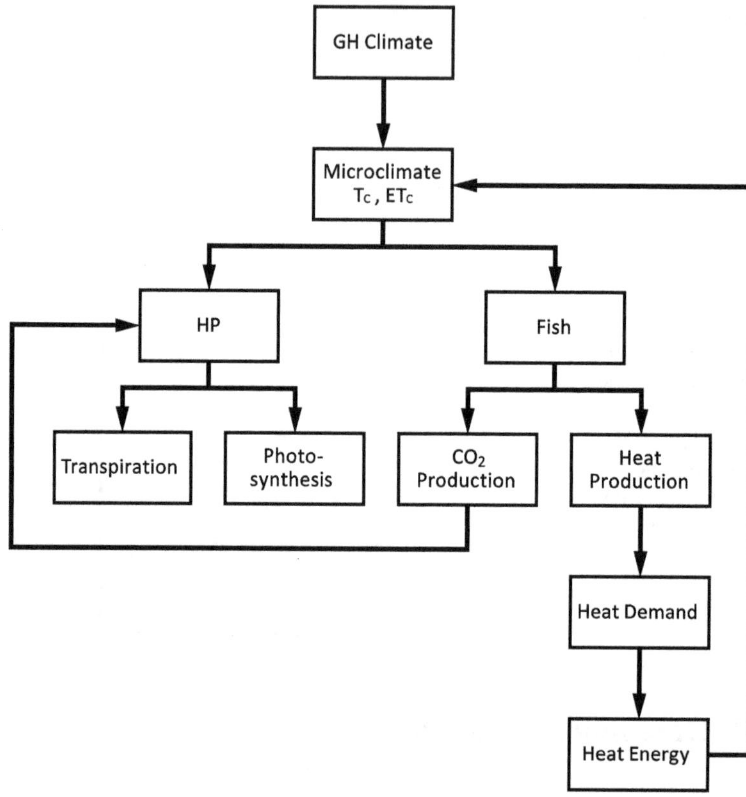

Fig. 11.15 Additional symbiotic behaviour of an aquaponic system

$$\begin{aligned} d_{O_2} &= \mathrm{f}_{\mathrm{fish}}\, \mathrm{f}_{O_2}\, \mathrm{w}_{O_2} \\ d_{CO_2} &= \frac{[\mathrm{CO_2}]}{[\mathrm{O_2}]}\, d_{O_2}\, \mathrm{Q}_{10,\mathrm{R}}{}^{(\mathrm{T_{H2O}}-\mathrm{T_{H2O,b}})/10} \end{aligned} \tag{11.16}$$

with feed amount for the fish (f_{fish}, g h^{-1}), oxygen consumption rate at base temperature (f_{O2}, kg [O_2] kg^{-1} [feed]), fraction of feed loss w_{O2} ($^{-}$) and mass balance of O_2/CO_2 ($^{-}$).

To calculate the basis of aquaponics, i.e. the process flow (indicated with arrows, →) greenhouse macroclimate → microclimate → evapotranspiration → nutrients uptake, various greenhouse simulators that were developed in the past can be used and combined with aquaculture to an aquaponic system. All greenhouse models include a crop growth model. The model quality, however, can vary a lot from simple empirical regression models, e.g. Boote and Jones (1987), via deterministic models, e.g. Heuvelink (1996), to functional structural plant models (FSPM), e.g. Buck-Sorlin et al. (2011). As current crop growth and development models are inaccurate and have limited predictive power (Poorter et al. 2013), models are occasionally employed in crop management, but then mainly for planning issues in greenhouse simulators, e.g. Vanthoor (2011) and Körner and Hansen (2011). Prediction accuracy is jeopardized by many sources of uncertainty, such as modelling

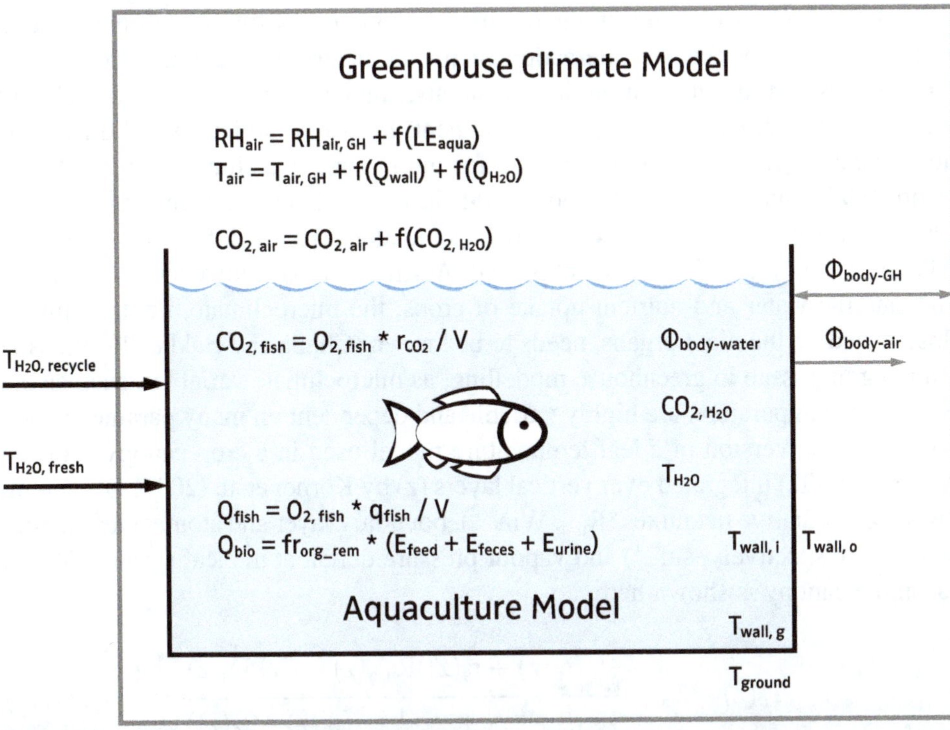

Fig. 11.16 Aquaculture system implemented in the greenhouse with humidity, temperature and CO2 concentrations of the air (RH_{air} T_{air}, $CO_{2,air}$), heat from (Q) fish environment (fish), biological breakdown (bio) and heat fluxes (ϕ), taken from Körner et al. (2017)

errors, variability between plants, variability between greenhouses and uncertain external climate conditions. As for predictions, accuracy also varies strongly per situation. However, online feeding of sensor information into the plant model can make plant model predictions considerably more reliable and useful to the grower.

Greenhouse simulators were developed in and for several places, e.g. the Virtual Grower (Frantz et al. 2010), KASPRO (De Zwart 1996), Greenergy Energy Audit Tool (Körner et al. 2008), The Virtual Greenhouse (Körner and Hansen 2011), The Adaptive Greenhouse (Vanthoor 2011), Hortex (Rath 1992, 2011) and the integrated aquaponic greenhouse model (Goddek and Körner 2019). At a research level, some models (i.e. simulation models in combination with certain greenhouse technologies) have been developed that potentially can be used to optimize investments and structural modifications to the production unit and production process. However, most systems entail closed software environments that can only be used by the developers, and many of them only exist in a research mode and lack further development and acceptance from the industry. However, there is yet no common basis for model sharing and collaborative model development. As a result, most modellers and modelling teams work in isolation developing their own models and codes. A shortcoming of that procedure is that greenhouse simulation models are developed in parallel in disparate research environments, which fail in cooperative growth and development.

All HP greenhouse model simulators are a compilation of sub-models that depend on the aim to integrate the interaction of plants and greenhouse equipment. The general two-part differentiation in greenhouse models and also in control and planning is the shoot and the root environment. Rather complicated and differentiated model approaches have been done for the greenhouse climate (Bot 1993; de Zwart 1996), and greenhouse crop growth has been intensively modelled in the 1990s for the main greenhouse crops such as tomatoes (Heuvelink 1996), cucumber (Marcelis 1994) and lettuce (Liebig and Alscher 1993). However, in order to calculate the water and nutrient uptake of crops, the microclimate, i.e. the climate close to and on the plant organs, needs to be known (Challa and Bakker 1999). This is an ongoing issue in greenhouse modelling, as microclimate variables, such as the central leaf temperature, are highly variable and dependent on many parameters and variables. One version of a leaf temperature model used in a crop canopy for crop temperature (T_c) integrated over vertical layers (z) by Körner et al. (2007) integrating absorbed irradiative net fluxes ($R_{n,a}$, Wm^{-2}), boundary layer and stomata resistances (r_b and r_s, respectively, sm^{-1}) and vapour pressure deficit at the leaf surface (VPD_s, Pa) in the canopy is shown here, i.e.

$$T_c(z) - T_a = \frac{\frac{1}{\rho_a c_p}(r_b(z) + r_s(z))R_{n,a}(z) - \frac{1}{\gamma}VPD_s(z)}{1 + \frac{\delta}{\gamma} + \frac{r_s(z)}{r_b(z)} + \frac{1}{\rho_a c_p/4\sigma T_a^3}(r_b(z) + r_s(z))} \tag{11.17}$$

with greenhouse air temperature (T_a, K), vapour pressure air density (ρ_a, g m^{-2}), Stefan-Boltzmann constant (σ, Wm^{-2} K^{-4}), specific heat capacity of the air (c_p, J g^{-1} K^{-1}), the psychrometric constant (γ, Pa K^{-1}) and the slope between saturated vapour pressure and greenhouse air temperature (δ, Pa K^{-1}).

Leaf temperature is the central part of the microclimate model, it has feedback loops to several input variables and especially stomata resistance (often also used as its reciprocal, the conductance), and the calculation needs several simulation steps for equilibrium. For HP, as part of the aquaponic system, however, modelling water and nutrient fluxes is most important. All water and nutrient balances in a closed multi-loop system are controlled based on the evapotranspiration rate of the crop ET_c (Chap. 8). Commonly ET_c is calculated as latent heat of evaporation, i.e. in energy terms (λE, Wm^{-2}), and can be in accordance to leaf temperature expressed in different canopy layers z

$$\lambda E(z) = \left(\frac{\frac{\delta}{\gamma}R_{n,a}(z) + \frac{\rho_a c_p}{\gamma}\left(\frac{1}{r_b(z)} + \frac{1}{\rho_a c_p/4\sigma T_a^3}\right)VPD_s(z)}{1 + \frac{\delta}{\gamma} + \frac{r_s(z)}{r_b(z)} + \frac{1}{\rho_a c_p/4\sigma T_a^3}(r_s(z) + r_b(z))} \right) \tag{11.18}$$

To calculate ET_c (L m^{-2}), λE needs to be multiplied with the constant L_w (heat of vaporization of water; 2454·103 J kg^{-1}) and the specific weight of water (9.789 kN·m^3 at 20 °C).

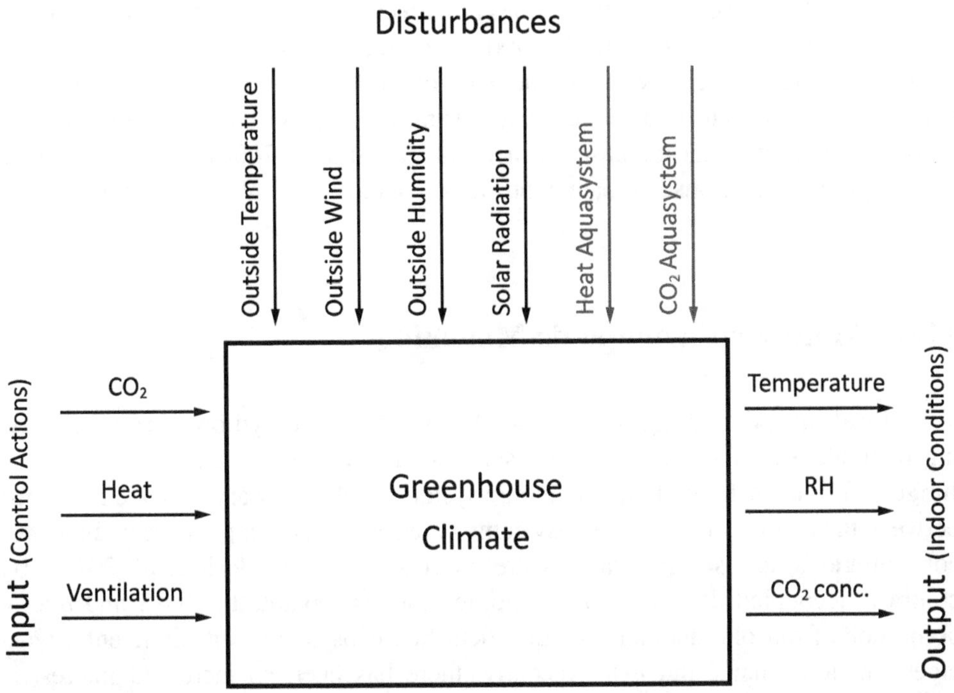

Fig. 11.17 Input-output system of a greenhouse

Equation (11.18), however, only calculates the water flux through the crop, while the easiest way to estimate nutrient uptake is the assumption that nutrients are taken up/absorbed as dissolved in irrigation water and assuming that no element specific chemical, biological or physical resistances exist. In reality uptake of nutrients is a highly complicated matter. Consequently, to maintain equilibrium, all nutrients taken up by the crop as contained in the nutrient solution need to be added back to the hydroponics system (see Chap. 8). However, Eq. (11.18) only calculates the potential ETc, while too high potential levels can result in a higher transpiration than plants can handle, and then potential water loss may exceed water uptake. For that, the simple assumption of nutrient uptake is not satisfying. As described in Chap. 10, the different nutrients can have different states and change states with, e.g. pH, while the plant availability strongly depends on pH and the relation of nutrients to each other. In addition, the microbiome in the root zone plays an important role, which is not implemented in models yet. Some models, however, differentiate between phloem and xylem pathways. The vast amount of nutrients, however, is not modelled in detail for aquaponics nutrient balancing and sizing of systems, while the easiest way to estimate nutrient uptake is the assumption that nutrients are taken up/absorbed as dissolved in irrigation water and apply the above explained ETc calculation approach.

For control purposes the greenhouse is typically considered as a black box, where outside climate conditions determine the disturbance inputs, CO_2 supply, heating and ventilation are the control inputs, and the greenhouse macro- and microclimate define the output of the system (Fig. 11.17).

To control the greenhouse, the actions are directed to minimize the fast impacts of the disturbances, i.e. being ahead of expected changes by smart control. For that, control actions such as feedback and feedforward are used (Chap. 8). The best control, however, can be achieved when using a complete greenhouse model and combine it with weather forecast (Körner and Van Straten, 2008) attaining a model-based optimal greenhouse climate control, as worked out by Van Ooteghem (2007).

11.6 Multi-loop Aquaponic Modelling

Traditional aquaponic designs comprise of aquaculture and hydroponic units involving recirculating water between both subsystems (Körner et al. 2017; Graber and Junge 2009). In such one-loop aquaponic systems, it is necessary to make trade-offs between the conditions of both subsystems in terms of pH, temperature and nutrient concentrations, as fish and plants share one ecosystem (Goddek et al. 2015). By contrast, decoupled double-loop aquaponic systems separate the RAS and hydroponic units from one another, creating detached ecosystems with inherent advantages for both plants and fish. Recently, there has been an increased interest in closing the loop in terms of nutrients as well as increasing the input-output efficiency. For that reason, remineralization (Goddek 2017; Emerenciano et al. 2017; Goddek et al. 2018; Yogev et al. 2016) and desalination loops (Goddek and Keesman 2018) have been incorporated into the overall system design. Such systems are called decoupled multi-loop aquaponic system (Goddek et al. 2016).

Sizing the respective subsystems is fundamental of having a functioning check-and-balance system. For sizing one-loop systems, a simple rule of thumb is generally used, determining the hydroponic cultivation area based on the daily feed input to the RAS (Knaus and Palm 2017; Licamele 2009). The higher degree of complexity of multi-loop systems does not allow this approach anymore, as it comes with inherent risks for making false assumptions for each subsystem. There is a growing body of literature that examines mass balances for aquaponic systems (Körner et al. 2017; Goddek et al. 2016; Reyes Lastiri et al. 2016; Karimanzira et al. 2016). While some research has been carried out in developing numerical models for one- and multi-loop aquaponic systems, no single study exists that integrates a multi-loop aquaponic model with a complemented full-scale deterministic greenhouse model. This is particularly relevant for sizing the system, since plant growth and nutrient uptake are location dependent with crop transpiration as major driver. In concrete terms, this means that the climate within a greenhouse – which is highly dependent on the external weather conditions – has a high impact on plant growth given environmental factors such as relative humidity (RH), light irradiation, temperature, carbon dioxide (CO_2) levels, etc. that were incorporated in greenhouse microclimate modelling (Körner et al. 2007; Janka et al. 2018).

11.7 Modelling Tools

In aquaponics, flow charts or stock and flow diagrams (SFD) and causal loop diagrams (CLDs) are commonly used to illustrate the functionality of the aquaponic system. In the following, flow chart and CLDs will be described.

11.7.1 Flow Charts

To get a systemic understanding of the aquaponics, flow charts with the most important components of the aquaponics are a good tool to show how material flows in the system. This can help, for example, in finding missing components and unbalanced flows and mainly influencing determinants of the subprocesses. Figure 11.18 shows a simple flow chart in aquaponics. In the flow chart, fish food and water are added to the fish tank, where the feed is taken by the fish for growth, the water is enriched with the fish waste and the nutrient-enriched water is added to the hydroponics system to produce plant biomass. From the flow chart, a CLD shown in Fig. 11.19 can be easily constructed.

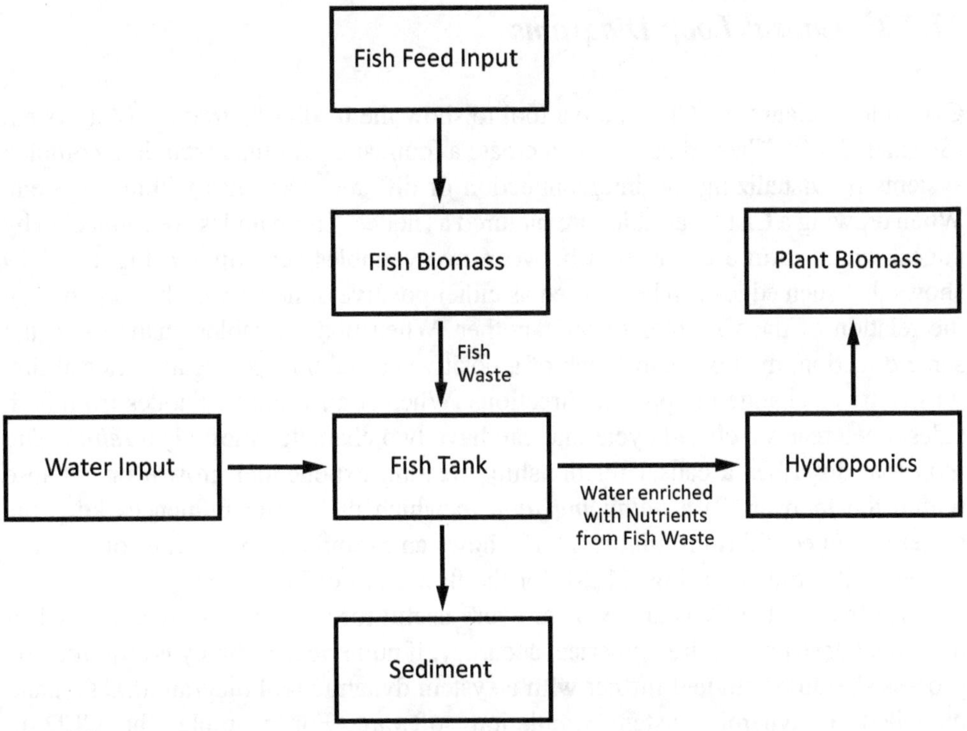

Fig. 11.18 Example of a flow chart in aquaponics (only RAS and HP exchange)

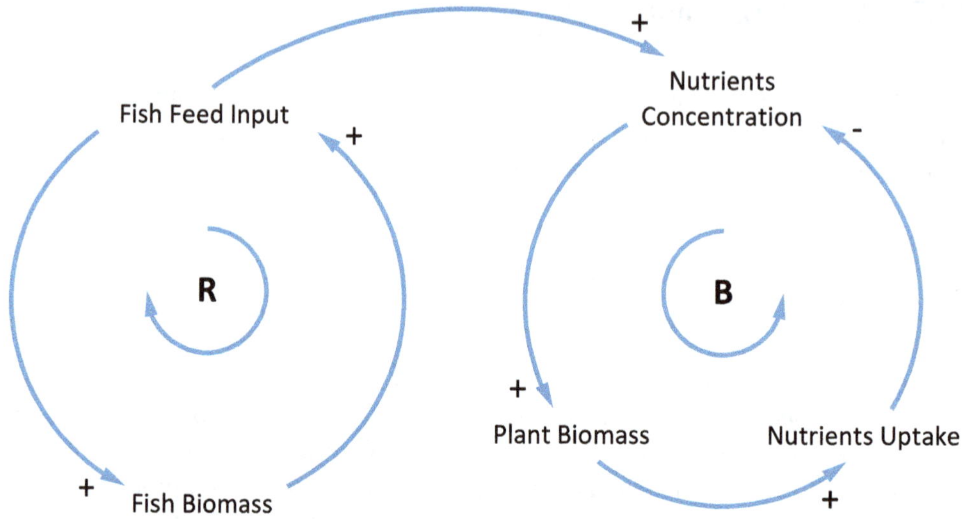

Fig. 11.19 Causal loop diagram (CLD) illustrating examples of a reinforcing and a balanced loop within aquaponic systems. The reinforcing loop (R) is one in which an action produces a result which influences more of the same action and consequently resulting in growth or decline, where as a balancing loop (B) attempts to bring things to a desired state and keep them there (e.g. temperature regulation in the house)

11.7.2 Causal Loop Diagrams

Causal loop diagrams (CLDs) are a tool to show the feedback structure of a system (Sterman 2000). These diagrams can create a foundation for understanding complex systems by visualizing the interconnection of different variables within a system. When drawing a CLD, variables are pictured as nodes. These nodes are connected by edges, which form a connection between two variables accordingly. Figure 11.19 shows that such edges can be marked as either positive or negative. This depends on the relation of the variables to one another. When both variables change into the same direction, then one can speak of a positive causal link. A negative causal link thus causes a change in opposite directions. When connecting two nodes from both sides, one creates a closed cycle that can have two characteristics: (1) a *reinforcing loop* that describes a causal relationship, creating exponential growth or collapse within the loop or (2) a *balancing loop* in which the causal influences keep the system in an equilibrium. Figure 11.19 shows an example of both types of loops.

Let us illustrate this (Fig. 11.20) for the flow chart of Fig. 11.18.

It is obvious that CLD and SFD are very useful for system understanding, when the model does not require numerical accuracy. If numerical accuracy is required, the process should be studied further with a system dynamic tool diagram (SDTD) and modelled in dynamic system simulation software. For example, the CLD in Fig. 11.20 can be augmented with differential equations to a SDTD (Fig. 11.21).

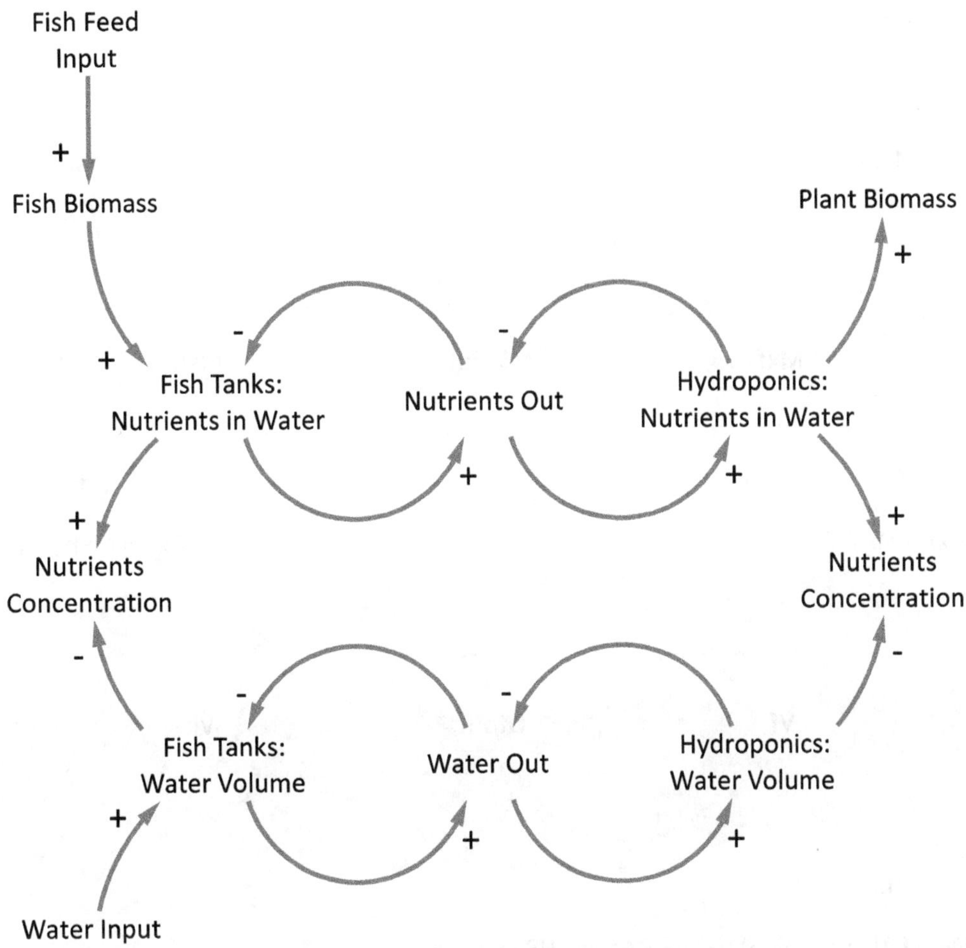

Fig. 11.20 Example CLD for RAS and HP exchange

From the SDTD, we can now see how the differential equations for the nutrients balance in the tank look like. We know that the nutrient flow out of the fish tank (M_{xfout}) must be the water flow (Q_{fout}) times the concentration in the out stream (C_{xf}):

$$M_{xfout} == C_{xf}\, Q_{fout}$$

Assuming a stirred tank gives the nutrient concentration of the fink tank to:

$$C_{\text{xf}} = M_{\text{xf}}/V_f$$

The differential equations of the RAS part can be derived to:

$$\begin{aligned} dV_f/dt &= Q_{\text{fin}} - Q_{\text{fout}} \\ dM_{\text{xf}}/dt &= M_{\text{xfin}} - M_{\text{xfout}}, \end{aligned}$$

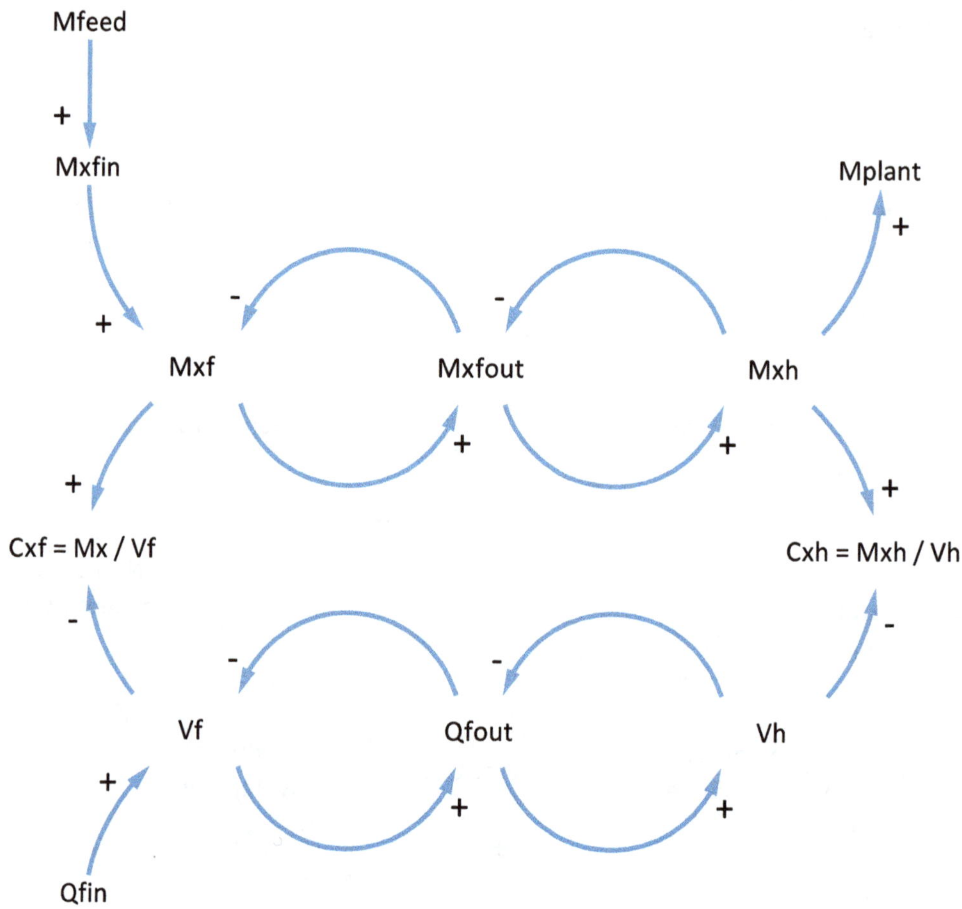

Fig. 11.21 Example SDTD for RAS and HP exchange

and for the concentration

$$dc/dt = (Q_{\text{fin}} C_{\text{xfin}} - Q_{\text{fout}} C_{\text{xf}})/V_f$$

11.7.3 Software

In addition to basic computer languages, such as Fortran, C++ and Python, for fast computation and fully user-specific implementation, all kinds of advanced software tools are available. These advanced software tools offer a variety of environments, concepts and options. We can model state variables, differential equations, connections and loops. In addition, we can use the model for simulations, stability analysis, optimization and control.

The main reasons for modelling of a system are to understand and control it. Therefore, the model helps to predict the system dynamics or behaviour. The software applications could allow us to do three consequent tasks: (a) the modelling itself, (b) the simulations of the model(s) and (c) optimization of the model and/or simulation.

Mathematica software is for functional analysis of mathematically described problems (Wolfram 1991). The concept is based on the LISP approach (McCarthy and Levin 1965.), a very effective functional programming language. The syntax is reasonably simple, and this software is popular in mathematics, physics and systems biology. Especially, the Ndsolve module helps to solve ordinary differential equations, plot the solution and find specific values.

Very similar tools for solving ODEs are offered by the *Maple*. This software is very powerful; between its features belong boundary problems solution, exact solutions and mathematical approximations. *Copasi* (complex pathway simulator) is a software tool for simulation and analysis of biochemical networks via ordinary differential equations.

SageMath is a free open-source mathematics software system. The software is Python-based and facilitates the simulation of ODE models. *Data2Dynamics* software is a collection of numerical methods for quantitative dynamic modelling and is a comprehensive model and data description language. The software allows the analysis of noise, calibration and uncertainty predictions and has libraries of biological models.

Probably the best simulation language is Simula (probably not in use anymore) and *Simula 67*, considered at the beginning as a package for Algol 60. These were the first fully object-oriented languages, introducing classes, inheritance, subclasses, garbage collector and others. In the beginning of the twenty-first century, the creators Ole-Johan Dahl and Kristen Nygaard were awarded the IEEE John von Neumann Medal and the A. M. Turing Award (Dahl and Nygaard 1966).

The idea behind Simula was that objects have life; they start to exist, do their being and cease. The objects are defined as general classes (template code), and each instance of such object has a 'life' in the simulation. The language was quite difficult to learn. However, it offered the possibility to model processes object by object and run simulation of their lives. The simulation runs on the basis of discrete events, and it is possible to simulate objects in co-routine. More tasks can start, run, detach, resume and complete in overlapping time periods in quasiparallel processes. Today's hardware allows us modelling and simulation in fully parallel threads. However, many of the Simula concepts were already used for development of other languages, namely, Java, C/C++/C# and persistent objects libraries like DOL (Soukup and Machacek 2014). Current successor of Simula is *BETA*, extending and featuring the possibilities of inheritance in concepts of nested (sub)classes (with nested local time) and patterns (Madsen et al. 1993).

It is always an option to use any of the object-oriented languages and specific libraries and program all the necessary code for a specific model. On the other hand, already existing graphical programming environments allow to design and link the structure of the modelled system from libraries of objects (signal generator, sum, integrator, etc.), parametrize them and run the simulation in virtual time.

Another popular software for simulation is MathWorks *Simulink*, describing itself as a model-based design tool. The environment allows to combine and parametrize predefined blocks (from wide range of libraries) and diagrams into subsystems. The programming is done using graphical blocks and their connections into functional parts with feedback loops. The environment is widely used for control, automation and signal processing. Another possibility is to integrate own code from MathWorks *Matlab* language or use various toolboxes (Jablonsky et al. 2016). One of them, *SensSB*, is focused on the sensitivity analysis and allows to import other models using the Systems Biology Markup Language. For just the visualization of existing models in Simulink, it is also possible to use the very quick models viewer *DiffPlug*. *PottersWheel* supports the modelling of time-dependent dynamic systems, parameter calibration, analysis and prediction. Interesting tool is the experimental design for model verification.

For the modelling and analysis of system dynamics, similar strategy is used by the *Stella* Architect isee software application, where the model is composed of blocks, which are connected by relations. Stella allows modelling and simulation of very different types of applications, ranging from medical needs through building construction to the airplanes. Stella is sometimes marketed as *iThink* software. *Powersim* software was designed originally for economic purposes. However, it developed into more sophisticated tool, including electronic, solar power or drug treatment simulations. The former developer of *Powersim* is currently producing a similar software for more complex tasks *Dynaplan Smia*. *Vensim* is a system for modelling big data relationships of real systems. The power of *Vensim* is that it allows causal tracking, sensitivity analysis, calibration and intensive simulation. However, the software is also capable of dealing with wide range of simple and complex real systems (Hassan et al. 2016). *True-World* system dynamics software facilitates complex multibody dynamic simulations in discrete and continuous time. The modelling basically starts from balances.

Completely different approaches to modelling and simulation are cellular automata or agent-based modelling approaches, popularized by Stephen Wolfram (Wolfram 1991) as a new kind of science. The approach is sometimes also called the game of life. The modelling is implemented via interactions of autonomous individuals (Macal and North 2005). The simulations show emergent behaviour and therefore are very popular in systems biology for population dynamics. A simple tool for basic (and advanced, as well) agent-based modelling and simulation is the *NetLogo* software, where simple descriptions and parametrization create powerful models. The software allows visualization of the time development and noise induction (Stys et al. 2015). The application is written in Java, which sometimes limits the available memory. Probably the biggest effort in the multi-agent modelling was done by development of *Wolfram* software, which is the continuation of popular Mathematica, with extended tools for modelling and simulation. It has put Simulink-like modelling in a more attractive suit and also creates possibilities of agent-based modelling and much more tools for other mathematical disciplines (multivariate statistics, data mining, global optimization).

AnyLogic is very interesting software for problems of flow – information, money, traffic, logistic and mining. The simulation solves the problem of optimal flow in the designed system with minimal effort and maximal efficiency. The used concepts are system dynamics and agent-based and discrete-event modelling. It also offers hybrids between different modelling concepts. The software is useful, for example, in the epidemic spread simulations (Emrich et al. 2007).

Another agent-based modelling tool is *Insight Maker*, for simulation of population interacting in geographic or network space. The software supports graphical model construction, usage of multiple paradigms, embedded scripting and optimization toolset (Fortmann-Roe 2014).

For the modelling itself, description of the state variables, solution of the ODEs, parametrization and time dependency analysis, the first group of software, from *Mathematica* to *Matlab*, could be used without hesitation. They represent powerful tools for the modelling purposes. In case of more complex analysis, such as related to big data, simulation, noise induction, optimization, sensitivity and stochastics, more advanced tools are required, with object-oriented approach, involving also higher induction in the programming languages syntax.

11.8 Discussion and Conclusions

Aquaponics are complex technical and biological systems. For example, possible explanations for fish not growing properly can be small food rations, adverse water quality, technical problems causing stress, etc. Due to the inherently slow biology, scientific investigations of the validity of these explanations would be tedious and require several experimental trials to get all important factors and their interactions, demanding a lot of facilities, expertise, research time and financial assets. Therefore, the issue of modelling aquaponic systems was addressed in this chapter. In aquaponics, modelling is required for different objectives: (i) insight/understanding, (ii) analysis, (iii) estimation and (iv) management and control. For all these objectives, appropriate models are required. For example, to achieve objectives (ii) and (iii), an empirical approach can be utilized which uses statistical models to analyse data from previous experimental trials with the objective of extracting as much information as possible without conducting new experiments. Statistical models can reveal the most important factors affecting fish and crop production in the aquaponic systems. Future experiments could concentrate on these factors, thus making the utilization of costly research assets more effective.

The complexity of aquaponic systems, due to their feedback character and the interactions between RAS and hydroponic system, water treatment and fish growth, implies that in order to fulfil objectives (i) and (iv), i.e. to understand or optimize a plant (configuration, size, fish, feed, flows, etc.) with respect to cost, stability, robustness and water quality, non-trivial theoretical models of most of the system components described in this chapter are required. The advantage of these theoretical models presented over statistical models is their stronger ability to analyse the process underlying the aquaponics and the possibility to model the time aspect (dynamics). Statistical models just confirm or refute a hypothesis and to what extent variables covary but give no evidence of the underlying processes. On the other hand, theoretical models allow us to simulate the processes according to a hypothesis, compare simulated with observed data, evaluate both the hypothesis and the model and make adaptions. The validity of statistical models may not be beyond the operational range they were trained for, whereas theoretical models can be defined and used for a wide range of environments, provided that the models are validated for these ranges before application. For example, the multiple regression model used to assess relationships between fish growth with *Oreochromis niloticus* as fish species and environmental variables in an aquaponics facility in Germany cannot be easily applied to Spain with *Cyprinus carpio*, whereas a theoretical model describing the underlying processes (e.g. fish behaviour, aquaculture, freshwater ecology) as mathematical equations can be adjusted relatively easily because the fish and ecological process underlying that model are basically the same for the two sites.

Nevertheless, theoretical models also require some parameters such as reaction constants and substance settling velocity in settling tank to be determined. This is achieved commonly based on empirical study of one facility or very few facilities or in most cases from previously published studies (secondary sources). Studies based on secondary sources have limitations imposed by the given structure and amount of the available data, which is not existent when the data come from an experimental setup designed ad hoc for the study. However, estimating model parameters using experimental data from one aquaponics facility only can have problems regarding generalizability and replication of the results due to particular conditions present in the study. The data scarcity sometimes imposes strong restrictions to models that limit their practicality. The development of studies for parameter estimation with primary data that use a larger number of aquaponics facilities than earlier studies does help to overcome the present limitations and provide better and reliable results. This, however, is not an easy challenge for aquaponics researchers.

Simulation of aquaponics with the mathematical models under a wide range of management conditions will improve the understanding of aquaponics, verify different aquaponics configurations and point the way to the most promising strategies for improving aquaponics facilities. Again, this can lead to a more efficient way of conducting experiments.

Some modelling tools were also presented in this chapter. Traditionally, stock and flow diagrams (SFD) have been used for understanding processes as support tools for quantitative analysis. They are used to comprehend the flow and fluxes of quantities but lack the ability to illustrate the information associated to the flow and fluxes. Causal loop diagram (CLD) can be used to transfer complex SFD system into understandable simplified feedback structures. Together, the SFDs and the CLDs fully define the differential equation system. If only a simple qualitative understanding of the system is required, then CLD and SFD may be enough, but if the answer requires a numerical accuracy, then the problem can be investigated further with system dynamic tool diagrams (SDTD) and subsequently be modelled in a software tool for numerical simulation.

References

Badiola M, Mendiola D, Bostock J (2012) Recirculating Aquaculture Systems (RAS) analysis: main issues on management. Aquac Eng 51:26–35. https://doi.org/10.1016/j.aquaeng.2012.07.004

Batstone DJ, Keller J, Angelidaki I, Kalyuzhnyi SV, Pavlostathis SG, Rozzi A, Sanders WTM, Siegrist H, Vavilin VA (2002) The IWA Anaerobic Digestion Model no 1 (ADM1). Water Sci Technol 45:65–73

Boote KJ, Jones JW (1987) Equations to define canopy photosynthesis from given quantum efficiency, maximum leaf rate, light extinction, leaf area index, and photon flux density. In: Biggins J (ed) Progress in photosynthesis research. Martinus Nijhoff, Dordrecht, pp 415–418

Bot GPA (1993) Physical modelling of greenhouse climate. In: Hashimoto Y, Bot GPA, Day W, Tantau HJ, Nonami H (eds) The computerized greenhouse. Academic Press, San Diego, pp 51–74

Buck-Sorlin G, De Visser PHB, Henke M, Sarlikioti V, Ven der Heijden G, Marcelis LFM, Vos J (2011) Towards a functional–structural plant model of cut-rose: simulation of light environment, light absorption, photosynthesis and interference with the plant structure. Ann Bot 108:1121–1134

Challa H, Bakker M (1999) Potential production within the greenhouse environment. In: Stanhill G, Enoch HZ (eds) Ecosystems of the world 20 – Greenhouse ecosystems. Elsevier, pp 333–347

Colt JEK (2013) Impact of aeration and alkalinity on the water quality and product quality of transported tilapia—a simulation study. *Aquac Eng*:46–58

Corominas L, Riegler L, Takács I (2010) New framework for standardized notation in wastewater. J Int Assoc Water Pollut Res 61(4):S841–S857

Dahl O-J, Nygaard K (1966) SIMULA: an ALGOL-based simulation language. Commun ACM 9 (9):671–678. https://doi.org/10.1145/365813.365819

de Zwart HF (1996) Analyzing energy-saving options in greenhouse cultivation using a simulation model. Wageningen Agricultural University, Wageningen, p 236

Delaide B, Goddek S, Keesman, KJ, Jijakli MH (2018). A methodology to quantify the aerobic and anaerobic sludge digestion performance for nutrient recycling in aquaponics. https://popups.uliege.be:443/1780-4507 22, 12

Drayer GE, Howard AM (2014) Modeling and simulation of an aquatic habitat for bioregenerative life support research. *Acta Astronaut* 93:S.138–S.147. https://doi.org/10.1016/j.actaastro.2013.07.013

El-Mashad H (2003) Solar Thermophilic Anaerobic Reactor (STAR) for renewable energy production PhD thesis Wageningen University. ISBN: 9058089533-238

Emerenciano M, Carneiro P, Lapa M, Lapa K, Delaide B, Goddek S (2017) Mineralizacão de sólidos. Aquac Bras 21–26

Emrich S, Suslov S, Judex F (2007) Fully agent based. Modellings of epidemic spread using anylogic. In: Proceedings of the EUROSIM

FAO (2016) *The State of World Fisheries and Aquaculture 2016. Contributing to food security and nutrition for all.* Food and Agriculture Organization of the United Nations, Rome

Ficara E, Hassam S, Allegrini A, Leva A, Malpei F, Ferretti G (2012) Anaerobic digestion models: a comparative study. In: Proceedings of the 7th Vienna international conference on mathematical modelling 2012, p 1052

Fortmann-Roe S (2014) Insight maker: a general-purpose tool for web-based modeling & simulation. *Simul Model Pract Theory* 47:28–45

Frantz JM, Hand B, Buckingham L, Ghose S (2010) Virtual grower: software to calculate heating costs of greenhouse production in the United States. HortTechnology 20:778–785

Galí A, Benabdallah T, Astals S, Mata-Alvarez J (2009) Modified version of ADM1 model for agro-waste application. Bioresour Technol 100(11):2783–2790

Goddek S (2017) Opportunities and challenges of multi-loop aquaponic systems. Wageningen University. https://doi.org/10.18174/412236

Goddek S, Delaide BPL, Joyce A, Wuertz S, Jijakli MH, Gross A, Eding EH, Bläser I, Reuter M, Keizer LCP, Morgenstern R, Körner O, Verreth J, Keesman KJ (2018) Nutrient mineralization and organic matter reduction performance of RAS-based sludge in sequential UASB-EGSB reactors. Aquac Eng 83:10–19. ISSN: 0144-8609

Goddek S, Keesman KJ (2018) The necessity of desalination technology for designing and sizing multi-loop aquaponics systems. Desalination 428:76–85. https://doi.org/10.1016/j.desal.2017.11.024

Goddek S, Körner O (2019) A fully integrated simulation model of multi-loop aquaponics: A case study for system sizing in different environments. Agric Syst

Goddek S, Delaide B, Mankasingh U, Ragnarsdottir K, Jijakli H, Thorarinsdottir R (2015) Challenges of sustainable and commercial aquaponics. Sustainability 7:4199–4224. https://doi.org/10.3390/su7044199

Goddek S, Espinal CA, Delaide B, Jijakli MH, Schmautz Z, Wuertz S, Keesman KJ (2016) Navigating towards decoupled aquaponic systems: a system dynamics design approach. Water (Switzerland) 8:303. https://doi.org/10.3390/W8070303

Graber A, Junge R (2009) Aquaponic systems: nutrient recycling from fish wastewater by vegetable production. Desalination 246:147–156

Halamachi I, Simon Y (2005) A novel computer simulation model for design and management of re-circulating aquaculture systems. Aquac Eng 32(3–4):S443–S464. https://doi.org/10.1016/j.aquaeng.2004.09.010

Hassan J et al (2016) Transient accumulation of NO2-and N2O during denitrification explained by assuming cell diversification by stochastic transcription of denitrification genes. PLoS Comput Biol 11(1):e1004621

He E, Wurtsbaugh W (1993) An empirical model of gastric evacuation rates for fish and an analysis of digestion in piscivorous brown trout. Trans Am Fish Soc *122(5)*:S.717–S.730

Henze M, Willi G, Takashi M, Mark L (2002) Activated sludge models ASM1, ASM2, ASM2d AND ASM3. IWA Publishing in its Scientific and Technical Report series, UK. ISBN: 1-900222-24-8

Heuvelink E (1996) Tomato growth and yield: quantitative analysis and synthesis. Department of Horticulture. Wageningen Agricultural University, Wageningen, The Netherlands, p 326

Jablonsky J, Papacek S, Hagemann M (2016) Different strategies of metabolic regulation in cyanobacteria: from transcriptional to biochemical control. Sci Rep 6:33024

Janka E, Körner O, Rosenqvist E, Ottosen CO (2018) Simulation of PSII-operating efficiency from chlorophyll fluorescence in response to light and temperature in chrysanthemum (Dendranthema grandiflora) using a multilayer leaf model. Photosynthetica 56:633–640

Karimanzira D, Keesman KJ, Kloas W, Baganz D, Rauschenbach T (2016) Dynamic modeling of the INAPRO aquaponic system. Aquac Eng 75:29–45. https://doi.org/10.1016/j.aquaeng.2016.10.004

Keesman KJ (2011) System identification: an introduction. Springer, London

Knaus U, Palm HW (2017) Effects of fish biology on ebb and flow aquaponical cultured herbs in northern Germany (Mecklenburg Western Pomerania). Aquaculture 466:51–63. https://doi.org/10.1016/j.aquaculture.2016.09.025

Körner O, Hansen JB (2011) An on-line tool for optimising greenhouse crop production. Acta Hortic 957:147–154

Körner O, Van Straten G (2008) Decision support for dynamic greenhouse climate control strategies. Comput Electron Agric 60:18–30

Körner O, Aaslyng JM, Andreassen AU, Holst N (2007) Modelling microclimate for dynamic greenhouse climate control. HortScience 42:272–279

Körner O, Warner D, Tzilivakis J, Eveleens-Clark B, Heuvelink E (2008) Decision support for optimising energy consumption in European greenhouses. Acta Hortic 801:803–810

Körner O, Gutzmann E, Kledal PR (2017) A dynamic model simulating the symbiotic effects in aquaponic systems. Acta Hortic 1170:309–316

Licamele JD (2009) Biomass production and nutrient dynamics in an aquaponics system. The University of Arizona

Liebig HP, Alscher G (1993) Combination of growth models for optimized CO_2- and temperature-control of lettuce. Acta Hortic 328:155–162

Lugert V, Thaller G, Tetens J, Schulz C, Krieter J (2014) A review on fish growth calculation: multiple functions in fish production and their specific application. Rev Aquac 8(1):30–42

Lupatsch I, Kissil GW (1998) Predicting aquaculture waste from gilthead seabream (*Sparus aurata*) culture using nutritional approach. Aquat Living Resour 11(4):265–268. https://doi.org/10.1016/S0990-7440(98)80010-7

Lupatsch I, Kissil GW, Sklan D (2003) Comparison of energy and protein efficiency among three fish species gilthead sea bream (Sparus aurata), European sea bass (Dicentrarchus labrax) and white grouper (Epinephelus aeneus): energy expenditure for protein and lipid deposition. *Aquaculture*:175–189

Macal CM, North MJ (2005) Tutorial on agent-based modeling and simulation. In: Simulation conference, 2005 proceedings of the winter. IEEE

Madsen LO, Møller-Pedersen B, Nygaard K (1993) Object-oriented programming in the BETA programming language. Addison-Wesley. ISBN 0-201-62430-3

Marcelis LFM (1994) Fruit growth and dry matter partitioning in cucumber. Department of Horticulture. Wageningen Agricultural University, Wageningen, p 173

McCarthy J, Levin MI (1965) LISP 1.5 programmer's manual. MIT Press, Cambridge, MA

Orellana JUW (2014) Culture of yellowtail kingfish (Seriola lalandi) in a marine recirculating aquaculture system (RAS) with artificial seawater. Aquac Eng:20–28

Pagand P, Blancheton JP, Casellas C (2000) A model for predicting the quantities of dissolved inorganic nitrogen released in effluents from a sea bass (Dicentrarchus labrax) recirculating water system. *Aquac Eng* 22(1–2):S137–S153

Pavlostathis SG, Giraldo-gomez E (1991) Kinetics of anaerobic treatment: A critical review. Crit Rev Environ Control 21:411–490

Poorter H, Anten NP, Marcelis LFM (2013) Physiological mechanisms in plant growth models: do we need a supra-cellular systems biology approach. Plant Cell Environ 36:1673–1690

Rath T (1992) Einsatz wissensbasierter Systeme zur Modellierung und Darstellung von gartenbautechnischem Fachwissen am Beispiel des hybriden Expertensystems HORTEX. University of Hannover, Germany

Rath T (2011) Softwaresystem zur Planung von Heizanlagen von Gewächshäusern. Fachgebiet Biosystem- und Gartenbautechnik. Leibniz University Hannover, Germany

Reyes Lastiri D, Slinkert T, Cappon HJ, Baganz D, Staaks G, Keesman KJ, (2016) Model of an aquaponic system for minimised water, energy and nitrogen requirements. Water Sci Technol. wst2016127. https://doi.org/10.2166/wst.2016.127

Richie M, Haley D, Oetker M (2004) Effect of feeding frequency on gastric evacuation and the return of appetite in tilapia Oreochromis niloticus (L.). Aquaculture 234(1–4):S657–S673. https://doi.org/10.1016/j.aquaculture.2003.12.012

Rusten BE (2006) Design and operations of the Kaldnes moving bed biofilm reactors. Aquac Eng:322–331

Sánchez-Romero A, Miranda-Baeza A, Rivas-Vega M (2016) Development of a model to simulate nitrogen dynamics in an integrated shrimp–macroalgae culture system with zerowater exchange. *J World Aquacult Soc 47(1):129–138*

Sinha NK, Kuszta B (1983) Modelling and identification of dynamic systems. Von-Nostrand Reinhold, New York

Soukup J, Macháček P (2014) Serialization and persistent objects. Springer. https://doi.org/10.1007/978-3-642-39323-5

Sterman J (2000) Business dynamics: systems thinking and modeling for a complex world. McGraw Hill, Boston

Štys D, Stys D Jr, Pecenkova J, Stys KM, Chkalova M, Kouba P, Pautsina A, Durniev D, Nahlık T, Cısa P (2015) 5iD Viewer-observation of fish school behaviour in labyrinths and use of semantic and syntactic entropy for school structure definition. World Acad Sci Eng Technol Int J Comput Electr Autom Control Inf Eng 9(1):281–285

van Ooteghem RJC (2007) Optimal control design for a solar greenhouse. Wageningen University, Wageningen, p 304

Vanthoor B (2011) A model-based greenhouse design method. Wageningen University, Wageningen, p 307

Waller U, Buhmann AK, Ernst A et al (2015) Integrated multi-trophic aquaculture in a zero-exchange recirculation aquaculture system for marine fish and hydroponic halophyte production. *Aquac Int* 23:1473

Weatherley LR, Hill RG, Macmillan KJ (1993) Process modelling of an intensive aquaculture system. *Aquac Eng*:215–230

Wik TEI, Lindén BT, Wramner PI (2009) Integrated dynamic aquaculture and wastewater treatment modelling for recirculating aquaculture systems. *Aquaculture* 287(3/4):361–370

Willems JC, Polderman JW (1998) Introduction to mathematical systems theory: a behavioral approach. Springer. ISBN: 978-1-4757-2953-5

Wolfram S (1991) Mathematica: a system for doing mathematics by computer. Wolfram Research, Champagne

Yogev U, Barnes A, Gross A (2016) Nutrients and energy balance analysis for a conceptual model of a three loops off grid, aquaponics. Water 8:589. https://doi.org/10.3390/W8120589

Zeigler BP, Praehofer H, Kim TG (2000) Theory of modeling and simulation, 2nd edn. Elsevier, London

12

Aquaponics and its Allied Methods of Food Production

Benz Kotzen, Maurício Gustavo Coelho Emerenciano, Navid Moheimani, and Gavin M. Burnell

Abstract Whilst aquaponics may be considered in the mid-stage of development, there are a number of allied, novel methods of food production that are aligning alongside aquaponics and also which can be merged with aquaponics to deliver food efficiently and productively. These technologies include algaeponics, aeroponics, aeroaquaponics, maraponics, haloponics, biofloc technology and vertical aquaponics. Although some of these systems have undergone many years of trials and research, in most cases, much more scientific research is required to understand intrinsic processes within the systems, efficiency, design aspects, etc., apart from the capacity, capabilities and benefits of conjoining these systems with aquaponics.

Keywords Aquaponics alternatives · Algaeponics · Aeroponics · Aquaeroponics · Biofloc Technologies · Digeponics · Haloponics · Maraponics · Vermiponics · Vertical aquaponics

---

B. Kotzen (✉)
School of Design, University of Greenwich, London, UK
e-mail: b.kotzen@greenwich.ac.uk

M. G. C. Emerenciano
Santa Catarina State University (UDESC), Aquaculture Laboratory (LAQ), Laguna, SC, Brazil

CSIRO Agriculture and Food, Aquaculture Program, Bribie Island Research Centre, Bribie Island, QLD, Woorim, Australia
e-mail: mauricio.emerenciano@csiro.au

N. Moheimani
Algae R&D Centre Director, School of Veterinary and Life Sciences, Murdoch University, Murdoch, WA, Australia
e-mail: n.moheimani@murdoch.edu.au

G. M. Burnell
School of Biological, Earth and Environmental Sciences, University College Cork, Cork, Ireland
e-mail: g.burnell@ucc.ie

12.1 Introduction

This chapter discusses a number of key allied and alternative technologies that either expand or have the potential to expand the functionality/productivity of aquaponic systems or are associated/stand-alone technologies that can be linked to aquaponics. The creation and development of these systems have at their core the ability, amongst other things, to increase production, reduce waste and energy and in most cases reduce water usage. Unlike aquaponics, which may be seen to be in a mid/teenage stage of development, the novel approaches discussed below are in their infancy. This, however, does not mean that they are not technologies valuable in their own right and have the potential to deliver future food, efficiently and sustainably. The methods discussed below include aeroponics, aeroaquaponics, algaeponics, biofloc technology for aquaponics, maraponics and haloponics and vertical aquaponics.

12.2 Aeroponics

12.2.1 Background

The US National Aeronautics and Space Administration (NASA) describes aeroponics as *the process of growing plants suspended in air without soil or media* providing *clean, efficient, and rapid food production*. NASA furthermore notes that *crops can be planted and harvested year-round without interruption, and without contamination from soil, pesticides, and residue* and that *aeroponic systems also reduce water usage by 98%, fertilizer usage by 60% percent, and eliminate pesticide usage altogether. Plants grown in aeroponic systems have been shown to absorb more minerals and vitamins, making the plants healthier and potentially more nutritious* (NASA Spinoff). Other advantages of aeroponics are seen to be that:

- The growing environment can be kept clean and sterile.
- This reduces the chances for plant diseases and the spread of infection.
- Seedlings do not stretch or wilt during root formation.
- Seedlings are easily removed for transplanting without transplant shock.
- Seedling growth is accelerated, which leads to increased crop cycles and thus more produce per annum.

For Weathers and Zobel (1992), aeroponics is defined as *the culture of whole plants and/or tissues with their roots or the whole tissue fed by an air/water fog (as opposed to immersion in/on water, soil, nutrient agar or other substrates).* For them, plants that are grown only partially with their roots in air and part in nutrient solutions or are grown for part of the time in air and part of the time in nutrient solution are grown through a process of aero-hydroponics and not areoponics.

Aeroponic systems thus function by spraying or misting the root zone area with nutrient solution. The roots of the plants are thus suspended in air and are subjected to a continuous or intermittent/periodic spray/misting of nutrient-rich water droplets, in the form of droplets or very fine mists, with droplet sizes from 5 to 50 μm (microns). It is usual to find 'hobby/domestic' kit with spray droplet sizes of 30–80 μm. Ultrasonic or dry-fog atomizers produce a droplet size <5 μm, but these require compressed air and very fine nozzles, or it may be possible to use ultrasonic transducers to produce these mists.

In aeroponics, as with hydroponics, nutrient supply can be optimized and in a comparison between hydroponics and aeroponics, Hikosaka et al. (2014) note that no difference was found between growth and harvest quality in lettuce using dry-fog aeroponics. However, there was a significant increase in root respiration rates and photosynthesis rates of leaves. They also note that this system also uses less water and that it can be more efficient and easier to manage than conventional hydroponics (Hikosaka et al. 2014). In a review paper on modern plant cultivation technologies in agriculture under controlled environments, Lakhiar et al. (2018) note that aeroponics 'is considered the best plant growing method for food security and sustainable development'.

12.2.2 Origin of Aeroponics

Richard J. Stoner II is considered the father of aeroponics. The NASA review of aquaponics (Clawson et al. 2000) notes that the origin of aquaponics is largely in the study of root morphology, but originates in nature, e.g. with plants, for example, orchids growing in tropical areas where mists occur naturally. Clawson et al. (2000) note the development of aeroponics from B. T. Barker, who 'succeeded in growing apple trees with a spray', and F. W. Went, who in 1957 grew tomatoes and coffee plants in mists and termed the process 'aeroponics'. With regard to the study of root morphology, Carter in 1942 used aeroponics as a way of investigating pineapple roots, and Klotz in 1944 investigated the roots of avocado and citrus, and then numerous others including Hubick and Robertson; Barak, Soffer, and Burger; Yurgalevitch and Janes; and Dutoit, Weathers, and Briggs all undertook various experiments in aeroponics (refer to Clawson et al. 2000 for details).

12.2.3 Aeroponics Growing Issues

Clawson et al. (2000) report the tests by Tibbits et al. (1994) that continuous misting can 'contribute to fungal and bacterial growth in the vicinity of or on the plants', and furthermore some researchers have found that due to fine droplets and with continuous fogging systems, there can be difficulties 'in delivering nutrients to all the plants where there is a high density of plants'. In this respect it has been shown that misting at intervals delivers a healthier system and healthier roots compared to continuous fogging and hydroponic techniques. Using intervals also makes the plants more resistant to any interruptions in misting, conditioning the plants to thrive longer on lower moisture levels, with a likely reduction in pathogen levels. For effective misting, 'droplet size and velocity are also important aeroponic parameters. The root's mist collection efficiency depends on its filament size, drop size, and velocity' (Clawson et al. 2000).

12.2.4 Combining Aquaponics and Aeroponics

Whilst a number of entrepreneurs and keen hobbyists are promoting combining aquaponics with aeroponics, there are a number of issues that need to be solved if considering this combo-technology for future farming. One issue that needs to be resolved is a name for this system, and it is suggested here that we call this combo-system 'aquaeroponics'.

Whilst there are numerous videos and discussion threads on the web, on combining aeroponics and aquaponics, the field is void of scientific literature. The web-based discussions raise the issues of clogging of mist sprayers and the need for fine filtration of aquaponic solutions. Another issue with aquaeroponics is the potential for pathogens to grow in the airy wet environment and research will be required to ascertain this. One solution to solving the problem of misters is to use ultrasonic vibration to create the mists but this does not solve any problems there may be with the growth of pathogens.

12.3 Algaeponics

12.3.1 Background

Microalgae are unicellular photoautotrophs (ranging from 0.2 μm up to 100 μm) and are classified in various taxonomic groups. Microalgae can be found in most environments but are mostly found in aquatic environments. Phytoplankton are responsible for over 45% of world's primary production as well as generating over 50% of atmospheric O_2. In general, there is no major difference in photosynthesis of microalgae and higher plants (Deppeler et al. 2018). However, due to their smaller size and the reduction in a number of internally competitive physiological organelles, microalgae can grow much faster than higher plants (Moheimani et al. 2015). Microalgae can also grow under limited nutrient conditions and have the ability to adapt to a wider range of environmental conditions (Gordon and Polle 2007). Most importantly, microalgal culture does not compete with food crop production regarding arable land and freshwater (Moheimani et al. 2015). Furthermore, microalgae can efficiently utilize inorganic nutrients from waste effluents (Ayre et al. 2017). In general, microalgal biomass contains up to 50% carbon making them a perfect candidate for bioremediating atmospheric CO_2 (Moheimani et al. 2012).

The increase in extensive worldwide agriculture and animal farming has resulted in significant increases in biologically available nitrogen and phosphorus entering the terrestrial biosphere (Galloway et al. 2004). Crop and animal farming and sewage systems contribute significant amounts to these nutrient loads (Schoumans et al. 2014). The infiltration of these nutrients into water streams can cause massive environmental issues such as harmful algal blooms and mass fish mortality. For instance, in the USA, nutrient pollution from agriculture is acknowledged as one of the major sources of eutrophication (Sharpley et al. 2008). Controlling the flow of nutrients from farming operations into the surrounding environment results in both technical and economic challenges that must be overcome to reduce such effects. There have been various successful processes developed to treat waste effluent with high organic loads. However, almost all of these methods are not very effective in removing inorganic elements from water. Furthermore, some of these methods are rather expensive to operate. One simple method for treating organic waste is anaerobic digestion (AD). The AD process is well understood and when operated efficiently, it can convert over 90% of the wastewater organic matters to bio-methane and CO_2 (Parkin and Owen 1986). The methane can be used to generate electricity and the generated heat can be used for various additional purposes. However, the AD process results in creating an anaerobic digestion effluent (ADE) which is very rich in inorganic phosphate and nitrogen as well as high COD (carbon oxygen demand). In certain locations, this effluent can be treated using microalgae and macroalgae (Ayre et al. 2017).

12.3.2 Algal Growth Systems

Since the United Nations committee recommended that conventional agricultural crops be supplemented with high-protein foods of unconventional origin, microalgae have become natural candidates (Richmond and Becker 1986). The first microalgal cultivation was achieved though in 1890 by culturing *Chlorella vulgaris* (Borowitzka 1999). Due to the fact that microalgae normally divide at a certain time of the day, the term cyclostat was developed in order to introduce a light/dark (circadian) cycle to the culture (Chisholm and Brand 1981). The large-scale culturing of microalgae and the partial use of its biomass especially as a base for certain products such as lipids was probably started seriously as early as 1953 with the aim of producing food from a large-scale culture of *Chlorella* (Borowitzka 1999). Typically, algae can be cultured in liquid using open ponds (Borowitzka and Moheimani 2013), closed photobioreactors (Moheimani et al. 2011), or a combination of these systems. Alga can also be cultured as biofilms (Wijihastuti et al. 2017).

Closed Photobioreactors (after Moheimani et al. 2011): Closed algal cultures (photobioreactors) are not exposed to the atmosphere but are covered with a transparent material or contained within transparent tubing. Photobioreactors have the distinct advantage of preventing evaporation. Closed and semi-closed photobioreactors are mainly used for producing high-value algal products. Due to the overall cost of operating expenditure (OPEX) and capital expenditure (CAPEX), closed photobioreactors are less economical than open systems. On the other hand, there is less contamination and less CO_2 losses, and by creating reproducible cultivation conditions and flexibility in technical design, this makes them a good substitute for open ponds. Some of the closed systems' weaknesses can be overcome by (a) reducing the light path, (b) solving shear (turbulence) complexity, reducing oxygen concentration, and (c) a temperature control system. Closed photobioreactors are mainly divided into (a) carboys, (b) tubular, (c) airlift and (d) plate photobioreactors.

Open Ponds (after Borowitzka and Moheimani 2013): Open ponds are most commonly used for large-scale outdoor microalgal cultivation. Major algal commercial production is based in open channels (raceways) which are less expensive, easier to build and operate when compared to closed photobioreactors. In addition, the growth of microalgae meets less difficulties in open than closed cultivation systems. However just a few species of microalgae (e.g. *Dunaliella salina*, *Spirulina* sp., *Chlorella* sp.) have been grown successfully in open ponds. Commercial microalgal production costs are high, approximated to be between 4 and 20 $US/g^{-1}. Large-scale outdoor open pond commercial microalgal culture has developed over the last 70 years, and both still (unstirred) and agitated ponds have been developed and have been used on a commercial basis. The very large unstirred open ponds are simply constructed from natural water ponds with open beds that are usually less than 0.5 m in depth. In some smaller ponds the surface may be lined with plastic lining sheets. Unstirred open ponds represent the most economical and least technical of all commercial culture methods and have been commercially used for *Dunaliella salina* β-carotene production in Australia. Such ponds are mainly limited to growing microalgae which are capable of surviving in poor conditions or have a competitive advantage that allows them to outgrow contaminants such as protozoa, unwanted microalgae, viruses, and bacteria. Agitated ponds on the other hand have the advantage of a mixing regime. Most agitated ponds are either (a) circular ponds with rotating agitators or (b) single or joined raceway ponds.

Circular cultivation ponds have primarily been used for the large-scale cultivation of microalgae especially in South East Asia. The circular ponds up to 45 m in diameter and usually 0.3–0.7 m in depth are uncovered, but there are some examples which are covered by glass domes. The low shear stresses that are required for microalgae production are produced in these systems particularly in the centre of the pond, and this is a distinct advantage of these kinds of systems. Some disadvantages include expensive concrete structures, inefficient land use with large footprints, difficulties in controlling the movement of the agitating device and the added cost in supplying CO_2.

Paddlewheel-driven raceways are the most common commercial microalgal cultivation system. Raceways are usually constructed in either a single channel or as linked channels. Raceways are usually shallow (0.15 to 0.25 m deep), are constructed in a loop and normally cover an area of approximately 0.5 to 1.5 ha. Raceways are mostly used and recommended for the major commercial culturing of three species of microalgae including *Chlorella*, *Spirulina* and *Dunaliella*. A high risk of contamination and low productivity, resulting mainly from poor mixing regimes and light penetration, are the main disadvantages of these open systems. In raceways, biomass concentrations of up to 1000 mg dry weight.L^{-1} and productivities of 20 g dry weight.m^{-2}.d^{-1} have been shown to be possible.

The price of microalgal production makes economic achievement highly dependent on the marketing of expensive and exclusive products, for which demand is naturally restricted. Raceways are also the most used cultivation system used for treating wastewater (Parks and Craggs 2010).

Solid Cultivation (after Wijihastuti et al. 2017): An alternative microalgal cultivation method is immobilizing the cells in a polymer matrix or attaching them to the surface of a solid support (biofilm). In general, the biomass yield of such biomass cultures are at least 99% more concentrated than liquid-based cultures. Dewatering is one of the most expensive and energy-intensive parts of any mass algal production. The main advantage of biofilm growth is the potential of reducing the dewatering process and the related energy consumption and thus costs. Biofilm cultivation can also increase cellular light capture, reduce environmental stress (e.g. pH, salinity, metal toxicity, very high irradiance), reduce the cost of production and reduce nutrient consumption. Solid-based cultivation methods can be used for treating wastewater (nutrient and metal removal). There are three main methods for biofilm cultivation: (a) 100% directly submerged in medium, (b) partially submerged in medium and (c) using a porous substrate to deliver the nutrients and moisture from the medium to the cells.

12.3.3 Algal Growth Nutrient Requirements

A number of inhibitory physical, chemical and biological factors can inhibit high microalgal production. These are described in Table 12.1.

A basic knowledge of the critical growth limitations is probably the most essential factor before applying any microalgae to any process. Light is by far the most important limiting factor affecting the growth of any alga. Temperature is also a critical factor for mass algal production (Moheimani and Parlevliet 2013). However, these variables are difficult to control (Moheimani and Parlevliet 2013). Next to light and temperature, nutrients are the most important limiting factor affecting the growth of any alga (Moheimani and Borowitzka 2007) and each microalgal species tends to have its own optimum nutrient requirements. The most important nutrients are nitrogen, phosphorus and carbon (Oswald 1988). Most

Table 12.1 Limits to growth and productivity of microalgae (Moheimani and Borowitzka 2007)

Abiotic factors	Light (quality, quantity)
	Temperature
	Nutrient concentration
	O_2
	CO_2 and pH
	Salinity
	Toxic chemicals
Biotic factors	Pathogens (bacteria, fungi, viruses)
	Competition by other algae
Operational factors	Shear produced by mixing
	Dilution rate
	Depth
	Harvest frequency
	Addition of bicarbonate

algae respond to N-limitation by increasing their lipid content (Moheimani 2016). For example, Shifrin and Chisholm (1981) reported that in 20 to 30 species of microalgae that they examined, the algae increased their lipid content under N-deprivation. Phosphorus is also an important nutrient required for microalgal growth as it plays an essential role in cell metabolism and regulation, being involved in the production of enzymes, phospholipids and energy-supplying compounds (Smith 1983). Brown and Button's (1979) studies on green alga *Selenastrum capricornutum* showed an apparent growth limitation when the phosphate concentration of the medium was lower than 10 nM. CO_2 is also a critical nutrient for achieving high algal productivity (Moheimani 2016). For example, if additional CO_2 is not added to the algal culture, the average productivity can be reduced by up to 80% (Moheimani 2016). However, the addition of CO_2 to algal ponds is rather costly (Moheimani 2016). The most economical way for introducing CO_2 to a culture media is the direct transfer of the gas into the media by bubbling through sintered porous stones or using pipes under submerged plastic sheets as CO_2 injectors (Moheimani 2016). Unfortunately, in all of these methods there is still high loss of CO_2 to the atmosphere because of the short retention time of the gas bubbles in the algal suspension.

Although adding N, P and C is critical, other nutrients also affect microalgal growth and metabolism. A lack of other nutrients, such as manganese (Mn) and various other cations (Mg^{2+}, K^+ and Ca^{2+}), is also known to reduce algal growth (Droop 1973). Trace elements are also critical for microalgal growth and some microalga also require vitamins for their growth (Croft et al. 2005). One effective and inexpensive way of supplying nutrients is by combining algal culture and wastewater treatment which is discussed immediately below.

12.3.4 *Algae and Wastewater Treatment*

With an increase in environmental deterioration and a greater necessity to generate alternative food and energy sources, there is the impetus to explore the feasibility of biological wastewater treatments coupled with resource recovery. Microalgal wastewater treatments have been particularly attractive, due to algal photosynthetic activities, where light is transferred into profitable biomass. Under certain conditions, wastewater-grown microalgal biomass can be equivalent or superior in biomass production to higher plant species. Thus, the process can transform a waste product into useful products (e.g. animal feed, aquaculture feed, bio-fertilizer and bioenergy). Thus, the waste effluent is no longer a negative waste product, but it becomes a valuable substrate for producing important substances and successful microalgal wastewater bioremediation has been reported for over half a century (Oswald and Gotass 1957; Delrue et al. 2016). Algal phytoremediation indeed provides an environmentally favourable solution for the treatment of wastewater as it can utilize organic and inorganic nutrients efficiently (Nwoba et al. 2017). Microalgal cultures hold an enormous potential for the later steps of wastewater treatment, especially for reducing 'N', 'P' and 'COD' (Nwoba et al. 2016). Moreover, the added ability of microalgae to grow via different nutritional conditions such as photoautotrophic, mixotrophic and heterotrophic conditions also enhances its capabilities in removing various different types of pollutants and chemicals from aqueous matrices. The ability of microalgae in sequestrating carbon (CO_2) allows CO_2 bioremediation. The synchronized algal-bacteria relationship established is also ideally synergetic for the bioremediation of wastewater (Munoz and Guieysse 2006). Through photosynthesis, microalgae provide oxygen required by aerobic bacteria for the mineralization of organic matter as well as the oxidation of NH_4^+ (Munoz and Guieysse 2006). In return, the bacteria supply carbon dioxide for the growth of microalgae, significantly reducing the amount of oxygen required for the overall wastewater treatment process (Delrue et al. 2016). In general, waste effluents with low carbon to nitrogen ratios are fundamentally suited to the growth of photosynthetic organisms. Most importantly, the microalgal domestic and agricultural wastewater treatments is an attractive option since the technology is relatively easy and they require very low energy compared to the standard of effluent treatment. Optimization of microalgal wastewater treatment in large-scale raceway ponds is appealing since it combines the effective treatment of a harmful waste product and the production of potentially valuable protein-rich algal biomass. Figure 12.1 summarizes a closed loop system for treating any organic waste by combination of anaerobic digestion and algal cultivation.

12.3.5 *Algae and Aquaponics*

Microalgae in aquaculture and in aquaponic systems is most often seen to be a nuisance as they can restrict water flows by clogging up pipes, consume oxygen,

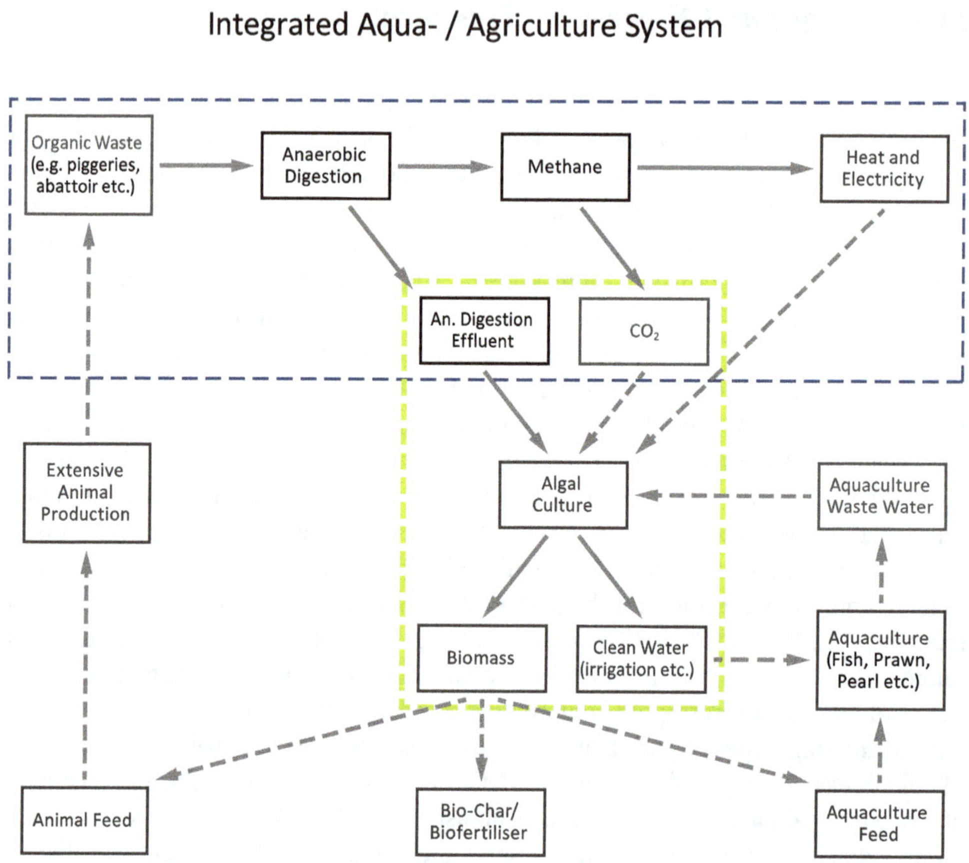

Fig. 12.1 Integrated process system to use algal culture for treating organic waste and potential end users. (The process is designed based on information from Ayre et al. 2017 and Moheimani et al. 2018)

may attract insects, reduce water quality and when decomposing can deplete oxygen. However, an experiment by Addy et al. (2017) shows that algae can improve water quality in an aquaponic system, help control pH drops related to the nitrification process, generate dissolved oxygen in the system, 'produce polyunsaturated fatty acids as a value-added fish feed and add diversity and improve resilience to the system'. One of the 'holy grails' of aquaponics is to produce at least part of the food that is fed to the fish as part of the system and it is here that research is required in producing algae that could be grown with part of the aquaponics water, most probably in a separate loop, which can then be fed as part of the diet to the fish.

12.4 Maraponics and Haloponics

Although freshwater aquaponics is the most widely described and practiced aquaponic technique, resources of freshwater for food production (agriculture and aquaculture) are becoming increasingly limited and soil salinity is progressively increasing in many parts of the world (Turcios and Papenbrock 2014). This has led to an increased interest and/or move towards alternative water sources (e.g. brackish to highly saline water as well as seawater) and the use of euryhaline or saltwater fish, halophytic plants, seaweed and low salt-tolerant glycophytes (Joesting et al. 2016). It is interesting to note that whilst the amount of saline in underground water is only estimated as 0.93% of world's total water resources at 12,870,000 km^3, this is more than the underground freshwater reserves (10,530,000 km^3) which makes up 30.1% of all freshwater reserves (Appelbaum and Kotzen 2016).

The use of saline water in aquaponics is a relatively new development and as with most new developments the terms used to describe the range/hierarchy of types needs to be established on a firm footing. In its short history, the term maraponics (i.e. marine aquaponics) has been coined for seawater aquaponics (SA), in other words, systems that use seawater as well as brackish water (Gunning et al. 2016). These systems are mainly located on-land, in coastal locations and in the case of SA, close to a seawater source. But there are fish as well as plants that grow and can be used in aquaponic units where water salinity levels vary. Thus whilst it makes etymological sense to use the term 'maraponics' for seawater aquaponics, it makes less sense to term brackish water aquaponics using this term. We thus suggest that a new term needs to be added to the aquaponic lexicon and this is 'haloponics', deriving from the Latin word *halo* meaning salt and combining this with suffix ponics. Thus maraponics is an on-land integrated multitrophic aquaculture (IMTA) system combining the aquacultural production of marine fish, marine crustaceans, marine molluscs, etc. with the hydroponic production of marine aquatic plants (e.g. marine seaweeds, marine algae and seawater halophytes) using oceanic strength seawater (approximately 35,000 ppm [35 g/L]). However aquaponic systems utilizing saline water below oceanic levels in a range of salinities should be termed haloponics (slightly saline water –1000 to 3000 ppm [1–3 g/L], moderately saline 3000–10,000 ppm [3–10 g/L] and high salinity 10,000–35,000 ppm [10–35 g/L]). These systems are also on-land IMTA systems combining aquacultural production with the hydroponic production of aquatic plants, but both the fish and plants are adapted to or grow well in what may be termed brackish water.

Although the concept of maraponics is very new, an interest in on-land seaweed-based integrated mariculture began to appear in the 1970s, starting from a laboratory-scale and then expanding to outdoor pilot-scale trials. In some of the earliest experimental studies, Langton et al. (1977) successfully demonstrated the growth of the red seaweed, *Hypnea musciformis*, cultured in tanks with shellfish culture effluent. Alternatively, crops that would usually be classed as glycophytes, such as the common tomato (*Lycopersicon esculentum*), the cherry tomato (*Lycopersicon esculentum* var. Cerasiforme) and basil (*Ocimum basilicum*), can achieve remarkably successful production levels at up to 4 g/L (4000 ppm) salinity and are often referred to as having low-moderate levels of salt tolerance (not to be confused with true halophytes, which are resistant to high salinities). Other crops that are tolerant of low-moderate salinities include turnip, radish, lettuce, sweet potato, broad bean, corn, cabbage, spinach, asparagus, beets, squash, broccoli and cucumber (Kotzen and Appelbaum 2010; Appelbaum and Kotzen 2016). For example, Dufault et al. (2001) and Dufault and Korkmaz (2000) experimented with shrimp waste (shrimp faecal matter and decomposed feed) as a fertilizer for broccoli (*Brassica oleracea italica*) and bell pepper (*Capsicum annuum*) production, respectively. Although their studies did not use maraponic techniques, they involved plants that are commonly grown using aquaponic (freshwater) techniques. Therefore, due to their salinity tolerance levels, these crops have enormous potential as candidate species for production in haloponic systems using low to medium salinities.

Recently, a number of studies have shown that halophytes can be successfully irrigated with aquacultural wastewater from marine systems using hydroponic techniques or as part of a recirculating aquaculture system (RAS). Waller et al. (2015) demonstrated the feasibility of nutrient recycling from a saltwater (16 psu salinity [16,000 ppm]) RAS for European sea bass (*D. labrax*) through the hydroponic production of three halophytic plants: *Tripolium pannonicum* (sea aster), *Plantago coronopus* (buck's horn plantain) and *Salicornia dolichostachya* (long spiked glasswort).

The majority of the maraponic work conducted so far involves the integration of two trophic levels – plants/algae and fish. However, an example of a system incorporating more than two trophic levels can be seen in an experiment conducted by Neori et al. (2000), who designed a small system for the intensive land-based culture of Japanese abalone (*Haliotis discus hannai*), seaweeds (*Ulva lactuca* and *Gracilaria conferta*) and pellet-fed gilthead bream (*Sparus aurata*). This system consisted of unfiltered seawater (2400 L/day) pumped to two abalone tanks and drained through a fish tank and finally through a seaweed filtration/production unit before being discharged back to the sea. Filter feeding molluscs could also be used in such a system. Kotzen and Appelbaum (2010) and Appelbaum and Kotzen (2016) compared the growth of common vegetables using potable water and moderately saline water (4187–6813 ppm) and found that basil (*Ocimum basilicum*), celery (*Apium graveolens*), leeks (*Allium ampeloprasum porrum*), lettuce (*Lactuca sativa* – various types), Swiss chard (*Beta vulgaris. 'cicla'*), spring onions (*Allium cepa*) and watercress (*Nasturtium officinale*) performed extremely well.

Maraponics (SAs) and haloponics offer a number of advantages over traditional crop and fish production methods. Because they use saline water (marine to brackish), there is a reduced dependence on freshwater, which in some parts of the world has become a very limited resource. It is typically practiced in a controlled environment (e.g. a greenhouse; controlled flow-rate tanks) giving better opportunities for intensive production. Many maraponic and haloponic systems are closed RAS with organic and/or mechanical biofilters and subsequently, water reuse is high, wastewater pollution is vastly reduced or eliminated, and contaminants are removed or treated. Even systems that are not RAS can significantly reduce the excess nutrients in the wastewater prior to discharge. Additionally, the occurrence of contaminants in non-RAS maraponic and haloponic systems can be reduced or eliminated through the use of water containing low levels of naturally occurring contaminants and the use of alternatives aquafeeds that do not contain dioxins or PCDs (e.g. novel feeds made from macroalgae). This improvement in water quality reduces the potential for disease occurrence and the need for antibiotic use is therefore vastly reduced. Due to their versatile configuration and low water requirements, maraponics and haloponics can be successfully implemented in a wide variety of settings, from fertile coastal areas to arid deserts (Kotzen and Appelbaum 2010), as well as in urban or peri-urban settlements. Another potential benefit is that many of the species that are suitable for these systems have a high commercial value. For example, the euryhaline European sea bass (*Dicentrarchus labrax*) and gilthead sea bream (*Sparus aurata*) can fetch a market price of €9/kg and €6/kg, respectively. Additionally, edible halophytes tend to have a high market price, with sea-agretti (*Salsola soda*), for example, having a market price of €4–€4.5/kg and marsh samphire (*Salicornia europaea*) selling at €18/kg in supermarkets.

The evidence is therefore compelling. Maraponics and haloponics provide a dynamic and rapidly growing field that has the potential to provide a number of services to communities, many of which are explored elsewhere in this publication.

12.5 Vertical Aquaponics

12.5.1 Introduction

Whilst aquaponics can be seen as part of a global solution to increase food production in more sustainable and productive ways and where growing more food in urban areas is now recognized as part of the solution to food security and a global food crisis (Konig et al. 2016), aquaponic systems can themselves become more productive and sustainable by adopting alternative growing technologies and learning from emerging technologies such as vertical farming and living walls [LWs] (Khandaker and Kotzen 2018). Additionally by being space-efficient, they can be better integrated into urban areas.

In the developed world most aquaponic systems are placed in greenhouses in order to control temperature; in northern Europe and North America for example, winter temperatures are too cold in winter and in Mediterranean areas such as Spain, Italy, Portugal, Greece and Israel, summer temperatures are too warm. There are of course many additional advantages in growing food in controlled greenhouses, such as the ability to regulate relative humidity and control air movement, to quarantine fish as well as plants from diseases as well as pests and potentially being able to add CO_2, to aid plant growth. However, growing produce in a greenhouse can readily raise costs through (a) the capital costs of the greenhouse (a broad estimate of US $\$350/m^2$ Arnold 2017) and (b) allied infrastructure such as microclimate controls which include heating and cooling systems and lighting. On top of the initial infrastructure costs, there are also the specific greenhouse production costs which include the energy/power supply for heating and cooling as well as lighting.

Most aquaponic systems such as the University of the Virgin Island (UVI) system (Fig. 12.1), designed by Dr. James Rakocy and his colleagues, use horizontal grow tanks or beds, emulating traditional land-based arable growing patterns to produce vegetables (Khandaker and Kotzen 2018). In other words, the system relies on horizontal rows/arrays of plants usually elevated to around waist level so that plant-related management tasks can be readily undertaken. Parallel developments in living wall and vertical farming technologies have arisen at almost the same time that aquaponics has evolved and are similarly in the adolescent stage of development. Similarly as in aquaponics, as more people become involved there is a concomitant increase in systems and technological development to increase productivity and reduce costs. The coupling of vertical growing systems (vertical farming systems and living walls) rather than horizontal beds to the fish and filtration tanks is potentially one key way of increasing productivity as it should be possible to increase the number of vegetables grown compared to the numbers produced in typical horizontal bed aquaponics. The UVI aquaponic systems (Fig. 12.2) produce approximately 32 plants per square metre (Al-Hafedh et al. 2008), depending on the species and cultivar that is grown, but as Khandaker and Kotzen (2018) note, approximately 96 plants can be grown per square metre 'using back-to-back elements of the *Terapia Urbana*[1] LW system which is more than three times the density

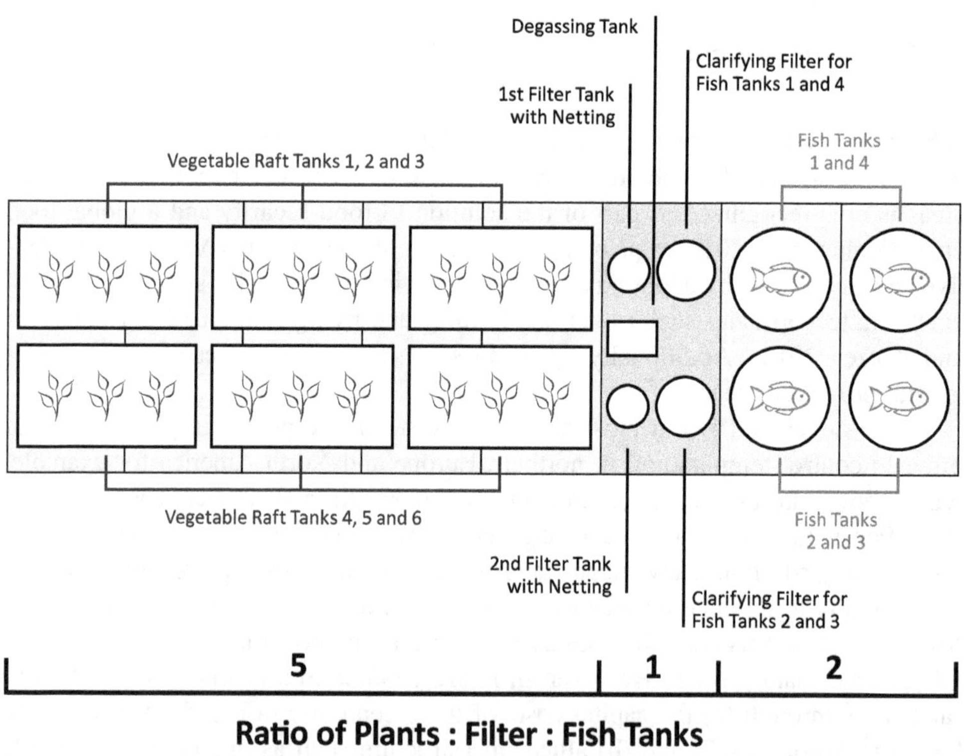

Fig. 12.2 Schematic diagram of a typical UVI system illustrating the ratio of fish tanks/filters/plant growing tanks which is 2:1:5. This shows that the greatest area is subsumed by the plants and it is in this area that space savings may be considered. (Khandaker and Kotzen 2018)

[1]Terapia Urbana S.L. produces a felt pocket type of living wall in Seville, Spain.

compared to the UVI horizontal growing system'. A conservative estimate should at least double the maximum amount grown in horizontal beds to 64 plants/m^2. In an experiment with lettuce (*Lactuca sativa* L. cv. 'Little Gem') using horizontal beds and planted columns, planted at similar densities, Touliatos et al. (2016) suggest that the 'Vertical Farming System (VFS) presents an attractive alternative to horizontal hydroponic growth systems (and) that further increases in yield could be achieved by incorporating artificial lighting in the VFS'.

Vertical Farming Systems (VFS)

Before we discuss the specific requirements for vertical systems we need to discuss the types of systems that are available. In VFS there are three main generic types (Fig. 12.3):

1. Stacked Horizontal Beds: Instead of only having one horizontal grow bed, the beds are stacked like shelves in tiers. This arrangement means that in a greenhouse, only the upper bed will be facing direct natural light and supplementary light needs to be provided at all levels. This is usually provided from directly under the grow bed above. In principle this could mean that the growing beds could be stacked as high as the greenhouse allows, but of course growing things at

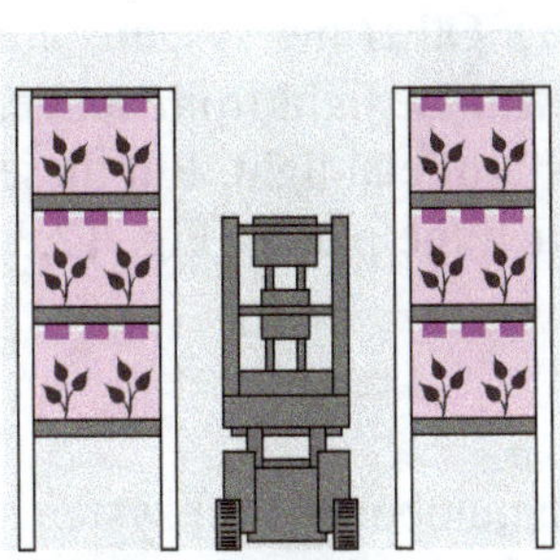

A. Stacked System (Elevation View)

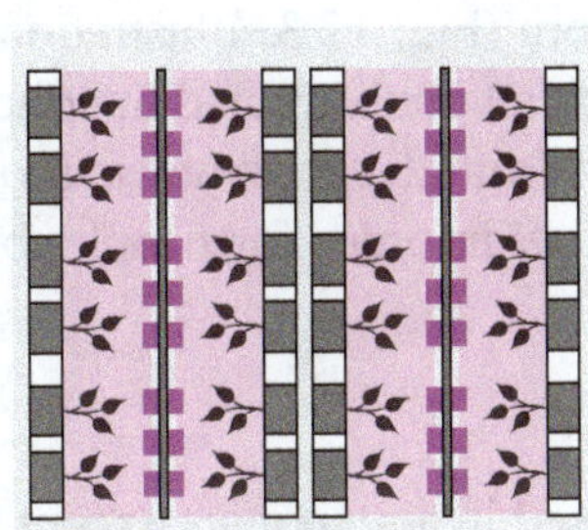

B1. Vertical System Zipgrow Type (Plan View)

B2. Vertical System Stacking Type (Plan View)

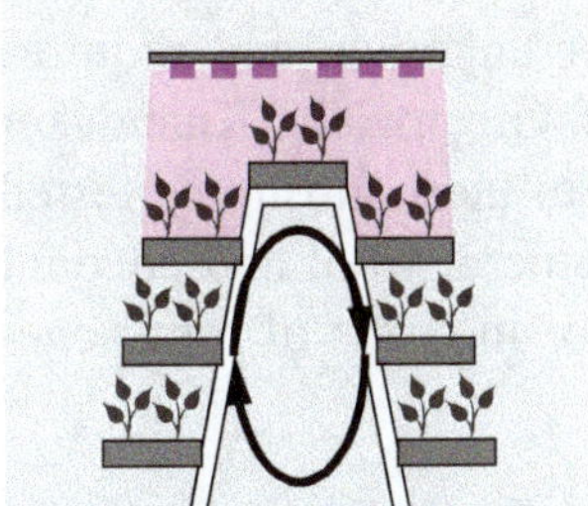

C1. Stacked Tier System with moving Troughs (Elevation View)

C2. Stepped Tier System with rigid Troughs (Elevation View)

Fig. 12.3 Vertical farming systems and their lighting arrangements

height means greater difficulty in the system's management including planting, maintenance and harvesting, requiring scissor lifts and additional energy to pump nutrient-rich water to all levels. According to Bright Agrotech (Storey 2015), up to four tiers is profitable and anything above that is unprofitable. Storey (2015) further notes that labour increases by 25% at the second, third and fourth levels when a scissor lift is required (Fig. 12.3, Illustration A).

2. Vertical Tower Systems (VTS): Vertical Tower Systems comprise systems which grow plants in vertical arrays within a container or series of stacked modules. Depending on the system, plants are grown facing one direction or if, for example, they are planted in a tube-like form, then they can be arranged facing any direction. An example of a vertical array system, where plants are grown facing in a single direction is the ZipGrow™ which are either hung or supported in rows (Fig. 12.3, Illustration B1). The rows between are approximately 0.5 metres (20 inches). Growing in a more three-dimensional way occurs with stacked systems or in tubular systems which allow more plants to be grown, but lighting is more complex (Fig. 12.3, Illustration B2).
3. Stepped Tiers: These systems contain rigid or moving plant troughs. The Sky Greens VFS in Singapore uses a rotating trough system which moves the troughs upwards and into the light. Additional natural light is more significant towards the top and less so at the bottom (Fig. 12.3, Illustration C1). Other tier systems are stepped so that each tier has an unobstructed interface with the light from above, whether this is natural light from the greenhouse roof or artificial light. But these systems have to be quite low in order for people to reach the plants (Fig. 12.3, Illustration C2).

Living Walls

Living walls have yet to be used in aquaponics except in a number of trial systems such as at the University of Greenwich, London (Khandaker and Kotzen 2018). Whereas most VFS use nutrient film technique (NFT) grow channels or encapsulated mineral wool blocks, LWs sometimes also use soil type substrates in pots or troughs, which provide the rooting medium. Whilst this is fine for growing ornamental plants as well as vegetables and herbs, when coupled with fish tanks, any addition of soil to the system may complicate the microbial character of the system and be detrimental to the fish. This is however unknown and requires research. Experiments undertaken at the University of Greenwich (Khandaker and Kotzen 2018) indicate that from a number of single, inert substrates tested (including hydroleica, perlite, straw, sphagnum moss, mineral wool and coconut fibre), coconut fibre and then mineral wool were superior in terms of root penetration and root growth in lettuce (*Lactuca sativa*).

Vertical v. Horizontal: Factors to Be Considered

There are four key aspects which need to be taken into account when comparing the benefits (productivity and sustainability) of vertical growing, compared to horizontal growing. These are (1) space, (2) lighting, (3) energy and (4) life cycle costs.

1. *Space*

The benefits of being able to grow produce vertically, back to back, need to be balanced with the amount of space that is required to provide an even spread of lighting as well as the row space required for management and maintenance. The width of a row in hydroponic systems varies. As noted the standard ZipGrow™ system is approximately 0.5 metres, whereas the usual row width for growing tomatoes and cucumbers hydroponically varies from 0.9 to 1.2 metres (Badgery-Parker and James 2010). Growing smaller plants such as lettuce and herbs such as basil, may allow for narrower rows, but of course row width must ensure that produce is not compromised by moving items such as trolleys and scissor lifts. A key issue with growing vertically is the conflict that occurs between having fixed rows and fixed lighting, which needs to be located in the rows between the planting facades. These lights will impede people movements and thus either the lights need to be (i) part of the growing structure or (ii) retractable or movable, so that workers can readily undertake tasks, or (iii) the planting structures are movable and the lights remain static.

2. *Lighting*

Greenhouse production of vegetables and other plants rely on specific spatial arrangements which allow for planting, management through growth and then harvesting. The spatial arrangement will depend on the types of plants and the types of mechanization that is installed. Additionally, growing efficiently relies on the supplement of additional light of different types, which have their own pros and cons. In general what these lights do is provide specific wavelengths for plant growth and for fruit or flower production. Whereas it is relatively simple and more common to evenly light plants grown horizontally, it is more of a challenge to evenly light a vertical surface.

With regard to types of lighting, many producers have moved to or are tempted to install LEDs (light-emitting diodes), due to their long lifespan, up to 50,000 hours or more (Gupta 2017), their low power requirements and their recent reduction in cost. Virsile et al. in Gupta (2017) note that *most applications of LED lighting in greenhouses choose the combinations of red and blue wavelengths with high photon efficiency* but that green and white light *containing substantial amounts of green wavelengths has a positive physiological impact on plants*. However, *the combination of blue and red lights creates a purplish-grey image*, and this hampers the visual evaluation of plant health. The type of wavelengths chosen is complex and can have benefits at different stages in the plant's life and even according to the cultivars of, for example, lettuce. Red-leafed lettuces, for example, respond to blue LED lighting, increasing their pigmentation (Virsile et al. in Gupta 2017). Additionally, blue LED lighting can improve the nutritional quality of green vegetables, reducing nitrate content, increasing antioxidants and phenolic and other beneficial compounds. The light spectra also affect taste, shape and texture (Virsile et al. in Gupta 2017). The costs of LEDs have dropped significantly and as the efficacy of LEDs has increased so the break-even return time on investment has decreased (Bugbee in Gupta 2017).

Other lighting of course exists and this includes fluorescent lighting, metal halide (MH) lighting and high pressure sodium (HPS) lighting. The type of lighting that is used in vertical farming and with living walls varies considerably depending on the scale and location. Compact fluorescent lamps (CFLs) are relatively thin and can easily fit into small spaces, but they require an inductive ballast to regulate current through the tubes. CFLs use only 20–30% of an incandescent bulb and they last six to eight times longer but they are almost 50% less efficient than LEDs. They are by far the cheapest of the three major types of grow lights. HPS grow light technology is over 75 years old and is well established for growing under glass, but they produce a lot of heat and are thus not suitable for vertical farming and living walls, where light needs to be delivered quite close to the plants. The heat produced by LED grow lights, on the other hand, is minimal. The cost however is higher than other two types, and eye protection is needed for longer-term exposure to LEDs as the long-term exposure to the light spectra can be damaging to the eyes. The arrangement of VFS units will dictate the lighting arrangement but on the whole these are lit by LEDs. The method of lighting living walls will depend on the height of the wall. The taller the wall the more difficult it is to apply an even spread across the surface, although it should be noted that the number of lights used should be no different to those used in horizontal grow beds and if the wall is tall then the lights may need to be staggered. As most living walls are located for aesthetic purposes, lighting needs to be kept as far as possible, out of the way and the lighting has to not only provide adequate light for plant growth and health, but also so that the plants look good (Fig. 12.4).

Fig. 12.4 A 4-metre-tall, 5-metre-long living wall can be adequately lit with six high-efficiency discharge lamps. Note these were chosen not only to provide adequate light for growth but also so that the plants in the living wall would look good. (University of Greenwich Living Wall. Source: Benz Kotzen)

The advances in LED technology, where lighting frequencies and intensity can be engineered to suit individual species and cultivars as well as their various life cycles means that LEDs will become the technology of choice in the near future. This will additionally be enhanced by reductions in costs.

3. *Energy*

More energy for lighting is likely to be required for VFS as well as LWs as even natural lighting cannot be achieved over vertical surfaces. Additionally more pumping power for irrigation will be required and this will be relative to the height of the VFS or LWs.

4. *Comparative Life Cycle Analysis (LCA)*

Whilst there are numerous studies undertaken on life cycle analysis of aquaponics and various aspects of aquaponic systems, there are no comparative studies that compare vertical versus horizontal aquaponics. This has yet to be done. We are getting to a point where vertical aquaponics is likely to warrant further testing and research and in time vertical aquaponics, which couples vertical farming systems or living wall systems with the fish tanks and filtration units, is likely to become more mainstream, as long as these can be profitable and sustainable.

12.6 Biofloc Technology (BFT) Applied for Aquaponics

12.6.1 Introduction

Biofloc technology (BFT) is considered the new 'blue revolution' in aquaculture (Stokstad 2010) since nutrients can be continuously recycled and reused in the culture medium, benefited by the in situ microorganism production and by the minimum or zero water exchange (Avnimelech 2015). These approaches might face some serious challenges in the sector such as competition for land and water and the effluents discharged to the environment which contain excess of organic matter, nitrogenous compounds and other toxic metabolites.

BFT was first developed in the early 1970s by the Aquacop team at Ifremer-COP (French Research Institute for Exploitation of the Sea, Oceanic Center of the Pacific) with different shrimp penaeid species including *Litopenaeus vannamei*, *L. stylirostris* and *Penaeus monodon* (Emerenciano et al. 2011). In the same period, Ralston Purina (a private US company) in connection with Aquacop applied the technology both in Crystal River (USA) and Tahiti, which lead to the greater understanding of the benefits of biofloc for shrimp culture. Several other studies enabled a comprehensive approach to BFT and researched the inter-relationships between water, animals and bacteria, comparing BFT with an 'external rumen' but now applied for shrimp. In the 1980s and at the beginning of the 1990s, both Israel and the USA (Waddell Mariculture Center) started R&D in BFT with tilapia and Pacific white shrimp *L. vannamei*, respectively, in which environmental concerns, water limitation and land costs were the main causative agents that promoted the research (Emerenciano et al. 2013).

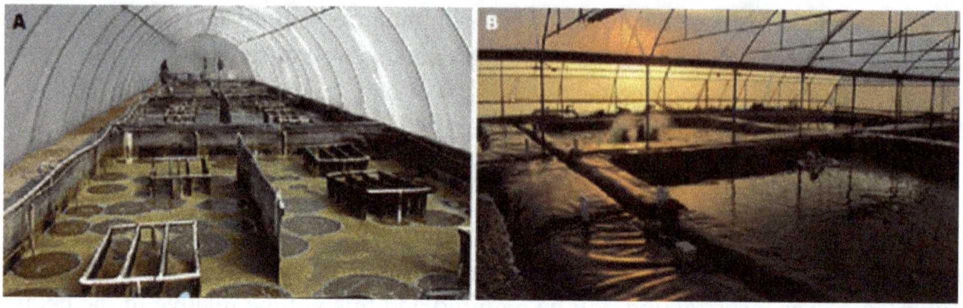

Fig. 12.5 Biofloc technology (BFT) applied for marine shrimp culture in Brazil (**a**) and for tilapia culture in Mexico (**b**) (Source: EMA-FURG, Brazil and Maurício G. C. Emerenciano)

The first commercial BFT operations and probably the most famous commenced in the 1980s at the 'Sopomer' farm in Tahiti, French Polynesia, and in the early 2000s at the Belize Aquaculture farm or 'BAL', located in Belize, Central America. The yields obtained using 1000 m^2 concrete tanks and 1.6 ha lined grow-out ponds were approximately 20–25 ton/ha/year with two crops at Sopomer and 11–26 ton/ha/cycle at BAL, respectively. More recently, BFT has been successfully expanded in large-scale shrimp farming in Asia, in South and Central America as well as in small-scale greenhouses in the USA, Europe and other areas. At least in one phase (e.g. nursery phase) BFT has been used with great success in México, Brazil, Ecuador and Peru. For commercial-scale tilapia culture, farms in Mexico, Colombia and Israel are using BFT with productions around 7 to 30 kg/m^3 (Avnimelech 2015) (Fig. 12.5b). Additionally, this technology has been used (e.g. in Brazil and Colombia) to produce tilapia juveniles (~30 g) for further stock in cages or earthen ponds (Durigon et al. 2017). BFT has mainly been applied to shrimp culture and to some extent with tilapia. Other species have been tested and show promise, as noted for silver catfish (*Rhamdia quelen*) (Poli et al., 2015), carp (Zhao et al., 2014), piracanjuba (*Brycon orbignyanus*) (Sgnaulin et al., 2018), cachama (*Colossoma macropomum*) (Poleo et al., 2011) and other crustacean species such as *Macrobrachium rosenbergii* (Crab et al., 2010), *Farfantepenaeus brasiliensis* (Emerenciano et al., 2012), *F. paulensis* (Ballester et al., 2010), *Penaeus semisulcatus* (Megahed, 2010), *L. stylirostris* (Emerenciano et al., 2011) and *P. monodon* (Arnold et al., 2006). The interest in BFT is evident by the increasing number of universities and research centres carrying out research particularly in the key fields of grow-out management, nutrition, reproduction, microbial ecology, biotechnology and economics.

12.6.2 How does BFT Work?

Microorganisms play a key role in BFT systems (Martinez-Cordoba et al. 2015). The maintenance of water quality, mainly by the control of the bacterial community over autotrophic microorganisms, is achieved using a high carbon to nitrogen ratio (C:N) since nitrogenous by-products can be easily taken up by heterotrophic bacteria. In the beginning of the culture cycles a high carbon to nitrogen ratio is required to guarantee optimum heterotrophic bacteria growth, using this energy for its maintenance and growth (Avnimelech 2015). Additionally, other microorganism groups are crucial in BFT systems. The chemoautotrophic bacterial community (i.e. nitrifying bacteria) stabilizes after approximately 20–40 days and might be responsible for two-thirds of the ammonia assimilation in the system (Emerenciano et al. 2017). Thus, the addition of external carbon should be reduced and alkalinity consumed by the microorganisms must be replaced by different carbonate/bicarbonate sources (Furtado et al. 2011). The stability of zero or minimal water exchange depends on the dynamic interaction amongst communities of bacteria, microalgae, fungi, protozoans, nematodes, rotifers, etc. that will occur naturally (Martinez-Cordoba et al. 2017). The aggregates (bioflocs) are a rich protein-lipid natural source of food that become available 24 h per day due to a complex interaction between organic matter, physical substrate and large range of microorganisms (Kuhn and Boardman 2008; Ray et al. 2010). The natural productivity in a form of microorganisms' production plays three major roles in the tanks, raceways or lined ponds: (1) in the maintenance of water quality, by the uptake of nitrogen compounds generating in situ microbial protein; (2) in nutrition, increasing culture feasibility by reducing feed conversion ratios and a decrease in feed costs; and (3) in competition with pathogens (Emerenciano et al. 2013).

Regarding the water quality for the culture organisms, besides oxygen, excess of particulate organic matter and toxic nitrogen compounds are the major concern in the biofloc systems. In this context, three pathways occur for the removal of ammonia nitrogen: at a lesser rate (1) photoautotrophic removal by algae and at a higher rate (2) heterotrophic bacterial conversion of ammonia nitrogen directly to microbial biomass and (3) autotrophic bacterial conversion from ammonia to nitrate (Martinez-Cordoba et al. 2015). The nitrate available in the systems plus other minor and major nutrients accumulated over the cycle could be used as substrate for plant growth in aquaponic systems (Pinho et al. 2017).

12.6.3 BFT in Aquaponics

The application of BFT in aquaponic systems is relatively new, although Rakocy (2012) mentions a commercial pilot-scale project with tilapia. Table 12.2 summarizes key recent studies that have used BFT in aquaponic systems.

Overall, the results demonstrate that biofloc technology can be used and integrated in a fish or shrimp-plant production. BFT when compared to other conventional aquaculture systems (such as RAS) actually improved the plant and fish yields and promoted better plant visual quality (Pinho et al. 2017), but not in all cases (Rahman 2010; Pinho 2018). Pinho et al. (2017) observed that lettuce yields with the BFT system were greater compared to the clear-water recirculation system (Fig. 12.6). This is possibly due to the higher nutrient availability provided by the

Table 12.2 Recent studies around the world applying the BFT in aquaponic systems for different aquatic and plant species

Aquatic species	Plant species	Main results	References
Tilapia	Lettuce	Biofloc technology did not improve lettuce production as compared to conventional hydroponic solution	Rahman (2010)
Tilapia	Lettuce	Yield and visual quality of lettuce was improved using BFT as compared to clear-water recirculation system	Pinho et al. (2017)
Tilapia (nursery)	Lettuce	Plant performance (lettuce) using tilapia in a nursery phase (1–30 g) was negatively influenced by biofloc wastewater as compared to RAS wastewater after two plant cycles (13 days each). Plant visual aspects were better in RAS as compared to BFT	Pinho (2018)
Tilapia	Lettuce	The presence of filtering elements (mechanical filter and biological filter) positively affected the lettuce production in aquaponic systems as compared to treatment without filters using BFT	Barbosa (2017)
Tilapia	Lettuce	Low salinity (3 ppt) can be performed in aquaponics using BFT. Visual and performance parameters indicated that the purple variety had better performance than the smooth and crisped varieties	Lenz et al. (2017)
Silver catfish	Lettuce	The use of bioflocs in the aquaponic system may improve the productivity of lettuce in an integrated culture with silver catfish	Rocha et al. (2017)
Litopenaeus vannamei	*Sarcocornia ambigua*	The performance of marine shrimp *L. vannamei* was not affected by the *S. ambigua* integrated aquaponics production and also improve the use of nutrients (e.g. nitrogen) in the culture system	Pinheiro et al. (2017)

Fig. 12.6 Experimental aquaponics greenhouse comparing biofloc technology and RAS wastewater at Santa Catarina State University (UDESC), Brazil. (Source: Pinho et al. 2017)

higher microbial activity. However, this trend was not observed in the study by Rahman (2010), who compared effluent from fish culture in a BFT system to a conventional hydroponic solution in a lettuce production. In addition, Pinho

Fig. 12.7 High salinity halophyte *Sarcocornia ambigua* aquaponics production integrated with Pacific white shrimp *Litopenaeus vannamei* successfully applying biofloc technology at Santa Catarina Federal University (UFSC), Brazil. (Source: LCM-UFSC, Brazil)

Fig. 12.8 Aquaponics lettuce production integrated with tilapia using biofloc technology (left) and accumulation of suspended solids in lettuce roots (right). Barbosa (2017)

(2018) in a recent study observed that productive performance of lettuce in aquaponic system using tilapia in a nursery phase (1–30 g) was negatively influenced by biofloc wastewater as compared to RAS wastewater over 46 days. The variation in results identifies the need for additional studies in this area.

BFT can be used with low salinity water, e.g. with some varieties of lettuce (Lenz et al. 2017), and higher salinity waters can be used, e.g. with halophyte plant species such as *Sarcocornia ambigua* co-culture with Pacific white shrimp *Litopenaeus vannamei* (Pinheiro et al. 2017) (Fig. 12.7). Silver catfish *Rhamdia quelen* has also shown good potential for the integration of aquaponics with BFT (Rocha et al. 2017).

With BFT, the concentration of solids can severely affect the roots and impact nutrient absorption and oxygen availability. As a result, yields can be affected but also the visual quality of the plants (e.g. lettuces) which is an important criterion for consumers. With this in mind, solids management is an important subject for further studies where the impact of solids (particulate fraction and also dissolved fraction) in aquaponic systems when applying BFT is considered (Fig. 12.8). In addition, economic studies need to be undertaken to compare the costs involved of the various aquaculture and plant growing systems and to identify appropriateness relative to different locations and conditions.

12.7 Digeponics

Anaerobic processing of purposely cultivated biomass, as well as residual plant material from agricultural activity, for biogas production is a well-established method. The bacterially indigestible digestate is returned to the fields as a fertilizer and for building humus. Whilst this process is widespread in agriculture, the application of this technology in horticulture is relatively new. Stoknes et al. (2016) claim that within the 'Food to waste to food' (F2W2F) project, an efficient method for the utilization of digestate as substrate and fertilizer has been developed for the first time. The research team coined the term 'digeponics' for this circular system. Digeponics, in contrast to aquaponics, replaces the aquaculture part with an anaerobic digester, or, when comparing it to a three loop aquaponic system that includes an anaerobic, the aquaculture part is removed from the system, leaving two main loops, the digestion loop and the horticultural loop.

The required organic input that is provided in the form of the fish food to an aquaponic system is replaced with food waste from human food production for digeponics. The varying composition of nutrients in the input stream opposed to the well-known, constant and probably nutritionally optimized nutrient stream resulting from the fish feed will most likely call for a more strict nutrient analysis and management regime than that required in aquaponics.

The produced biogas, which mainly contains methane and carbon dioxide, can be utilized within the facility for electricity and heat production. The resulting carbon dioxide-rich exhaust gas can be used as a fertilizer directly in the greenhouse reducing emissions in comparison to classical biogas plants used in agriculture.

Since the 'fresh and untreated digestate in anaerobic liquid slurry (contains) plant toxic substances, a very high electrical conductivity (EC) and chemical oxygen demand (COD)' (Stoknes et al. 2016), it has to be treated to make it suitable for plant fertilization. Several methods of moderation have been examined within the F2W2F project. The relatively high EC of the digestate and the operational flexibility of a digester fed with low-cost food waste alleviate some of the tight coupling issues often attributed to coupled aquaponic systems (see Chap. 7). Thus digeponics may serve as an interesting alternative to aquaponics in situations where the aquaculture part represents a challenge. With respect to a three loop aquaponic system that already comprises a loop with an anaerobic digester, the inclusion of a food waste stream for organic input might represent an interesting future direction. The methane yield of aquaculture sludge is rather limited. A targeted inclusion of residual agricultural biomass with the aim of methane yield optimization could enhance overall performance.

12.8 Vermiponics and Aquaponics

It would be remiss in this chapter not to mention earthworms and their introduction into aquaponics, and thus this chapter concludes with a brief résumé of these detritivore invertebrates and their abilities to convert organic waste into fertilizer. It is said that worms and the way that they digest matter were of interest to Aristotle and Charles Darwin as well as the philosophers Pascal and Thoreau (Adhikary 2012) and they were protected by law under Cleopatra. Earthworms are valued in agriculture and horticulture as they are 'vital to soil health because they transport nutrients and minerals from below to the surface via their waste, and their tunnels aerate the ground' (National Geographic).

Modern vermiculture is attributed to Mary Appelhof, who in the early 1970s and 1980s produced a number of publications on composting with worms. Contemporary vermicomposting occurs on large and small scales with the objective of getting rid of organic waste and producing fertilizer in the forms of compost and 'worm tea'. Worm tea can be produced by soaking worm casts or by leaching the nutrients from the compost through wetting or natural wetting leachate from precipitation.

Vermiponics uses the worm casts of mainly red wriggler worms also known as tiger worms (*Eisenia fetida*) or (*E. foetida*) to provide nutrients in a hydroponic system. When worms are introduced into an aquaponic system, we suggest that the system is termed 'vermi-aquaponics' to differentiate the systems. It is thus the introduction of worms into the growing beds of the plant parts of an aquaponic system. It should be noted that vermi-aquaponics is in its infancy and mainly practiced by hobbyists and in research laboratories. The worms are introduced mainly into the plant growing media, usually gravel beds, where they can help to break down any solid waste from the fish and any detritus from the plants and additionally provide additional nutrients for the plants, and they can also be fed to carnivore fish. In most instances the beds are of a flood and drain type, so that the worms are not constantly under water.

Acknowledgements The authors thank National Council for Scientific and Technological Development-CNPq (Project Number 455349/2012-6) and Scientific and Technological Research Foundation of Santa Catarina State-FAPESC (Project Number 2013TR3406 and 2015TR453).

References

Addy MM, Kabir F, Zhang R, Lu Q, Deng X, Current D, Griffith R, Ma Y, Zhou W, Chen P, Ruan R (2017) Co-cultivation of microalgae in aquaponic systems. Bioresour Technol 245 (2017):27–34

Adhikary S (2012) Vermicompost, the story of organic gold: a review. Agricult Sci 3(7). https://doi.org/10.4236/as.2012.37110

Al-Hafedh YS, Alam A, Beltagi MS (2008) Food production and water conservation in a recirculating Aquaponic system in Saudi Arabia at different ratios of fish feed to plants. J World Aquacult Soc 39(4):510–520

Appelbaum S, Kotzen B (2016) Further investigations of Aquaponics using brackish water resources of the Negev Desert. Ecocycles Scientific J Eur Ecocycl Soc 2(2):26–35. ISSN 2416-2140 https://doi.org/10.19040/ecocycles.v2i2.53

Arnold J (2017) Greenhouse business: start-up costs, profits and labor, 19 April 2017. https://blog.brightagrotech.com/author/jason-arnold

Arnold SJ, Sellars MJ, Crocos PJ, Coman GJ (2006) An evaluation of stocking density on the intensive production of juvenile brown tiger shrimp (*Penaeus esculentus*). Aquaculture 256 (1):174–179

Avnimelech Y (2015) Biofloc technology: a practical guide book, 3rd edn. The World Aquaculture Society, Baton Rouge

Ayre JM, Moheimani NR, Borowitzka MA (2017) Growth of microalgae on undiluted anaerobic digestate of piggery effluent with high ammonium concentrations. Algal Res 24:218–226

Badgery-Parker J, James L (2010) Commercial greenhouse cucumber production. NSW Agriculture, Orange

Ballester ELC, Abreu PC, Cavalli RO, Emerenciano M, De Abreu L, Wasielesky W Jr (2010) Effect of practical diets with different protein levels on the performance of *Farfantepenaeus paulensis* juveniles nursed in a zero exchange suspended microbial flocs intensive system. Aquac Nutr 16 (2):163–172

Barbosa M (2017) Biofloc technology: do filtering elements might affects lettuce aquaponics production integrated with tilapia? A thesis presented at animal science postgraduate program, Santa Catarina State University (Degree of Master of Science), Chapecó, Santa Catarina, Brazil, December 2017

Borowitzka MA (1999) Commercial production of microalgae: ponds, tanks, tubes and fermenters. J Biotechnol 70(1–3):313–321

Borowitzka MA, Moheimani NR (2013) Open pond culture systems. In: Algae for biofuels and energy. Springer, Dordrecht

Brown EJ, Button DK (1979) Phosphate-limited growth kinetics of *Selenastrum capricornutum* (CHLOROPHYCEAE). J Phycol 15(3):305

Bugbee B (2017) Economics of LED lighting. In: Gupta D (ed) Light emitting diodes for agriculture – smart lighting. Springer, Singapore

Chisholm SW, Brand LE (1981) Persistence of cell division phasing in marine phytoplankton in continuous light after entrainment to light: dark cycles. J Exp Mar Biol Ecol 51(2–3):107–118

Clawson JM, Hoehn A, Stodieck LS, Todd P, Stoner RJ (2000) NASA – review of aeroponics – aeroponics for spaceflight plant growth. Aeroponics DIY. https://aeroponicsdiy.com/nasa-review-of-aeroponics

Crab R, Chielens B, Wille M, Bossier P, Verstraete W (2010) The effect of different carbon sources on the nutritional value of biofloc, a feed for *Macrobrachium rosenbergii* postlarvae. Aquac Res 41:559–567

Croft MT, Lawrence AD, Raux-Deery E, Warren MJ, Smith AG (2005) Algae acquire vitamin B 12 through a symbiotic relationship with bacteria. Nature 438(7064):90

Delrue F, Álvarez-Díaz PD, Fon-Sing S, Fleury G, Sassi JF (2016) The environmental biorefinery: using microalgae to remediate wastewater, a win-win paradigm. Energies 9(3):132

Deppeler S, Petrou K, Schulz KG, Westwood K, Pearce I, McKinlay J, Davidson A (2018) Ocean acidification of a coastal Antarctic marine microbial community reveals a critical threshold for CO_2 tolerance in phytoplankton productivity. Biogeosciences 15(1):209–231

Droop MR (1973) Some thoughts on nutrient limitation in algae. J Phycol 9(3):264–272

Dufault RJ, Korkmaz A (2000) Potential of biosolids from shrimp aquaculture as a fertiliser in bell pepper production. Compost Sci Util 3:310–319

Dufault RJ, Korkmaz A, Ward B (2001) Potential of biosolids from shrimp aquaculture as a fertiliser for broccoli production. Compost Sci Util 9:107–114

Durigon EG, Sgnaulin T, Pinho SM, Brol J, Emerenciano MGC (2017) Bioflocos e seus benefícios nutricionais na pré-engorda de tilápias. Aquacult Brasil 8:50–54

Emerenciano M, Cuzon G, Goguenheim J, Gaxiola G (2011) Floc contribution on spawning performance of blue shrimp *Litopenaeus stylirostris*. Aquac Res 44(1):75–85

Emerenciano MGC, Ballester ELC, Cavalli RO, Wasielesky W (2012) Biofloc technology application as a food source in a limited water exchange nursery system for pink shrimp *Farfantepenaeus brasiliensis* (Latreille, 1817). Aquac Res 43(3):447–457

Emerenciano M, Gaxiola G, Cuzon G (2013) Biofloc technology (BFT): a review for aquaculture application and animal food industry. In: Biomass now-cultivation and utilization. InTech, Rijeka

Emerenciano MGC, Martínez-Córdova LR, Martínez-Porchas M, Miranda-Baeza A (2017) Biofloc technology (BFT): a tool for water quality management in aquaculture, water quality, In: Hlanganani Tutu (ed). InTech

Furtado PS, Poersch LH, Wasielesky W (2011) Effect of calcium hydroxide, carbonate and sodium bicarbonate on water quality and zootechnical performance of shrimp Litopenaeus vannamei reared in bio-flocs technology (BFT) systems. Aquaculture 321:130–135

Galloway JN, Dentener FJ, Capone DG, Boyer EW, Howarth RW, Seitzinger SP, Asner GP (2004) Nitrogen cycles: past, present, and future. Biogeochemistry 70(2):153–226

Gordon JM, Polle JE (2007) Ultra-high bioproductivity from algae. Appl Microbiol Biotechnol 76 (5):969–975

Gunning D, Maguire J, Burnell G (2016) The development of sustainable saltwater-based food production systems: a review of established and novel concepts. Water 8(12):598. https://doi.org/10.3390/w8120598

Gupta D (ed) (2017) Light emitting diodes for agriculture – smart lighting. Springer, Singapore

Hikosaka Y, Kanechi M, Uno Y (2014) A novel aeroponic technique using dry-fog spray fertigation to grow leaf lettuce (*Lactuca sativa* L. var. crispa) with water-saving hydroponics. Adv Hortic Sci 28(4):184–189

Joesting HM, Blaylock R, Biber P, Ray A (2016) The use of marine aquaculture solid waste for nursery of salt marsh plants Spartina alterniflora and *Juncus roemerianus*. Aquac Rep 3:108–114

Khandaker M, Kotzen B (2018) The potential for combining living wall and vertical farming systems with aquaponics with special emphasis on substrates. Aquacult Res 23 January 2018. https://doi.org/10.1111/are.13601

Kotzen B, Appelbaum S (2010) An investigation of aquaponics using brackish water resources in the Negev Desert. J Appl Aquacult 22(4):297–320. ISSN 1045-4438 (print), 1545-0805 (online). https://doi.org/10.1080/10454438.2010.527571

König B, Junge R, Bittsanszky A, Villarroel M, Komives T (2016) On the sustainability of aquaponics. Ecocycles 2(1):26–32

Kuhn DD, Boardman GD (2008) Use of microbial flocs generated from tilapia effluent as a nutritional supplement for shrimp, Litopenaeus vannamei, in recirculating aquaculture systems. J World Aquacult Soc 39:72–82

Lakhiar IA, Gao J, Naz ST, Chandio FA, Buttar NA (2018) Modern plant cultivation technologies in agriculture under controlled environment: a review on aeroponics. J Plant Interact 13 (1):338–352. https://doi.org/10.1080/17429145.2018.1472308

Langton RW, Haines KC, Lyon RE (1977) Ammonia nitrogen produced by the bivalve mollusc *Tapes japonica* and its recovery by the red seaweed *Hypnea musciformis* in a tropical mariculture system. Helgol Wiss Meeresunters 30:217–229

Lenz GL, Durigon EG, Lapa KG, Emerenciano MGC (2017) Produção de alface (*Lactuca sativa*) em efluentes de um cultivo de tilápias mantidas em sistema BFT em baixa salinidade. Bol Inst Pesca 43:614–630

Martínez-Córdova LR, Emerenciano M, Miranda-Baeza A, Martínez-Porchas M (2015) Microbial-based systems for aquaculture of fish and shrimp: an updated review. Rev Aquac 7(2):131–148

Martínez-Córdova LR, Martínez-Porchas M, Emerenciano MG, Miranda-Baeza A, Gollas-Galván T (2017) From microbes to fish the next revolution in food production. Crit Rev Biotechnol 37:287–295

Megahed M (2010) The effect of microbial biofloc on water quality, survival and growth of the green tiger shrimp (*Penaeus semisulcatus*) fed with different crude protein levels. J Arab Aquacult Soc 5:119–142

Moheimani NR (2016) *Tetraselmis suecica* culture for CO_2 bioremediation of untreated flue gas from a coal-fired power station. J Appl Phycol 28(4):2139–2146

Moheimani NR, Borowitzka MA (2007) Limits to productivity of the alga Pleurochrysis carterae (Haptophyta) grown in outdoor raceway ponds. Biotechnol Bioeng 96(1):27–36

Moheimani NR, Parlevliet D (2013) Sustainable solar energy conversion to chemical and electrical energy. Renew Sust Energ Rev 27:494–504

Moheimani NR, Isdepsky A, Lisec J, Raes E, Borowitzka MA (2011) Coccolithophorid algae culture in closed photobioreactors. Biotechnol Bioeng 108(9):2078–2087

Moheimani NR, Webb JP, Borowitzka MA (2012) Bioremediation and other potential applications of coccolithophorid algae: a review. Algal Res 1(2):120–133

Moheimani NR, Parlevliet D, McHenry MP, Bahri PA, de Boer K (2015) Past, present and future of microalgae cultivation developments. In: Biomass and biofuels from microalgae. Springer, Cham, pp 1–18

Moheimani NR, Vadiveloo A, Ayre JM, Pluske JR (2018) Nutritional profile and in vitro digestibility of microalgae grown in anaerobically digested piggery effluent. Algal Res 35:362–369

Munoz R, Guieysse B (2006) Algal–bacterial processes for the treatment of hazardous contaminants: a review. Water Res 40(15):2799–2815

NASA Spinoff, Experiments Advance Gardening at Home and in Space, https://spinoff.nasa.gov/Spinoff2008/ch_3.html

National Geographic, About the common earthworm, https://www.nationalgeographic.com/animals/invertebrates/c/common-earthworm

Neori A, Shpigel M, Ben-Ezra D (2000) A sustainable integrated system for culture of fish, seaweed, and abalone. Aquaculture 186:279–291

Nwoba EG, Ayre JM, Moheimani NR, Ubi BE, Ogbonna JC (2016) Growth comparison of microalgae in tubular photobioreactor and open pond for treating anaerobic digestion piggery effluent. Algal Res 17:268–276

Nwoba EG, Moheimani NR, Ubi BE, Ogbonna JC, Vadiveloo A, Pluske JR, Huisman JM (2017) Macroalgae culture to treat anaerobic digestion piggery effluent (ADPE). Bioresour Technol 227:15–23

Oswald WJ (1988) Role of microalgae in liquid waste treatment and reclamation. In: Lembi CA, Robert Waaland J (eds) Algae and human affairs. Sponsored by the Phycological Society of America, Inc

Oswald WJ, Gotaas HB (1957) Photosynthesis in sewage treatment. Trans Am Soc Civ Eng 122 (1):73–105

Park JBK, Craggs RJ (2010) Wastewater treatment and algal production in high rate algal ponds with carbon dioxide addition. Water Sci Technol 61(3):633–639

Parkin GF, Owen WF (1986) Fundamentals of anaerobic digestion of wastewater sludges. J Environ Eng 112(5):867–920

Pinheiro I, Arantes R, Santo CME, Seiffert WQ (2017) Production of the halophyte Sarcocornia ambigua and Pacific white shrimp in an aquaponic system with biofloc technology. Ecol Eng 100:261–267

Pinho SM (2018) Tilapia nursery in aquaponics systems using bioflocs technology. A thesis presented at aquaculture Centre of Sao Paulo State University. Degree of Master of Science, Jaboticabal, Sao Paulo, Brazil, February 2018

Pinho SM, Molinari D, De Mello GL, Fitzsimmons KM, Emerenciano MGC (2017) Effluent from a biofloc technology (BFT) tilapia culture on the aquaponics production of differente lettuce varieties. Ecol Eng 103:146–153

Poleo G, Aranbarrio JV, Mendoza L, Romero O (2011) Cultivo de cachama blanca en altas densidades y en dos sistemas cerrados. Pesq Agrop Brasileira 46(4):429–437

Poli MA, Schveitzer R, Nuñerr APO (2015) The use of biofloc technology in a South American catfish (*Rhamdia quelen*) hatchery: effect of suspended solids in the performance of larvae. Aquac Eng 66:17–21

Rahman SSA (2010) Effluent water characterization of intensive Tilapia culture units and its application in an integrated lettuce aquaponic production facility. A thesis submitted to the graduate faculty. Auburn University, Degree of Master of Science, Auburn Alabama, December 13 2010

Rakocy JE (2012) Aquaponics – integrating fish and plant culture. In: Tidwell JH (ed) Aquaculture Production Systems, 1st edn. Wiley-Blackwell, Oxford, pp 343–386

Ray AJ, Lewis BL, Browdy CL, Leffler JW (2010) Suspended solids removal to improve shrimp (Litopenaeus vannamei) production and an evaluation of a plant-based feed in minimal-exchange, superintensive culture systems. Aquaculture 299:89–98

Richmond A, Becker EW (1986) Technological aspects of mass cultivation, a general outline. In: CRC handbook of microalgal mass culture, CRC Press, Boca Raton, pp 245–263

Rocha AF, Biazzetti Filho ML, Stech MR, Silva RP (2017) Lettuce production in aquaponic and biofloc systems with silver catfish *Rhamdia quelen*. Bol Inst Pesca 44:64–73

Schoumans OF, Chardon WJ, Bechmann ME, Gascuel-Odoux C, Hofman G, Kronvang B et al (2014) Mitigation options to reduce phosphorus losses from the agricultural sector and improve surface water quality: a review. Sci Total Environ 468:1255–1266

Sgnaulin T, Mello GL, Thomas MC, Esquivel-Garcia JR, Oca GARM, Emerenciano MGC (2018) Biofloc technology (BFT): an alternative aquaculture system for Piracanjuba *Brycon orbignyanus*? Aquaculture 485:119–123

Sharpley AN, Kleinman PJ, Heathwaite AL, Gburek WJ, Folmar GJ, Schmidt JP (2008) Phosphorus loss from an agricultural watershed as a function of storm size. J Environ Qual 37 (2):362–368

Shifrin NS, Chisholm SW (1981) Phytoplankton lipids: interspecific differences and effects of nitrate, silicate and light-dark cycles. J Phycol 17(4):374–384

Smith VH (1983) Low nitrogen to phosphorus ratios favor dominance by blue-green algae in lake phytoplankton. Science 221(4611):669–671

Stoknes K, Scholwin F, Krzesiński W, Wojciechowska E, Jasińska A (2016) Efficiency of a novel "food to waste to food" system including anaerobic digestion of food waste and cultivation of vegetables on digestate in a bubble-insulated greenhouse. Waste Manag 56:466–476. https://doi.org/10.1016/j.wasman.2016.06.027

Stokstad E (2010) Down on the shrimp farm. Science 328:1504–1505

Storey, A., 2015, Vertical farming costs and the math behind them, posted 6 October 2015 https://blog.brightagrotech.com/vertical-farming-costs-and-the-math-behind-them/

Tibbitts TW, Cao W, and Wheeler RM (1994) Growth of potatoes for CELSS. NASA Contractor Report 177646

Touliatos D, Dodd IC, McAinsh M (2016) Vertical farming increases lettuce yield per unit areacompared to conventional horizontal hydroponics. Food and Energy Security 5(3):184–191

Turcios AE, Papenbrock J (2014) Sustainable treatment of aquaculture effluents – what can we learn for the past for the future? Sustainability 6:836–856

Virsile A, Olle M, Duchovskis P (2017) LED lighting in horticulture. In: Gupta D (ed) Light emitting diodes for agriculture – smart lighting. Springer, Singapore

Waller U, Buhmann AK, Ernst A, Hanke V, Kulakowski A, Wecker B, Orellana J, Papenbrock J (2015) Integrated multi-trophic aquaculture in a zero-exchange recirculation aquaculture system for marine fish and hydroponic halophyte production. Aquac Int 23:1473–1489

Weathers PJ, Zobel RW (1992) 1992, Aeroponics for the culture of organisms. Tissues Cells Biotechnol Adv 10(1):93–115

Wijihastuti RS, Moheimani NR, Bahri PA, Cosgrove JJ, Watanabe MM (2017) Growth and photosynthetic activity of *Botryococcus braunii* biofilms. J Appl Phycol 29(3):1123–1134

Zhao Z, Xu Q, Luo L, Wang CA, Li J, Wang L (2014) Effect of feed C/N ratio promoted bioflocs on water quality and production performance of bottom and filter feeder carp in minimum-water exchanged pond polyculture system. Aquaculture 434:442–448

Permissions

All chapters in this book were first published in AFPSCAHPTF, by Springer; hereb published with permission under the Creative Commons Attribution License or equivalen Every chapter published in this book has been scrutinized by our experts. Their significanc has been extensively debated. The topics covered herein carry significant information for comprehensive understanding. They may even be implemented as practical applications o may be referred to as a beginning point for further studies.

The contributors of this book come from diverse backgrounds, making this book a trul international effort. We would like to thank all the contributing authors for lending thei expertise to make the book truly unique. They have played a crucial role in the developmen of this book. Without their invaluable contributions this book wouldn't have been possibl They have made vital efforts to compile up to date information on the varied aspects o this subject to make this book a valuable addition to the collection of many professional and students.

This book was conceptualized with the vision of imparting up-to-date and integrate information in this field. To ensure the same, a matchless editorial board was set up. Every individual on the board went through rigorous rounds of assessment to prove their worth After which they invested a large part of their time researching and compiling the mos relevant data for our readers.

The editorial board has been involved in producing this book since its inception. They have spent rigorous hours researching and exploring the diverse topics which have resulted ir the successful publishing of this book. They have passed on their knowledge of decade through this book. To expedite this challenging task, the publisher supported the team a every step. A small team of assistant editors was also appointed to further simplify the editing procedure and attain best results for the readers.

Apart from the editorial board, the designing team has also invested a significant amount of their time in understanding the subject and creating the most relevant covers. They scrutinized every image to scout for the most suitable representation of the subject and create an appropriate cover for the book.

The publishing team has been an ardent support to the editorial, designing and production team. Their endless efforts to recruit the best for this project, has resulted in the accomplishment of this book. They are a veteran in the field of academics and their pool of knowledge is as vast as their experience in printing. Their expertise and guidance has proved useful at every step. Their uncompromising quality standards have made this book an exceptional effort. Their encouragement from time to time has been an inspiration for everyone.

The publisher and the editorial board hope that this book will prove to be a valuable piece of knowledge for students, practitioners and scholars across the globe.

Index

www.ingramcontent.com/pod-product-compliance
Lightning Source LLC
LaVergne TN
LVHW080642210926
848485LV00003B/311